Lecture Notes in Mathematics

Volume 2396

This series reports on new developments in all areas of mathematics and their applications - quickly, informally and at a high level. Mathematical texts analysing new developments in modelling and numerical simulation are welcome. The type of material considered for publication includes:

1. Research monographs 2. Lectures on a new field or presentations of a new angle in a classical field 3. Summer schools and intensive courses on topics of current research.

Texts which are out of print but still in demand may also be considered if they fall within these categories. The timeliness of a manuscript is sometimes more important than its form, which may be preliminary or tentative. Please visit the LNM Editorial Policy (https://drive.google.com/file/d/1MOg4TbwOSokRnFJ3ZR3ciEeKs9hOnNX_/view?usp=sharing)

Titles from this series are indexed by Scopus, Web of Science, Mathematical Reviews, and zbMATH.

Alexandru Buium · Adrian Vasiu

Delta-Invariant Theory for Hecke Correspondences

Springer

Alexandru Buium
Department of Mathematics and Statistics
University of New Mexico
Albuquerque, NM, USA

Adrian Vasiu
Department of Mathematics and Statistics
Binghamton University
Binghamton, NY, USA

ISSN 0075-8434 ISSN 1617-9692 (electronic)
Lecture Notes in Mathematics
ISBN 978-3-032-20570-4 ISBN 978-3-032-20571-1 (eBook)
https://doi.org/10.1007/978-3-032-20571-1

Mathematics Subject Classification: 11F06, 11F46, 11F55, 11F99, 11G15, 14K02, 14L24, 14L35, 15A15, 15A24, 15A69, 20C20, 20G05, 11E04, 11F60, 11G10, 11G18, 14G35, 14K99, 14L30, 14R10, 14R20, 15A18, 15A72

This work was supported by Simons Foundation (311773, 615356).

This Springer imprint is published by the registered company Springer Nature Switzerland AG
The registered company address is: Gewerbestrasse 11, 6330 Cham, Switzerland

Preface

Our starting point is the following pathology present in all 'classical' geometries such as algebraic geometry, formal algebraic geometry, analytic geometry, differential geometry, etc. Assume X is an object of such a geometry and assume we are given a correspondence on X, i.e., another object X' of this geometry together with two morphisms from X' to X. Such a correspondence generates an equivalence relation on X and one would like to construct, in our geometry, an object that plays the role of the quotient of X by this equivalence relation. However, in many interesting cases, the orbits of the equivalence relation are dense in X in which case the categorical quotient is trivial. One way out of this puzzle would be to consider a refinement of the original geometry by allowing more functions on our objects in order to achieve the 'separation of the orbits'. An example of such a move is Alain Connes' invention of noncommutative geometry [Con] which led to a refinement of differential geometry and to the construction of new quotients in a series of remarkable examples. In a different direction, a refinement of algebraic geometry was proposed in a series of papers beginning with [Bu95a] (and partially summarized in [Bu05]); this refined geometry can be referred to as *δ-geometry* and is roughly obtained from the usual algebraic geometry by replacing unital commutative rings with δ-rings, i.e., with $\mathbb{Z}_{(p)}$-algebras that admit (and are equipped with) a Fermat quotient operator (which we denote by δ and we refer to as a *p-derivation*). The latter led in [Bu05] to the construction of quotients of various curves X (rational, elliptic, modular) by various remarkable correspondences.

The aim of the present monograph is twofold. First we revisit the foundations of δ-geometry by offering a version of it which is more flexible and more general than that of [Bu05]. Second, as our main application, we construct quotients for a remarkable series of higher dimensional examples that generalize the modular curve examples in [Bu05].

The method offered in this monograph is general in nature but we concentrate on a foundational case study which is that of quotients of moduli schemes of principally polarized abelian schemes of arbitrary fixed relative dimension by Hecke correspondences and which we hope it will be used for all future generalizations to closed subschemes of these moduli schemes that are defined by integral canonical models of Shimura varieties of Hodge type (cf. [V], Chap. 1, Sect. 1.6).

Our treatment here is independent of that in [Bu05] and, in fact, all our prerequisites here belong to 'usual' algebraic geometry: we mainly assume some familiarity with abelian schemes and their deformation theory for which we provide detailed references. A key technical role in our proofs here is played by some old and new results in the classical geometric invariant theory of tuples of quadrics. These are carefully reviewed and developed in the last chapter of our monograph. With some rare exceptions, all results in this monograph appear here for the first time.

It is interesting to note that some of the examples of quotient problems for which δ-geometry is effective are related to examples for which noncommutative geometry proves to be effective. We also note that a variant of δ-geometry was used in [Borg] as a candidate for the geometry over $\mathbb{F}_1$, the field with one element (see also [Bu05], Introd.); on the other hand, noncommutative geometry has also been proposed as a path towards $\mathbb{F}_1$. Finally note that δ-rings recently played a role in the foundation of prismatic cohomology (see [BS]). It would be interesting to interrelate these topics and the topic of the present monograph.

Albuquerque, USA — Alexandru Buium
Binghamton, USA — Adrian Vasiu

Acknowledgments The first author acknowledges support from the Simons Foundation through grant 311773, from the IHES (Bures sur Yvette), and from the MPI (Bonn). The second author would like to thank SUNY Binghamton for good working conditions and the opportunity to teach a graduate course in geometric invariant theory and Ofer Gabber for the example of Footnote 2 in Chap. 2 and for suggesting the proof of Lemma 3.2.5. We would like to thank the referees for many valuable comments and suggestions and Hari Asokan for a careful reading of Chap. 4 that helped us to fix few typos and to correct Theorem 4.2.22 (which pertains to characteristic 0) by restricting its scope to the case $g \leq 3$, the case when its proof would work for $g \geq 3$ and arbitrary characteristic being the topic of his Ph.D. thesis.

Competing Interests The authors have no competing interests to declare that are relevant to the content of this manuscript.

Contents

Notation

Let $\mathbb{Z}$ be the ring of integers and let $\mathbb{Z}_{\geq 0}$ be its semiring of non-negative integers. Let $\mathbb{N}$ be the set of positive integers. For a prime p, let $\mathbb{F}_p$ be the prime field with p elements. In Chap. 4, and hence for its notation, p is either a prime or 0. Let $\mathbb{C}$ be the field of complex numbers.

With S a unital commutative ring, let $\mathbb{A}_S^n$ and $\mathbb{P}_S^n$ be the affine space and the projective space (respectively) over $\mathrm{Spec}(S)$, let $S^\times$ be the group of units of S, and let $\bar{S} = S/pS$. If S is an integral domain, let $\mathrm{Frac}(S)$ be its field of fractions. If S is local, then $\mathfrak{m}_S$ is its maximal ideal. The reduction modulo p of X is denoted by $\bar{X}$, where X is either a scheme over $\mathrm{Spec}(S)$ or a formal scheme over the p-adic completion with structure ringed sheaf $\mathcal{O}$.

Axioms

- (DAX1) – (DAX3) the axioms defining p-derivations
- (GAX1) – (GAX5) the axioms defining δ-graded rings
- (GAX6) and (GAX7) the extra axioms defining filtered δ-graded rings
- (MAX1) – (MAX4) the axioms defining a moduli datum
- (NAX1) and (NAX2) the nondegeneration axioms for abelian schemes over R
- (UAX1) and (UAX2) the unitary axioms defining k-valued points for which one has good Serre–Tate expansions

Capital Letters

- A_0 ordinary abelian variety of dimension g over k
- $B_{\langle f \rangle}$ and $B_{((p))}$ localizations of an integral δ-graded ring B
- $C_p(T_1, T_2) = \frac{T_1^p + T_2^p - (T_1 + T_2)^p}{p} \in \mathbb{Z}[T_1, T_2]$

- $D(A)$ the multiplicative set of all $a \in A$, with A a δ-ring, such that the image of a in A/pA is a nonzero divisor
- $D(g,r) = (r+1)\frac{g(g+1)}{2} - g^2 + 1$ the dimension of either $\mathbb{H}^r_{g,\mathbb{K}}$ or $\mathbb{H}^r_{g,\mathrm{tot},\mathbb{K}}$
- $D_p(g,r,s)$ the dimension of the $\mathbb{K}$-vector space $\mathbb{H}^r_g(s)_{\mathbb{K}}$
- G a reductive group over $\mathrm{Spec}(\mathbb{K})$ and the connected component G^0 of its identity element
- $H^d(X_\bullet)$ the d-th cohomology group of a semi-simplicial object $X_\bullet$
- $H^d(X_\delta)$ the d-th cohomology group of an ss-δ-space X_δ
- $H^r_g(U_0,s)_B$ or $\mathbb{H}^r_g(s)$ (resp. $H^r_{gg}(U_0)_B$ or $H^r_{gg}(U_0)^\dagger_B$ or $H^r_{gg}(U_0)^\ddagger_B$) the B-module of polynomials (resp. matrices with entries polynomials) satisfying the appropriate functional equation, with B an integral domain that contains $\mathbb{Z}_p$ and $s \in \frac{1}{2}\mathbb{Z}$
- $J^r(X)$ the r-th p-jet space of an R-scheme X
- K the field of fractions of R; so $K = R[\frac{1}{p}]$
- M, often with lower right index in $\mathbb{N} \cup \{0\}$, matrix in Mat_g
- M^t and M^* the transpose and the adjugate of a square matrix M
- M_g the R-algebra of Siegel δ-modular functions of genus g; R-subalgebras M^r_g and R-submodules $M^r_g(w)$
- M_{gg} the R-algebra of Siegel δ-modular functions of genus g and size g; R-subalgebras M^r_{gq} and R-submodules $M^r_{gg}(w', w'')$
- $N_b = (f^\partial)^{\phi^{b-1}} \cdot (f^\partial)^{\phi^{b-2}} \cdots (f^\partial)^{\phi} \cdot f^\partial \in \mathbb{I}^b_{gg,\mathrm{ord}}(-\phi^b, 1)$ with $b \geq 1$ an integer
- Q, often with lower right index in $\mathbb{N} \cup \{0\}$, symmetric matrix in V_g
- R the ring $W(k)$ of p-typical Witt vectors with coefficients in k
- $S^{r+1}_2 := \prod_{l=0}^r S^{(l)}_2$ the singular locus of $\mathbf{V}^{r+1}_2$
- $S_{\text{big-for}}$ the larger target (codomain) of Serre–Tate expansions
- S_{can} the target (codomain) of Fourier expansions; R-subalgebras $S_{\text{small-can}}$, $S^r_{\text{small-can}}$ and S^r_{can}
- S_{for} the target (codomain) of Serre–Tate expansions; R-subalgebras S^r_{for}
- $T^{(l)}$ a $g \times g$ symetric matrix whose ij entry is $T^{(l)}_{ij}$, with $l \in \mathbb{N} \cup \{0\}$ and the index $^{(0)}$ often dropped
- $T_p(A_0)$ the p-adic Tate module of an ordinary abelian variety A_0 over k
- $T_\ell(A)$ the ℓ-adic Tate module of an abelian variety A over an algebraically closed field in which the prime ℓ is invertible
- $U_0 = U_{A_0,\theta_0,e}$ submonoid of $\mathbf{GL}_g(\mathbb{Z}_p)$ endowed with an involution † induced by the Rosati involution † of $\mathrm{End}(A_0)$ relative to the principal polarization θ_0
- V a finite dimensional vector space V over $\mathbb{K}$ with dual V^*, symmetric $\mathbb{K}$-algebra $\mathrm{Sym}(V)$, and moduli affine variety $\mathbf{V} = \mathrm{Spec}(\mathrm{Sym}(V^*))$
- V^{r+1} the direct sum of r copies of V with $r \in \mathbb{Z}_{\geq 0}$
- $W = \mathbb{Z}[\phi]$ the commutative ring of polynomials in an indeterminate ϕ with $\mathbb{Z}$-coefficients, with its subsets $W_+ = \{w \in W | w \geq 0\}$ and, for every $r \in \mathbb{Z}_{\geq 0}$, $W(r) = \{w \in W | \mathrm{ord}(w) \leq r\}$
- W_g the standard $\mathbf{SL}_g$ of dimension g
- $W_{\mathrm{even}}(r) = \{w \in W(r) | \deg(w) \in 2\mathbb{Z}\}$ set of weights
- $X(R)^{\mathrm{ss}}_{\mathbf{f}}$ the sets of $\mathbf{f}$-semistable points in $X(R)$, with X a smooth scheme over $\mathrm{Spec}(R)$

- $X(R)^{\mathrm{u}}_{\mathbf{f}}$ the sets of **f**-unstable points in $X(R)$, with X as above
- $X_\delta(\delta\text{-pts})$ the set of all δ-points of the ss-δ-space X_δ
- $Z^1(\mathfrak{a}) = \mathrm{Spf}(\mathcal{O}(J^1(X))/\mathfrak{a})$ formal scheme, with $\mathfrak{a}$ an ideal of $\mathcal{O}(J^1(X))$

Capital Letters Boldfaced

- **Aff** the category of affine schemes over some ground ring (usually $\mathbb{Z}_{(p)}$ or R)
- $\mathbf{Aff}_\delta$ the category of affine δ-spaces
- $\mathbf{FilRing}_\delta$ the category of filtered δ-rings
- $\widehat{\mathbf{FilRing}}_\delta$ the category whose objects are the filtered δ-rings S with each S^r a p-adically complete, flat R-algebra and with inclusions $S^r \subset S^{r+1}$
- $\mathbf{FilSpss}_\delta$ the category of filtered ss-δ-spaces
- $\mathbf{GrRing}_\delta$ the category of δ-graded rings
- **Ring** the category of rings over some ground ring (usually $\mathbb{Z}_{(p)}$ or R)
- $\mathbf{Ring}_\delta$ the category of δ-rings
- $\mathbf{GL}_g$ general linear group scheme over $\mathrm{Spec}(\mathbb{K})$ of invertible $g \times g$ matrices with the standard maximal torus $\mathbf{GT}_g$ and GL_g as group of $\mathbb{K}$-valued points
- $\mathbf{Mat}_g$ the ring scheme over $\mathrm{Spec}(\mathbb{Z})$ in Chaps. 2 and 3 and over $\mathrm{Spec}(\mathbb{K})$ in Chap. 4; the identity element of $\mathbf{Mat}_g(S)$ denoted by 1_g
- $\mathbf{M}'(S)$ the set of morphisms of the category $\mathfrak{M}(S)$
- $\mathbf{M}(S)$ the set of objects of the category $\mathfrak{M}(S)$
- $\mathbf{SL}_g$ special linear group scheme over $\mathrm{Spec}(\mathbb{K})$ of $g \times g$ matrices of determinant 1 with the standard maximal torus $\mathbf{T}_g$ and SL_g as group of $\mathbb{K}$-valued points
- $\mathbf{SO}_g$ the closed subgroup scheme of $\mathbf{SL}_g$ of orthogonal matrices of determinant 1
- $\mathbf{Spss}_\delta$ the category of ss-δ-spaces

Capital Letters Symbols

- (A, θ) a principally polarized abelian scheme of relative dimension g, often with lower right indices
- $(\mathfrak{M}, \mathfrak{D}, G, D)$ a quadruple that defines a moduli datum, i.e., satisfies axioms (MAX1) to (MAX4)
- $\mathrm{Diag}(\lambda_1, \ldots, \lambda_g)$ the diagonal $q \times g$ matrix whose ii entry is λ_i for each $i \in \{1, \ldots, g\}$
- $\mathrm{End}(A, \theta)^{(p)}$ the subring of $\mathrm{End}(A)$ generated by all polarized isogenies $u : A \to A$ of degree prime to p
- $L_g = \mathbb{K}1_g \times \mathrm{Z}_g$ a $\mathbb{K}$-vector subspace of both Z_g^2 and V_g^2
- Mat_g the $\mathbb{K}$-vector space of $g \times g$ matrices
- $\mathrm{Proj}_\delta(B, \mathbf{f})$ the ss-δ-space attached to a δ-graded ring B and a linear system $\mathbf{f}$
- $\mathrm{Spec}_\delta(B_\bullet) = (\mathrm{Spec}_\delta(B_d))_{d \geq 0}$ an ss-δ-space

- $\mathrm{Stab}_G(v)$ the stabilizer subgroup scheme with v an element of a G-module V
- Tr trace maps
- V_g the vector space over $\mathbb{K}$ of symmetric $g \times g$ matrices
- $\mathrm{V}_g^{\mathrm{alt}}$ the $\mathbb{K}$-vector subspace of V_g of alternating symmetric matrices and $\mathrm{V}_g^{\mathrm{Diag}} = \mathrm{V}_g^{\mathrm{alt}}$ when $p = 2$
- Z_g the vector space over $\mathbb{K}$ of diagonal $g \times g$ matrices
- $\check{A}$ the dual of an abelian scheme A
- $\widehat{L}$ the p-adic completion of an S-module L
- $\widehat{X}$ the p-adic formal scheme obtained by p-adically completing a noetherian scheme X

Small Letters

- b^w short notation for $b^{a_0}\phi(b)^{a_1} \cdots \phi^n(b)^{a_n} \in B$, where $w = \sum_{i=0}^n a_i \phi^i \in \mathbb{Z}[\phi]$ and $\phi : B \to B$ is an injective ring endomorphism
- $b_p(g, 2)$ additional invariant in $\mathbb{N}$
- $c_p(g, \mathrm{odd}, r, l)$ nondecreasing sequence with $l \in \{1, \ldots, \varrho_p(g, r)\}$ of integers ≥ 1 when $\mathbb{H}^r_{g,\mathrm{tot},\mathbb{K}}$ is assumed to be Cohen–Macaulay
- $c_p(g, r, l)$ nondecreasing sequence with $l \in \{1, \ldots, \varrho_p^-(g, r)\}$ of integers ≥ 2 when $\mathbb{H}^r_{g,\mathrm{tot},\mathbb{K}}$ is assumed to be Cohen–Macaulay
- $f^r_{g,\mathrm{crys}}$ the basic Hecke covariant δ-modular form of genus g, size g, order $\leq r$, and weight $(\phi^r, 1)$ characterized by the following basic and simple Serre–Tate expansion $\mathcal{E}_{A_0,\theta_0,e}(f^r_{g,\mathrm{crys}}) = \Psi^{\phi^{r-1}} + p\Psi^{\phi^{r-2}} + \cdots + p^{r-1}\Psi$
- $f^{<a>} = f^{\langle a \rangle}_{g,\mathrm{crys}} = N_a^t \cdot (f^1_{g,\mathrm{crys}})^{\phi^{a-1}} \cdot N_{a-1} \in \mathbb{I}^a_{gg}(1, 1)$ with transpose $f^{\langle a \rangle t}$ δ-modular forms of size g indexed by integers $a \geq 1$
- $f^{[a]} = f^{[a]}_{g,\mathrm{crys}} = N_a^t \cdot f^a_{g,\mathrm{crys}} \in \mathbb{I}^a_{gg}(1, 1)$ with transpose $f^{[a]t}$ δ-modular forms of size g indexed by integers $a \geq 1$
- $f^{\partial}_{g,\mathrm{crys}}$ the basic ordinary Hecke covariant Siegel δ-modular form of weight $(-\phi, 1)$ characterized by Serre–Tate expansion $\mathcal{E}_{A_0,\theta_0,e}(f^{\partial}_{g,\mathrm{crys}}) = 1_g$; often abbreviated as f^{∂} and its inverse abbreviated as f_{∂}
- $f^{ab} = \phi^b f^{a-b}_{g,\mathrm{crys}} \in \mathbb{I}^a_{gg}(\phi^a, \phi^b)$ and its transpose $f^{ba} = (f^{ab})^t \in \mathbb{I}^a_{gg}(\phi^b, \phi^a)$ Hecke covariant modular forms
- k the residue field of R; so $k = R/pR$ is an algebraic closure of $\mathbb{F}_p$
- $q_A : T_p(A_0) \times T_p(\check{A}_0) \to 1 + \mathfrak{m}_R = 1 + pR$ Serre–Tate bilinear map attached to an abelian scheme over $\mathrm{Spec}(R)$ whose reduction modulo p is an ordinary abelian variety A_0
- $q_{A,\theta} : T_p(A_0) \times T_p(A_0) \to 1 + \mathfrak{m}_R$ Serre–Tate bilinear map attached to a principally polarized abelian scheme (A, θ), with A_0 as above
- $q_{\mathrm{univ},A_0} : T_p(A_0) \times T_p(\check{A}_0) \to 1 + \mathfrak{m}_{R_g}$ Serre–Tate bilinear map attached to the formal deformation of A_0 over $\mathcal{M}_{A_0} = \mathrm{Spf}(R_g)$, with A_0 an ordinary abelian variety over $\mathrm{Spec}(k)$

- u polarized isogeny of degree prime to p between two principally polarized abelian schemes

Small Letters Symbols

- asy.dim(B) the asymptotic dimension of a filtered ring B
- asy.tr.deg(B) the asymptotic transcendence degree of a filtered ring B
- $\{b_1, b_2\}_\delta = \frac{b_1^p b_2^\phi - b_2^p b_1^\phi}{p} \in B((\phi + p)w)$, where $b_1, b_2 \in B(w)$ with B a δ-graded ring
- $\mathbf{f} = (f_0, \ldots, f_n)$ linear system, i.e., an $n-1$-tuple in $(B(w)\backslash pB(w))^{n+1}$ for an integral δ-graded ring B
- tr.deg$_{B_1}(B_2)$ the transcendental degree of Frac(B_2) over Frac(B_1), where $B_1 \subset B_2$ is an inclusion between integral domains
- $[u]_\omega$ invertible $g \times g$ matrix defined with respect to a column basis of 1-forms $\omega = (\omega_1 \cdots \omega_g)^{\mathrm{t}}$ by the identity $u^*\omega = [u]_\omega\omega$

Capital Greek Letters

- $\Delta(g, r)$ and $\Delta^{+1}(g, r)$ are subsets of $\mathbb{Z}_{\geq 0}^{r+1}$, for $r \geq 2$ the former contains $\Delta_3(g, r)$ and $\Delta_3^*(g, r)$
- $\Gamma_p(g, r)$ and $\Gamma_p(g, \mathrm{tot}, r)$ the smallest number of homogeneous generators of the $\mathbb{Z}_{\geq 0}$-graded $\mathbb{K}$-algebras $\mathbb{H}^r_{g,\mathbb{K}}$ and of the $\mathbb{W}_{\varepsilon_g}$-graded $\mathbb{K}$-algebra $\mathbb{H}^r_{g,\mathrm{tot},\mathbb{K}}$
- Λ, often with lower right index in $\mathbb{N} \cup \{0\}$, invertible matrix in GL_g
- Ω a comparison map
- $\overline{\Omega} : \overline{\mathbb{H}_g^{r-1}(s)} \times \overline{\mathbb{H}_g^{r-1}(s)} \to \mathbb{V}_g^{r+1}((p+1)s)[p]^\tau$ an antisymmetric k-bilinear map
- Ψ a $g \times g$ matrix with entries in S^r_{for} which is important in Serre–Tate expansions
- Θs the partial polarizations of the invariant polynomial det
- $\Upsilon_{q_1,\ldots,q_{\frac{g(g+1)}{2}}}$ in $\mathbb{H}^r_{g,\mathrm{tot},\mathbb{K}}$ with $r \geq \frac{g(g+1)}{2}$
- Ξ_ω the invariant polynomial in $\mathbb{H}^r_{2,\mathrm{tot},\mathbb{K}}$ uniquely determined by an m-cycle ω in the set Perm$(\{0, \ldots, r\})$

Small Greek Letters

- χ_w families of 'character' homomorphisms with $w \in \mathcal{W}$
- χ_M the characteristic polynomial of a square matrix M
- $\omega = (\omega_1 \cdots \omega_g)^t$ a column basis of 1-forms on an abelian scheme A of relative dimension g
- ϕ-tr.deg$_C(B)$ the ϕ-transcendence degree of B over C
- ρ the slope of a moduli datum

- σ the shift $\mathbb{K}$-monomorphism of $\mathbb{T}_g$ or $\mathbb{X}$
- τ_g the involutary automorphism of $\mathbf{SL}_g$ given by the rule $\Lambda \mapsto (\Lambda^t)^{-1}$
- $\varepsilon_g \in \{0, 1\}$ the remainder of the division of g by 2
- $\varpi_g^{r+1,l}$ morphisms of affine $\mathbb{K}$-varieties with $r \geq l \geq 2$
- $\varrho_p^-(g, r)$ non-negative integer related to the rank of $\mathbb{H}^r_{g,\mathbb{K}}$ when $\mathbb{H}^r_{g,\text{tot},\mathbb{K}}$ is assumed to be Cohen–Macaulay
- $\varrho_p(g, r) = 1 + \varrho_p^-(g, r) + D_p(g, r, 1) - D(g, r)$ natural number related to the rank of $\mathbb{H}^r_{g,\mathbb{K}}$ when $\mathbb{H}^r_{g,\text{tot},\mathbb{K}}$ is assumed to be Cohen–Macaulay

Blackboard Bold Letters

- $\mathbb{D}_g$ the multiplicative set of all homogeneous elements of $\mathbb{I}_g \backslash p\mathbb{I}_g$ whose weights have even degree
- $\mathbb{E}_g$ the $g \times g$ matrix with all entries 1
- $\mathbb{H}^n_{g,\text{conj}}$ an R-algebra of polynomials satisfying an appropriate functional equation
- $\mathbb{H}^r_{g,\text{tot},\mathbb{K}} = \oplus_{s\in\frac{1}{2}\mathbb{Z}}\mathbb{H}^r_{g,\mathbb{K}}(s)$ the $\mathbb{K}$-algebra of invariants of the congruence action of $\mathbf{SL_g}$ on V_g^{r+1} (resp. the R-algebra version $\mathbb{H}^r_{g,\text{tot}} = \oplus_{s\in\frac{1}{2}\mathbb{Z}}\mathbb{H}^r_g(s)$ of it); contains $\mathbb{H}^r_{g,\mathbb{K}}$ (resp. $\mathbb{H}^r_g$)
- $\mathbb{H}^r_{gg}(1)_B$ the B-module of homogeneous matrices of degree 1 satisfying the appropriate functional equation, with B an infinite integral domain
- $\mathbb{H}^{r,\mathbb{Z}}_{g,\text{tot}} = \mathbb{H}^r_{g,\text{tot},\mathbb{C}} \cap (\mathbb{Z}[T, \ldots, T^{(r)}])$ and $\mathbb{H}^{r,\mathbb{Z}}_g = \mathbb{H}^r_{g,\mathbb{C}} \cap (\mathbb{Z}[T, \ldots, T^{(r)}])$ are $\mathbb{Z}$-algebras of invariants which contain $\mathbb{H}^{r,\Theta}_g$
- $\mathbb{H}^{r,\perp}_{g,\mathbb{K}}$ an $\mathbb{H}^r_{g,\mathbb{K}}$-module which is a direct supplement of $\mathbb{H}^r_{g,\mathbb{K}}$ in $\mathbb{H}^r_{g,\text{tot},\mathbb{K}}$ when g is even
- $\mathbb{H}^{\text{torus}}_{g,\text{tot}}$ an $\mathbb{H}^r_{g,\text{tot}}$-algebra that contains $\mathbb{H}^{r,\text{torus}}_{g,\text{tot}}$, $\mathbb{H}^{\text{torus}}_g$, and $\mathbb{H}^{r,\text{torus}}_g$
- $\mathbb{I}^r_{gg}(w', w'')$ the R-submodule of $M^r_{gg}(w', w'')$ of Hecke covariant Siegel δ-modular forms
- $\mathbb{I}^{r,\mathbb{Z}}_{g,\text{ord}} = \oplus_{s\in\frac{1}{2}\mathbb{Z}}\mathbb{I}^r_{g,\text{ord}}(-2s)$ and R-subalgebra of $\mathbb{I}^r_{g,\text{ord}}$
- $\mathbb{I}_g$ the R-algebra of Hecke covariant Siegel δ-modular forms
- $\mathbb{I}_{g,\text{ord}}$ the ordinary version of the above
- $\mathbb{K}$ an algebraically closed field of characteristic p which is either 0 or a prime
- $\mathbb{K}_{g,\infty}$ the union of the $\mathbb{K}_{g,r} = \text{Frac}(\mathbb{H}^r_{g,\text{tot},\mathbb{K}})$s taken inside $\text{Frac}(\mathbb{T}_g)$; it contains $\mathbb{K}_{g,\text{even},\infty}$ and $\mathbb{K}_{g,\text{even},r}$
- $\mathbb{M}_g = \bigcup_{r\geq 0}\mathbb{M}^r_g$ and $\mathbb{M}'_g = \bigcup_{r\geq 0}\mathbb{M}'^r_g$ W-graded R-algebra which are particular cases of the general moduli data R-algebras $\mathbb{M} = \bigcup_{r\geq 0}\mathbb{M}^r$ and $\mathbb{M}' = \bigcup_{r\geq 0}\mathbb{M}'^r$
- $\mathbb{M}_X$ the ring (R-algebra) $\mathbb{S}_{X,L}$ of δ sections, with X an open subscheme of $\mathcal{X}$ with $\bar{X}$ connected and with L the Hodge bundle over X; contains the R-subalgebra $\mathbb{M}^r_X = \oplus_{w\in W(r)}\mathbb{M}^r_X(w)$
- $\mathbb{S}_n$ the R-algebra of δ-homogeneous polynomials
- $\mathbb{S}_{X,L}$ ring (R-algebra) of δ-sections
- $\mathbb{T}_g$ the R of $\mathbb{K}$-algebra generated by the $T^{(l)}_{ij}$s with $l \in \mathbb{N} \cup \{0\}$, $i, j \in \{1, \ldots, j\}$, $i \leq j$

- $\mathbb{V}_g^r(s) = \frac{\mathbb{I}^r_{g,\mathrm{ord}}(-2s)}{\diamondsuit(\mathbb{H}_g^{r-1}(s))}$ and $\mathbb{V}_g^r = \frac{\mathbb{I}^{r,\mathbb{Z}}_{g,\mathrm{ord}}}{\diamondsuit(\mathbb{H}^{r-1}_{g,\mathrm{tot}})}$ torsion R-modules
- $\mathbb{W}_0 = \frac{1}{2}\mathbb{Z}_{\geq 0}$ and $\mathbb{W}_1 = \mathbb{Z}_{\geq 0}$
- $\mathbb{X}$ polynomial $\mathbb{K}$-algebra in indeterminates $x_i^{(l)}$ with $i \in \{1, \dots, d\}$ and $l \in \mathbb{N} \cup \{0\}$; contains $\mathbb{X}_r$
- $\mathbb{Y}_r = \mathbb{K}[u_i v_i, v_i | i \in \{0, \dots, r\}]$ a $\mathbb{Z}_{\geq 0}$-graded $\mathbb{K}$-algebra;
- $\mathbb{Y}_r^{\text{all-even}} = \mathbb{K}[u_i^2, u_i v_i, v_i^2 | i \in \{0, \dots, r\}]$ a $\mathbb{K}$-subalgebra of $\mathbb{Y}_r$
- $\mathbb{Z}_{(p)}$ the localization of $\mathbb{Z}$ at the prime ideal $(p) = p\mathbb{Z}$

Calligraphic Letters

- $\mathcal{A}'_{g,1,N,R}$ the union of all prime to p Hecke correspondences on the scheme $\mathcal{A}_{g,1,N,R}$
- $\mathcal{A}'_{g,1,R}$ the stack corresponding to the union of all prime to p Hecke correspondences on $\mathcal{A}_{g,1,R}$
- $\mathcal{A}'_{g,1,R,\mathrm{ord}}$ the ordinary version of the above
- $\mathcal{A}_{g,1,N,R}$ the moduli scheme of principally polarized abelian schemes over R equipped with symplectic similitude level-N structure, often denoted by $\mathcal{X}$, with $g \in \mathbb{N}$ and $N \geq 3$ an integer relatively prime to p
- $\mathcal{A}_{g,1,R}$ the stack of principally polarized abelian schemes over R
- $\mathcal{A}_{g,1,R,\mathrm{ord}}$ the ordinary version of the above
- $\mathcal{C}_{r,m}$ the set of all m-cycles of the permutation group $\mathrm{Perm}(\{0, \dots, r\})$ and its subset $\mathcal{C}_{r,m}^{\mathrm{uni}}$ of increasing or unimodular m-cycles
- $\mathcal{D}^{a_1,\dots,a_s}$ nonzero δ-modular form $\mathbb{I}_g(-(g^2 - g + 1)\sum_{i=1}^{2s}\phi^{a_i})$ related to the commutativity of rings $\mathrm{End}(A, \theta)^{(p)}$
- $\mathcal{E}_{A_0,\theta_0,e} : M_g^r(w) \to S_{\mathrm{for}}^r$ and $\mathcal{E}_{A_0,\theta_0,e} : M_{gg}^r(w', w'') \to \mathbf{Mat}_g(S_{\mathrm{for}}^r)$ the Serre–Tate expansion R-linear maps
- $\mathcal{E}_{A_0,\theta_0,e} : M^r_{g,\mathrm{ord}}(w) \to S_{\mathrm{for}}^r$ and $\mathcal{E}_{A_0,\theta_0,e} : M^r_{gg,\mathrm{ord}}(w', w'') \to \mathbf{Mat}_g(S_{\mathrm{for}}^r)$ the ordinary version of the above
- $\mathcal{H}_g^r(U_0)$ (resp. $\mathcal{H}_{gg}^r(U_0)$, $\mathcal{H}_{gg}^r(U_0)^\dagger$ or $\mathcal{H}_{gg}^r(U_0)^\ddagger$) the R-module of series (resp. matrices with entries series) satisfying the appropriate functional equation
- $\mathcal{T}_0$ algebraically independent subset of $\mathbb{H}_g^r(1)_{\mathbb{K}}$ which is a transcendental bases of either $\mathbb{H}^r_{g,\mathbb{K}}$ or $\mathbb{H}^r_{g,\mathrm{tot},\mathbb{K}}$ over $\mathbb{K}$
- $\mathcal{U}_0$ submonoid of $\mathrm{End}(A_0)$
- $\mathcal{W}$ the multiplicative submonoid of W generated by -1 and ϕ

Special Characters

- $\clubsuit$ a comparison map
- $\diamondsuit$ a comparison map
- $\heartsuit$ a comparison map

- $\jmath$, also with lower index l, $\mathbb{K}$-homomorphism
- ♠ a comparison map
- □ a comparison map
- $\star$ left and right operations of $\mathbf{Mat}_g(\mathbb{Z}_p)$ on $\mathbf{Mat}_g(1 + \mathfrak{m}_R)$
- Δ a comparison map
- $\wp_p(g, r)(x)$ and $\wp_p(g, \mathrm{tot}, r)(x^{\frac{1}{2}})$ the Hilbert series of $\mathbb{H}^r_{g,\mathbb{K}}$ and $\mathbb{H}^r_{g,\mathrm{tot},\mathbb{K}}$
- $\wp_p(g, r; 1)(x) = \frac{1+[D_p(g,r,1)-D(g,r)]x}{(1-x)^{D(g,r)}}$ a Hilbert series that depends on (g, r, p) only; in particular, $\wp_0(g, r; 1)(x) = \frac{1+[\binom{g+r}{r}-D(g,r)]x}{(1-x)^{D(g,r)}}$
- † upper index for elements of certain submonoids that relate to a Rosati involution

Chapter 1
Introduction

1.1 Motivation

Recall that a *coequalizer*, alternatively *categorical quotient*, of a *correspondence* (i.e., pair of morphisms) $\pi_1, \pi_2 : X' \rightrightarrows X$ in a category[1] is a morphism $\pi : X \to Y$ of the category such that $\pi \circ \pi_1 = \pi \circ \pi_2$ and the following universal property is satisfied: for every morphism $\pi' : X \to Y'$ with $\pi' \circ \pi_1 = \pi' \circ \pi_2$ there is a unique morphism $\alpha : Y \to Y'$ such that $\alpha \circ \pi = \pi'$. We usually set $X/X' := Y$. If X/X' is a final object of the category, then we say that X/X' is *trivial*.

Consider now the category of schemes over a (unital commutative) ring R and a correspondence $X' \rightrightarrows X$ in this category. In many interesting examples of such correspondences the categorical quotient X/X' (exists and) is trivial. This is a basic pathology in algebraic geometry which typically persists if instead of the category of schemes one considers variants such as the category of formal schemes, the category of algebraic stacks, or the 'birational category' (whose objects are varieties and whose morphisms are the dominant rational maps). In order to 'rectify' this problem (i.e., in order to obtain interesting categorical quotients) one needs a more drastic departure from algebraic geometry. The departure we propose leads to what we refer to as *δ-geometry*. In this new framework we consider a class of typical examples for which the above pathology occurs, namely Hecke correspondences on moduli schemes of principally polarized abelian schemes. (For background on these basic concepts of moduli theory we refer to Appendix.)

Specifically, let $g \in \mathbb{N}$, let p be a prime, and let $N \geq 3$ be an integer relatively prime to p. If $\mathbb{Z}_{(p)}$ is the localization of the ring of integers $\mathbb{Z}$ at (p) and R is a $\mathbb{Z}_{(p)}$-algebra, let $\mathcal{A}_{g,1,N,R}$ be the moduli scheme of principally polarized abelian schemes of relative dimension g over R-schemes that are equipped with symplectic similitude level-N structures. Let $\mathcal{A}'_{g,1,N,R} \rightrightarrows \mathcal{A}_{g,1,N,R}$ be the correspondence obtained by

[1] Throughout this monograph we assume a Grothendieck universe is given such that each category **C** under consideration is small with respect to it, i.e., we have that Ob(**C**) and Mor(**C**) are sets in that universe.

A. Buium and A. Vasiu, *Delta-Invariant Theory for Hecke Correspondences*,
Lecture Notes in Mathematics 2396, https://doi.org/10.1007/978-3-032-20571-1_1

taking the union of all prime to p Hecke correspondences. The categorical quotient $\mathcal{A}_{g,1,N,R}/\mathcal{A}'_{g,1,N,R}$ in the category of schemes over R is easily seen to be trivial. We show that many natural analogs of this categorical quotient in δ-geometry are highly nontrivial and their rich structure is studied in detail.

In what follows we overview the framework of δ-geometry.

1.2 The World of δ

The δ-geometry (see [Bu95a, Bu05]) is an arithmetic analog of the differential algebraic geometry of Ritt and Kolchin [Ko] and of the difference algebraic geometry of Ritt and Cohn [Coh, Le]: the derivations in [Ko] and the difference operators in [Coh, Le] are replaced in the δ-geometry by *Fermat quotient operators* (*p-derivations*), see [Jo, Bu95a].

Recall that a *p-derivation* on a $\mathbb{Z}_{(p)}$-algebra S is a function $\delta : S \to S$ such that the following axioms hold (see Definition 2.1.3).

For all $x, y \in S$ we have $\delta(x + y) = \delta x + \delta y + C_p(x, y)$.

For all $x, y \in S$ we have $\delta(xy) = x^p \delta y + y^p \delta x + p\delta x \delta y$.

We have $\delta(1) = 0$.

Here C_p is the polynomial $C_p(T_1, T_2) := \frac{T_1^p + T_2^p - (T_1+T_2)^p}{p} \in \mathbb{Z}[T_1, T_2]$.

The ring $\mathbb{Z}_{(p)}$ has a unique p-derivations given by the rule: $\delta(x) := \frac{x - x^p}{p}$. If S is a non-zero $\mathbb{Z}_{(p)}$-algebra equipped with a p-derivation then the structure ring homomorphism $\mathbb{Z}_{(p)} \to S$ is automatically injective and commutes with the p-derivations (see Lemma 2.1.4).

We introduce the δ-geometry in Sect. 2.1; our version of it is related to but more expanded and flexible than that in [Bu05], Sect. 2.1. We start with the category of *δ-rings* whose objects are $\mathbb{Z}_{(p)}$-algebras equipped with a p-derivation (and thus, if non-zero, they are of characteristic 0) and whose morphisms are $\mathbb{Z}_{(p)}$-algebra homomorphisms that commute with the p-derivations. Next we consider the opposite of the above category, whose objects are referred to as *affine δ-spaces*. So $\mathbb{Z}_{(p)}$ (resp. $\mathrm{Spec}(\mathbb{Z}_{(p)})$) together with the unique p-derivation of $\mathbb{Z}_{(p)}$ is the initial (resp. final) object of the category of δ-rings (resp. affine δ-spaces). To go beyond the affine case we consider the category of semi-simplicial objects in the category of affine δ-spaces which we refer to as *semi-simplicial δ-spaces* (or simply *ss-δ-spaces*). Geometry in this latter category is what we refer to, in this monograph, as *δ-geometry*. Note that an alternative way to pass from the affine case to the not necessarily affine case would be 'sheafification'. We felt that the semi-simplicial setting offers a quicker path to our results than that of 'sheafification' so we opted for it in our presentation.
A special role in the theory is played by the ring $R := W(k)$ of p-typical Witt vectors with coefficients in an algebraic closure k of $\mathbb{F}_p$. The ring R is, up to isomorphism, the unique complete discrete valuation ring of mixed characteristic $(0, p)$ with residue field k and uniformizer p. Moreover, R has a unique p-derivation. Typically the

objects and morphisms of interest in the theory are 'over R'; so R (resp. Spec(R)) together with the p-derivation of R is the initial (resp. final) object of the category of δ-rings (resp. affine δ-spaces) over R.

We develop an analog in δ-geometry of geometric invariant theory (see [MFK]) which can be referred to as '*δ-invariant theory*'. The latter comes with δ analogs of all the paraphernalia of weights, covariance, modular functions and forms, homogeneous polynomials, power series, valued points, semistability, etc. (see Sect. 2.3). Our analogs in δ-geometry of the categorical quotient $\mathcal{A}_{g,1,N,R}/\mathcal{A}'_{g,1,N,R}$ appear as certain 'Proj_g' objects attached to 'linear systems' in the graded R-algebra $\mathbb{I}_g$ of *Hecke covariant Siegel δ-modular forms of genus g*.

Siegel δ-modular forms were introduced in [Bu00], Sect. 1.3 for genus $g = 1$ and in [BB] for all genera $g \geq 1$.[2] The present monograph can be viewed as an independent continuation of [Bu05, BB]: only few facts from [Bu05, BB] are needed here and we briefly review those facts in the body of the present monograph. Note that our theory here is worked out over the $\mathbb{Z}_p$-algebra R instead of, as in [BB], over $\mathbb{Z}_p$; moreover, in [Bu00, BB, Bu05] the prime p was assumed to be at least 5 but the methods of the present monograph work for $p \in \{2, 3\}$ as well.

We also note that, as in [BB], our definition of Siegel δ-modular forms does not involve symplectic similitude level-N structures. So we often implicitly work with the algebraic stacks $\mathcal{A}_{g,1,R}$ of principally polarized abelian schemes of relative dimension g over R-schemes and with their respective unions of correspondences $\mathcal{A}'_{g,1,R}$. Many of our proofs, however, crucially use the schemes $\mathcal{A}_{g,1,N,R}$, $\mathcal{A}'_{g,1,N,R}$.

No knowledge of stacks is required in our treatment as we only use valued points of stacks and no special properties or concepts; the reader familiar with stacks will recognize the 'stacky' nature of some of our concepts. It is conceivable that δ-geometry can be extended to the stack world but no attempt at such an extension is made here as it is not necessary.

The theory of the present monograph can be viewed as an arithmetic analog of the differential algebraic theory developed in [Bu95b]. In particular, for instance, the maps of sets in Diagram (2.48) of Sect. 2.3.7 defined by tuples of Hecke covariant Siegel δ-modular forms can be viewed as arithmetic analogs of the maps defined by 'Schwarzian type' differential equations considered in [Bu95b], Thm. 0.2. Our monograph is logically independent from [Bu95b] but some of our motivation and ideas have their origin in it.

Here is an informal description of the philosophy behind our monograph (and [BB, Bu05]). Classical modular forms and p-adic modular forms (see [Ka73a]) can be envisioned as sections of certain line bundles on the moduli stack $\mathcal{A}_{g,1,R}$; there exist some technical caveats in this description (related to nonrepresentability issues) which are not accounted for here. On the other hand, to each scheme X, which is either affine or of finite type over R, one attaches its p-jet spaces $J^r(X)$ indexed

---

[2] Over R, the Hecke covariance condition here (see Definition 2.3.13) is a priori weaker than the *isogeny covariance* condition in [BB], Sect. 1.3.1 (see Remark 2.3.15) for $g \geq 2$ and they are the same, of course, for $g = 1$. One expects that over R they are the same for all g. It seems that Hecke covariance is more natural and, for technical reasons, better suited for developing the higher dimensional theory.

by an integer $r \geq 0$ (see [Bu95a], Sect. 1.2, where $J^r(X)$ is denoted by X^r). Then Siegel δ-modular forms of orders $\leq r$ can be informally described as sections of certain line bundles on the 'p-jet stacks $J^r(\mathcal{A}_{g,1,R})$'; similar caveats as above apply in this description. Weights of Siegel δ-modular forms are no longer assumed to be integers but rather elements in the ring $\mathbb{Z}[\phi]$, where ϕ is an indeterminate thought of as a Frobenius lift. The δ-geometric approach to the study of Hecke correspondences is based on replacing classical or p-adic Siegel modular forms by Siegel δ-modular forms.

We emphasize that, as a limitation of δ-geometry, the non-zero δ-rings have characteristic 0 and all test objects for evaluating Siegel δ-modular functions (implicitly also those that have a weight and are called forms) involve flat (unions of) p-adically complete R-algebras, such as R-algebras that define p-jet spaces of flat affine schemes of finite type over R (cf. Definition 2.3.3). Hence, only those torsion, i.e., modulo powers of p, phenomena are captured via these evaluations which are liftable to flat contexts over R.

1.3 Strategy

Among Siegel δ-modular forms we single out certain forms that have a covariance property with respect to the Hecke correspondences on $\mathcal{A}_{g,1,R}$, to be referred to as the *Hecke covariance*. It is a property stronger than that of being a *Hecke eigenform*; in fact, no classical or p-adic modular forms can be Hecke covariant, i.e., our theory has no classical counterpart.

Besides the $\mathbb{Z}[\phi]$-graded R-algebra $\mathbb{I}_g$ of Hecke covariant Siegel δ-modular forms of genus g, we also study a larger R-algebra $\mathbb{I}_{g,\mathrm{ord}}$ which is the variant of $\mathbb{I}_g$ in which $\mathcal{A}_{g,1,R}$ is replaced by its open substack $\mathcal{A}_{g,1,R,\mathrm{ord}}$ whose fiber over Spec(k) is the ordinary locus of $\mathcal{A}_{g,1,k}$. The rings $\mathbb{I}_g$ and $\mathbb{I}_{g,\mathrm{ord}}$ are integral domains.

One can consider the 'Proj$_\delta$' objects Proj$_\delta(\mathbb{I}_g, \mathbf{f})$ and Proj$_\delta(\mathbb{I}_{g,\mathrm{ord}}, \mathbf{f})$ over Spec(R) attached to these $\mathbb{Z}[\phi]$-graded R-algebras and to 'linear systems' $\mathbf{f} = (f_0, \ldots, f_n)$ with $f_0, \ldots, f_n$ of the same weight in $\mathbb{I}_g$ and $\mathbb{I}_{g,\mathrm{ord}}$ (respectively). These 'Proj$_\delta$' objects are objects in the category of ss-δ-spaces. Then Proj$_\delta(\mathbb{I}_g, \mathbf{f})$ and Proj$_\delta(\mathbb{I}_{g,\mathrm{ord}}, \mathbf{f})$ are the categorical quotients, in the category of ss-δ-spaces, of suitable correspondences that depend on $\mathbf{f}$ and that play, in δ-geometry, the roles of the correspondences $\mathcal{A}'_{g,1,R} \rightrightarrows \mathcal{A}'_{g,1,R}$ and their restrictions $\mathcal{A}'_{g,1,R,\mathrm{ord}} \rightrightarrows \mathcal{A}_{g,1,R,\mathrm{ord}}$ (respectively).

This formalism is somewhat parallel to the formalism of (classical) geometric invariant theory as in [MFK]; instead of group actions we consider Hecke correspondences and instead of usual varieties or schemes we consider ss-δ-spaces.

The main tool in [BB] was a theory of Fourier expansions of Siegel δ-modular forms (at infinity). For $g = 1$ one has a Fourier expansion theory (at infinity) and also a Serre–Tate expansion theory (at ordinary k-valued points), see [Bu00], Sect. 7, [Bar], Sect. 5, [Bu05], Sects. 8.1 and 8.3. One of the somewhat surprising outcomes was that for Hecke covariant Siegel δ-modular forms the two types of expansions coincide up to a natural identification.

The main tool developed in this monograph is a Serre–Tate expansion theory for Siegel δ-modular forms of genus $g \geq 1$. The main new idea is that this tool allows us to construct a series of 'comparison maps' between the R-algebras $\mathbb{I}_g$ and $\mathbb{I}_{g,\text{ord}}$ (see Sects. 3.6 and 3.7) and certain R-algebras that appear in the geometric invariant theory of multiple quadratic forms and multiple endomorphisms over R; we then transfer properties of the latter R-algebras proved in Chap. 4 to the former R-algebras.

So invariant theory shows up in our monograph in three distinct ways as follows.

First, as an intrinsic δ-invariant theory, the R-algebras $\mathbb{I}_g$ and $\mathbb{I}_{g,\text{ord}}$ are two 'δ-geometric analogs' of the classical algebras of invariants of linear algebraic group actions.

Second, in Chap. 4 we use concrete instances of classical geometric invariant theory in the context of the congruent $\mathbf{SL}_g$-actions on vector spaces of multiple quadratic forms (i.e., of symmetric matrices) and of the conjugate $\mathbf{SL}_g$-actions on vector spaces of multiple endomorphisms over either k or an algebraic closure of the field of fractions $K := \text{Frac}(R)$ of R. For instance, we compute dimensions of spectra of finitely generated algebras of invariant homogeneous polynomials over either k or K, dimensions of certain vector spaces of invariant homogeneous polynomials over either k or K and smallest numbers of homogeneous generators that are important for computing or estimating Hilbert series of the mentioned algebras, and (for $g = 2$) relations (see beginning of Chap. 4 for precise references).

Third, the comparison maps we construct connect the intrinsic δ-invariant theory with the mentioned instances of classical geometric invariant theory.

1.4 Results

We now outline our main results that are described in Sect. 2.4.

First, we prove a Serre–Tate expansion principle similar to the classical Fourier expansion principle (see Theorem 2.4.1). The expansions are 'around' an arbitrary k-valued point of $\mathcal{A}_{g,1,R}$ to which the attached principally polarized abelian variety over k satisfies the axioms (UAX1) and (UAX2) introduced in Sect. 3.2. These axioms hold for products of g copies of an ordinary elliptic curve over k with the Abel–Jacobi map as its principal polarization (see Proposition 3.2.4 and Remark 3.2.6(b) and (c)). Generalizing the known result for $g = 1$, we prove that the Serre–Tate expansions of the *basic* Hecke covariant Siegel δ-modular forms $f^r_{g,\text{crys}}$ of genus g, size g, order r and weight $(\phi^r, 1)$ (introduced in [BB], Sect. 4.1 and reviewed in Sect. 3.4 using the language recalled in Sect. 2.3) coincide, up to a natural identification, again, with their Fourier expansions (see Theorem 2.4.2). For special generating properties of $f^1_{g,\text{crys}}$ see Theorem 2.4.4(a). Furthermore, generalizing a construction introduced in [Bar] in case $g = 1$, we define (see Sect. 3.4) for all $g \geq 1$ a remarkable ordinary Siegel δ-modular form $f^{\partial}_{g,\text{crys}}$ of genus g, size g, and order 1 (also referred to as *basic*) which is invertible and whose Serre–Tate expansion equals the identity $g \times g$ matrix (see Theorem 2.4.5(a)).

Second, using the Serre–Tate expansion theory and the basic δ-modular forms we show that the R-algebra $\mathbb{I}_{g,\mathrm{ord}}$ coincides, up to torsion, with a ring of Laurent polynomials in a countable number of indeterminates over the R-algebra of invariants of multiple quadratic forms (see Theorem 2.4.7). The analog of the latter R-algebra over an algebraically closed field is a classical object and is revisited and further studied in detail in Sects. 4.1–4.3. As a consequence of this link with geometric invariant theory, if we denote by $\mathbb{I}^r_{g,\mathrm{ord}}$ the R-subalgebra of elements of $\mathbb{I}_{g,\mathrm{ord}}$ of order $\leq r$, we obtain an exact formula for the transcendence degree of the fraction field $\mathrm{Frac}(\mathbb{I}^r_{g,\mathrm{ord}})$ over K (see Remark 2.4.14). This transcendence degree divided by r tends to a limit, referred to as the *asymptotic transcendence degree* of $\mathbb{I}_{g,\mathrm{ord}}$ and denoted by asy.tr.deg($\mathbb{I}_{g,\mathrm{ord}}$), which we prove to be equal to

$$\dim(\mathcal{A}_{g,1,K}) + 1 = \dim(\mathcal{A}_{g,1,N,K}) + 1 = \frac{g(g+1)}{2} + 1.$$

We also prove that $\mathrm{Frac}(\mathbb{I}_{g,\mathrm{ord}})$ is ϕ-generated over K by finitely many elements in $\mathbb{I}^4_{g,\mathrm{ord}}$ (in the sense of Definition 2.2.2(e)). Finally, we prove that the ϕ-transcendence degree of $\mathrm{Frac}(\mathbb{I}_{g,\mathrm{ord}})$ over K, i.e., the maximum number of ϕ-algebraically independent elements of $\mathrm{Frac}(\mathbb{I}_{g,\mathrm{ord}})$, equals $\frac{g(g+1)}{2} + 1$ (see Theorem 2.4.12 for the last three results).

Third, defining $\mathbb{I}^r_g$ similarly (so $\mathbb{I}^r_g = \mathbb{I}_g \cap \mathbb{I}^r_{g,\mathrm{ord}}$) and using the geometric invariant theory of multiple endomorphisms acted upon via conjugation, we show that the sequence obtained by dividing the transcendence degree of $\mathrm{Frac}(\mathbb{I}^r_g)$ over K by r tends to a limit asy.tr.deg($\mathbb{I}_g$) which we prove to be equal, again, to $\frac{g(g+1)}{2} + 1$ (see Theorem 2.4.17). Note that, by a general result in difference algebra (see [Le], Thm. 4.4.1, p. 292), the field $\mathrm{Frac}(\mathbb{I}_g)$ is also ϕ-finitely generated. We prove that ϕ-transcendence degree of $\mathrm{Frac}(\mathbb{I}_g)$ over K equals $\frac{g(g+1)}{2} + 1$ (see Theorem 2.4.17). The latter result was first obtained for $g \in \{1, 2\}$ in [BB], Thms. 1.13 and 1.14, and for $g \geq 3$ solves the 'main open problem' formulated in the paragraph after [BB], Thm. 1.13. We also prove that the transcendence degree of $\mathrm{Frac}(\mathbb{I}_{g,\mathrm{ord}})$ over $\mathrm{Frac}(\mathbb{I}_g)$ is $\leq \frac{g(g+1)}{2} + 2$ (see Remark 2.4.16). We can view the latter as saying that the R-subalgebra $\mathbb{I}_g$ of $\mathbb{I}_{g,\mathrm{ord}}$ is 'transcendentally close' to the algebra $\mathbb{I}_{g,\mathrm{ord}}$.

As an application of our structure theory for $\mathbb{I}_g$ we show that various Zariski dense loci in $\mathcal{A}_{g,1,N,R}(R)$ are contained in zero loci of (many) Hecke covariant Siegel δ-modular forms that are not constant on $\mathcal{A}_{g,1,N,R}(R)$. This is the case with the locus (set):

(i) of abelian schemes with complex multiplication (see Remark 2.4.11);

(ii) of a polarized isogeny prime to p class (see Remark 2.4.10);

(iii) of principally polarized abelian schemes with noncommutative 'prime to p' ring of endomorphisms (see Theorem 2.4.20);

(iv) of principally polarized abelian schemes whose polarized isogeny prime to p class is decomposable (see Theorem 2.4.21).

We also prove a 'p-adic approximation' result saying, informally speaking, that each element of $\mathbb{I}_{g,\text{ord}}$ is a p-adic limit of fractions of elements in $\mathbb{I}_g$ (see Theorem 2.4.22). This means that the R-subalgebra $\mathbb{I}_g$ of $\mathbb{I}_{g,\text{ord}}$ is also 'big' in the sense of a localization of it being a 'close p-adic approximation' of $\mathbb{I}_{g,\text{ord}}$.

Fourth, we prove a δ-*flow* structure theorem for the arithmetic differential equation defined by the restriction of the basic δ-modular form $f^1_{g,\text{crys}}$ to the ordinary locus (see Theorem 2.4.23). The concept of δ-flow appearing here is an arithmetic analog of the classical concept of flow on a manifold and played a role, for instance, in [BM], Def. 5.3, p. 117 and [BPII], Lem. 4.45. Using our δ-flow structure theorem plus the arithmetic analog (see [Bu97], Cor. 1.7) of Manin's Theorem of the Kernel (see [Man], Sect. 5, Thm. 2), we prove (see Theorem 2.4.25) an extension to arbitrary genus $g \geq 1$ and to all primes p the 'Reciprocity Theorem for CL (canonical lifts) points' in [BPII], Thm. 3.5; we do not obtain, however, the conclusion, in the latter theorem, that our 'reciprocity map' is nonconstant. Theorem 2.4.25 is a purely diophantine result about $\mathbb{Z}$-linear dependence in the group of K-rational points of an abelian variety over K of points arising from R-valued CL (canonical lifts) points on a Siegel modular variety. Recall that for $p \geq 3$, R-valued CL points are R-valued CM (complex multiplication) points with ordinary reduction (for instance, see [V], Thm. 1.6.3).

As an addition, we would like to mention that Propositions 3.11.1–3.11.6 compute cohomology groups of our ss-δ-spaces $\text{Proj}_\delta(\mathbb{I}_g, \mathbf{f})$ and $\text{Proj}_\delta(\mathbb{I}_{g,\text{ord}}, \mathbf{f})$ in some simple cases.

1.5 Open Problems

Our results suggest a number of problems and directions of investigations, including the following ones.

(a) Find generators and relations for the rings $\mathbb{I}_{g,\text{ord}}$ and $\mathbb{I}_g$.

(b) Compute the exact transcendence degree of $\text{Frac}(\mathbb{I}_{g,\text{ord}})$ over $\text{Frac}(\mathbb{I}_g)$ and of each $\text{Frac}(\mathbb{I}^r_g)$ with $r \in \mathbb{N}$.

(c) Compute the exact ranks of the R-submodules of $\mathbb{I}_{g,\text{ord}}$ and $\mathbb{I}_g$ of δ-modular forms of a given order and weight.

(d) Compute the ideals of all Siegel δ-modular forms in the rings $\mathbb{I}_{g,\text{ord}}$ and $\mathbb{I}_g$ that vanish on various loci (i) to (iv) of $\mathcal{A}_{g,1,N,R}(R)$.

(e) Compute the cohomology groups of all $\text{Proj}_\delta(\mathbb{I}_g, \mathbf{f})$ and $\text{Proj}_\delta(\mathbb{I}_{g,\text{ord}}, \mathbf{f})$.

(f) Develop an analog of our theory for more general Shimura varieties. (The case of Shimura curves was developed in [Bu03, Bu04, Bu05].)

(g) Recalling that in the case of $\mathcal{A}_{1,1,R}$ and of Shimura curves (see [Bar, Bu04, Bu05]) an important role was played by certain operators, called δ-*Serre operators*, that are analogs of the Serre differential operators acting on classical or p-adic modular forms, develop a theory of δ-Serre operators for stacks of higher dimensional Shimura varieties, in particular for $\mathcal{A}_{g,1,R}$.

For our partial results and more details and comments on questions (a)–(e) we refer to Theorems 2.4.4, 2.4.5, 2.4.15 and 2.4.17 and Remarks 2.4.3, 2.4.8, 2.4.10, 2.4.13, 2.4.14, 2.4.16, 2.4.18 and 2.4.19.

1.6 Plan of the Monograph

Section 2.1 reviews some notation, terminology, and concepts including p-jet spaces introduced in [Bu95a], p. 315.

Section 2.2 introduces some dimension concepts that are useful in the monograph.

The basic concepts on Siegel δ-modular forms [BB] are recalled and expanded in Sect. 2.3.

The main results are stated in Sect. 2.4 and are proved in Chap. 3. For a detailed description of the contents of Chap. 3 and, in particular, for our detailed strategy of proofs, we refer to the beginning of Chap. 3.

Chapter 4 is devoted to revisiting the classical geometric invariant theory of multiple quadratic forms and multiple endomorphisms and proving new results that are being often used and cited in this chapter to Chap. 3: it is organized as an independent appendix, i.e., it is completely a self-contained chapter (including the notation). For more details on Chap. 4 we refer to its beginning.

Chapter 2
Main Concepts and Results

2.1 Concepts of δ-Geometry

We begin by reviewing some ring theoretic terminology and notation. Unless otherwise stated, all rings are always assumed commutative with 1. We do not assume $1 \neq 0$ as often we need to use the zero ring.

By a *filtered* ring we mean a ring B equipped with an increasing exhausting filtration $(B^r)_{r\geq 0}$ of subrings; so we have $B^r \subset B^{r+1}$ for all integers $r \geq 0$ and $B = \bigcup_{r=0}^{\infty} B^r$. Filtered rings are the objects of a category whose morphisms are ring homomorphisms $u : B \to \tilde{B}$ that preserve the filtrations, i.e., we have $u(B^r) \subset \tilde{B}^r$ for all integers $r \geq 0$.

By a *graded* ring we mean a ring B equipped with a direct sum decomposition $B = \oplus_{w\in W} B(w)$ where W is an abelian semigroup and $B(w)$ are subgroups such that we have $B(w)B(v) \subset B(w+v)$ for all $w, v \in W$. The elements of $B(w)$ are referred to having *degree* w or *weight* w. We also say that B is a W-graded ring. The W-graded rings are the objects of a category whose morphisms are ring homomorphisms $u : B \to \tilde{B}$ that preserve the gradings, i.e., we have $u(B(w)) \subset \tilde{B}(w)$ for all $w \in W$.

Definition 2.1.1 A *filtered graded ring* is a ring B equipped with a filtered ring structure, given by a filtration $(B^r)_{r\geq 0}$, and also equipped with a graded ring structure, given by $B = \oplus_{w\in W} B(w)$ with W as an abelian semigroup, the two structures being compatible in the sense that for all integers $r \geq 0$ we have a direct sum decomposition $B^r = \oplus_{w\in W} B^r \cap [B(w)]$. We also call B a *filtered W-graded ring* and we set $B^r(w) := B^r \cap [B(w)]$.

Filtered W-graded rings are the objects of a category whose morphisms are ring homomorphisms that preserve the filtrations and the gradings.

Assume one is given a sequence of W-graded rings $B^r = \oplus_{w\in W} B^r(w)$, $r \geq 0$, such that B^r is a subring of B^{r+1} and $B^r(w) \subset B^{r+1}(w)$ for all integers $r \geq 0$ and $w \in W$; let us refer to this data as a *filtered W-graded system* or just as a *filtered graded system*. To each filtered graded system as above we can attach a filtered

A. Buium and A. Vasiu, *Delta-Invariant Theory for Hecke Correspondences*,
Lecture Notes in Mathematics 2396, https://doi.org/10.1007/978-3-032-20571-1_2

graded ring as follows: define $B := \bigcup_{r\geq 0} B^r$ and $B(w) := \bigcup_{r\geq 0} B^r(w)$. Then it is easy to check that $B = \oplus_{w\in W} B(w)$ and $B^r(w) = B^r \cap B(w)$ for all integers $r \geq 0$ and $w \in W$, so B with the above grading and with filtration $(B^r)_{r\geq 0}$ is a filtered graded ring. All filtered graded rings can be obtained in this way.

Throughout the monograph p denotes a field characteristic: so it is either 0 or a prime natural number. Until Chap. 4 we assume that $p > 0$.

If S is a ring and L is an S-module, let $\widehat{L}$ be its p-adic completion. We recall that L is called p-adically complete if the functorial S-linear map $L \to \widehat{L}$ is an isomorphism. We write $\overline{S} := S/(p) = S/pS$ and we denote by $S^\times$ the group of units of S. If S is local, we denote by $\mathfrak{m}_S$ its maximal ideal; we have $S^\times = S \setminus \mathfrak{m}_S$.

If X is a scheme, we denote by $\mathcal{O}_X$ its structure ringed sheaf and we use the shorter notation $\overline{X} := X \times_{\mathrm{Spec}(\mathbb{Z})} \mathrm{Spec}(\mathbb{F}_p)$ and $\mathcal{O}(X) := H^0(X, \mathcal{O}_X)$. If in fact $X = \mathrm{Spec}(S)$ is affine, then $\overline{X} = \mathrm{Spec}(\overline{S})$ and for a scheme Z over X we let $Z(S)$ be the set of all X-valued points of Z.

Similarly, if Y is a p-adic formal scheme (i.e., a formal scheme with ideal of definition generated by p), we denote by $\mathcal{O}_Y$ its structure ringed sheaf and we use the shorter notation $\overline{Y} := Y \times_{\mathrm{Spf}(\mathbb{Z}_p)} \mathrm{Spf}(\mathbb{F}_p)$ and $\mathcal{O}(Y) := H^0(Y, \mathcal{O}_Y)$. If $Y = \mathrm{Spf}(S)$ is affine, then $\overline{Y} = \mathrm{Spf}(\overline{S}) = \mathrm{Spec}(\overline{S})$ and for a p-adic formal scheme Z over Y we let $Z(S)$ be the set of all Y-valued points of Z.

For a noetherian scheme X we denote by $\widehat{X}$ the p-adic formal scheme obtained by p-adically completing X. If X is affine, then $\overline{X} = \overline{\widehat{X}}$.

Let $\mathbf{Mat}_g$ be the ring $\mathbb{Z}$-scheme of $g \times g$ matrices. Let $1_g \in \mathbf{Mat}_g(S)$ be the identity matrix. Let $\mathbb{E}_g \in \mathbf{Mat}_g(S)$ be the matrix with all entries equal to 1. For each matrix $M \in \mathbf{Mat}_g(S)$ let M^{t} and M^* denote the transpose and the adjugate of M (respectively); so $MM^* = \det(M) \cdot 1_g$.

Let $\rho : S \to S'$ be a ring homomorphism. We say that ρ is *stalk-preserving* (or *identifies local rings*) if for every prime ideal P' in S', setting $P := \rho^{-1}(P')$, we have that the induced homomorphism between the localizations $S_P \to S'_{P'}$ is an isomorphism. Alternatively we say that the induced morphism of schemes $f : X' := \mathrm{Spec}(S') \to X := \mathrm{Spec}(S)$ is stalk-preserving. We say that ρ (resp. f) is a *local isomorphism* if every point in X' has an open neighborhood whose induced morphism to X is an open immersion. Every local isomorphism is stalk-preserving. Every stalk-preserving homomorphism is flat. Every homomorphism of the form $\rho : S \to \Sigma^{-1}S$ (where $\Sigma \subset S$ is a multiplicative system) is stalk-preserving; it is a local isomorphism in case Σ consists of the powers of one element. Also, if $S \to S'_i$ are stalk-preserving (resp. local isomorphisms) for $i \in \{1, \ldots, n\}$, then the induced homomorphism $S \to \prod_{i=1}^n S'_i$ is stalk-preserving (resp. a local isomorphism). The properties of being stalk-preserving or a local isomorphism are stable under composition and base change.

Definition 2.1.2 Let $\rho : S_1 \to S_2$ be a ring homomorphism between two $\mathbb{Z}_{(p)}$-algebras and let $\bar{\rho} : S_1/pS_1 \to S_2/pS_2$ be its reduction modulo p. By a *Frobenius lift relative to* ρ we mean a homomorphism $\phi : S_1 \to S_2$ whose reduction modulo p maps $x \in S_1/pS_1$ to $\bar{\rho}(x)^p \in S_2/pS_2$. By a *Frobenius lift* on a $\mathbb{Z}_{(p)}$-algebra S we understand a Frobenius lift $\phi : S \to S$ relative to the identity 1_S.

Throughout the monograph we let k, R, and K be as in the Introduction; so:

- k is an algebraic closure of $\mathbb{F}_p$;
- $R = W(k)$ is the ring of p-typical Witt vectors with coefficients in k;
- $K = R[\frac{1}{p}] = \mathrm{Frac}(R)$.

Recall that there exists a unique Frobenius lift ϕ on R.

We next review the concepts of p-derivations [Jo, Bu95a] and of p-jet spaces [Bu95a]. Consider the polynomial

$$C_p(T_1, T_2) := \frac{T_1^p + T_2^p - (T_1 + T_2)^p}{p} \in \mathbb{Z}[T_1, T_2].$$

Definition 2.1.3 If $\iota : S \to B$ is a ring homomorphism between two $\mathbb{Z}_{(p)}$-algebras,[1] then by a *p-derivation (relative to* ι) we mean a function $\delta : S \to B$ such that the following axioms hold.

(DAX1) For all $x, y \in S$ we have $\delta(x + y) = \delta x + \delta y + \iota(C_p(x, y))$.

(DAX2) For all $x, y \in S$ we have $\delta(xy) = \iota(x)^p \delta y + \iota(y)^p \delta x + p\delta x \delta y$.

(DAX3) We have $\delta(1) = 0$.

If B is p-adically separated, then axioms (DAX1) and (DAX2) imply (DAX3).

If ι is the identity automorphism of a $\mathbb{Q}$-algebra S, then $\delta : S \to S$ satisfies axioms (DAX1) and (DAX2) iff the rule $x \mapsto x^p + p\delta(x)$ defines a ring endomorphism of S which is not necessarily unital; in particular, if $\delta(x) = -\frac{x^p}{p}$ for each $x \in S$, then axioms (DAX1) and (DAX2) hold but $\delta(1) = -\frac{1}{p} \neq 0$ and, in fact, if moreover $\mathrm{Spec}(S)$ is connected (i.e., the only idempotents of S are 0 and 1), then this function $\delta : S \to S$ is the only one for which axioms (DAX1) and (DAX2) hold and axiom (DAX3) fails.

If δ is a p-derivation relative to ι, then axioms (DAX1) to (DAX3) trivially imply that the homomorphism $\phi : S \to B$ defined by the rule

$$\phi(x) := \iota(x)^p + p\delta(x)$$

(with $x \in S$) is a Frobenius lift relative to ι, to be called the *Frobenius lift attached to* δ. Conversely, given a Frobenius lift $\phi : S \to B$ relative to $\iota : S \to B$, if p is a nonzero divisor in B, then $\delta : S \to B$, defined by the rule $\delta(x) := \frac{\phi(x) - \iota(x)^p}{p}$, is a p-derivation relative to ι.

By a *p-derivation* on a $\mathbb{Z}_{(p)}$-algebra S we mean a p-derivation $\delta : S \to S$ relative to the identity 1_S. In particular the $\mathbb{Z}_{(p)}$-algebra $\mathbb{Z}_{(p)}$ has a unique p-derivation. Note also that the $\mathbb{Z}_{(p)}$-algebra R has a unique p-derivation $\delta : R \to R$.

Here are a few more immediate consequences of the axioms (DAX1) to (DAX3).

---

[1] In practice ι is injective.

Lemma 2.1.4 *For each non-zero $\mathbb{Z}_{(p)}$-algebra S equipped with a p-derivation, the unique ring homomorphism $h : \mathbb{Z}_{(p)} \to S$ is automatically injective and commutes with the p-derivations on S and $\mathbb{Z}_{(p)}$.*

Proof. We denote by δ_0 and δ the p-derivation of $\mathbb{Z}_{(p)}$ and S (respectively). We have $\delta(0) = 0$ by axiom (DAX1) applied with $x = y = 0$ and $\delta(1) = 0$ by axiom (DAX3). Using axiom (DAX1), by induction on $n \in \mathbb{N} \cup \{0\}$ we get that we have an identity $\delta(h(n)) = h(\delta_0(n)) \in S$. So if there exists $m \in \mathbb{N}$ such that $h(p)^m = 0$ we choose the smallest such m, and from $\delta(h(p)^m) = 0$ and $\delta_0(p^m) \in p^{m-1}\mathbb{Z}_{(p)}^{\times}$ it follows that $h(p)^{m-1} = 0$; hence the minimality assumption gives that $m = 1$, so S is the zero ring, which contradicts our hypotheses. So no such m exists and hence h is injective. As $h(\delta_0(n)) = \delta(h(n))$ for each $n \in \mathbb{N}$, it easily follows using axiom (DAX2) that $h(\delta_0(x)) = \delta(h(x))$ for each $x \in \mathbb{Z}_{(p)}$. □

We have the following compatibilities of p-derivations with the basic ring operations. Assume S is a $\mathbb{Z}_{(p)}$-algebra equipped with a p-derivation δ and let $\phi : S \to S$ be the Frobenius lift attached to δ. Then the p-adic completion $\widehat{S}$ has a unique p-derivation compatible with the p-derivation δ on S. If $\Sigma \subset S$ is a multiplicative system such that $\phi(\Sigma) \subset \Sigma$ then the ring of fractions $\Sigma^{-1}S$ has a unique p-derivation (which we still denote here by δ) compatible with the p-derivation δ on S. It is given by the 'quotient rule':

$$\delta\left(\frac{x}{y}\right) := \frac{y^p\delta x - x^p\delta y}{y^p\phi(y)} \quad \text{for } x \in S,\ y \in \Sigma.$$

In particular if $P \subset S$ is a prime ideal that contains p, then one checks directly that $\phi(S \setminus P) \subset S \setminus P$ and hence the localization S_P has a unique p-derivation compatible with the p-derivation δ on S. If $I \subset S$ is an ideal such that $\delta(I) \subset I$, then the factor ring S/I has a unique p-derivation compatible with the p-derivation on S.

We have the following compatibility of p-derivations with étale homomorphisms. If $S \to S'$ is an étale homomorphism of $\mathbb{Z}_{(p)}$-algebras and B is a p-adically complete S'-algebra via a homomorphism $\iota' : S' \to B$ of S-algebras, then every p-derivation $S \to B$ relative to ι extends uniquely to a p-derivation $B \to C$ relative to ι'. (See [Bu95a], Lem. 1.3 which holds without assuming finite presentation.)

Definition 2.1.5 **(a)** By a *δ-ring* we mean a $\mathbb{Z}_{(p)}$-algebra S equipped with a p-derivation $\delta : S \to S$.

(b) By a *filtered δ-ring* we mean a δ-ring S which is also a filtered ring with filtration $(S^r)_{r \geq 0}$ such that we have $\delta(S^r) \subset S^{r+1}$ for all integers $r \geq 0$.

Note that each δ-ring has characteristic 0 by Lemma 2.1.4 and that a filtered δ-ring in the above sense is not a filtered object in the category of δ-rings; rather it is an object of the category of δ-rings and also an object of the category of filtered rings and the two structures are subject to an appropriate compatibility condition.

The next example shows that δ-rings are not necessarily flat $\mathbb{Z}_{(p)}$-algebras.

Example 2.1.6 Let $S = \mathbb{Z}_{(p)}[x, x', x'', \ldots]$ be endowed with the unique p-derivation δ that maps x to x', x' to x'', etc. Let I be the ideal of S generated by $\delta^i(px)$ with $i \in \mathbb{N} \cup \{0\}$ (with the convention $\delta^0(px) = px$). Axioms (DAX1) and (DAX2) imply that $\delta(I) \subset I$. Hence S/I with the p-derivation induced by δ is a δ-ring in which $x + I$ is a nonzero element annihilated by p. In particular S/I is not a flat $\mathbb{Z}_{(p)}$-algebra.

A morphism of δ-rings is a ring homomorphism that commutes with the corresponding p-derivations. The δ-rings and their morphisms form a category $\mathbf{Ring}_\delta$ which has products, fiber coproducts and equalizers. If **Ring** is the category of rings over $\mathbb{Z}_{(p)}$, then the forgetful functor $\mathbf{Ring}_\delta \to \mathbf{Ring}$, which forgets the p-derivations, commutes with products, fiber coproducts and equalizers.

A morphism of filtered δ-rings $u : S \to \tilde{S}$ is a morphism of δ-rings such that $u(S^r) \subset \tilde{S}^r$ for all integers $r \geq 0$. Filtered δ-rings and their morphisms form a category $\mathbf{FilRing}_\delta$ that has products, fiber coproducts and equalizers.

The opposite category $\mathbf{Ring}_\delta^{\mathrm{op}}$ will be denoted by $\mathbf{Aff}_\delta$ and its objects will be referred to as *affine δ-spaces*. The object in $\mathbf{Aff}_\delta$ that corresponds to a δ-ring $B \in \mathrm{Ob}(\mathbf{Ring}_\delta)$ will be denoted by $\mathrm{Spec}_\delta(B) \in \mathrm{Ob}(\mathbf{Aff}_\delta)$. We often drop mentioning the p-derivations δ when referring to objects or morphisms of either $\mathbf{Ring}_\delta$ or $\mathbf{Aff}_\delta$.

Remark 2.1.7 Note that in the definition above the word 'affine' is not an adjective as we did not introduce a concept of *δ-space*. One can introduce such a concept as follows, cf. [Borg], p. 7. We let **Ring** be the category of rings over $\mathbb{Z}_{(p)}$ and we let **Aff** be the category of affine schemes over $\mathbb{Z}_{(p)}$; we have an equivalence of categories $\mathrm{Spec} : \mathbf{Ring}^{\mathrm{op}} \to \mathbf{Aff}$. We equip the category **Aff** with the étale topology and we denote by **Sp** the category of sheaves of sets on **Aff**. The infinite-length p-typical Witt vector functor $W^* : \mathbf{Sp} \to \mathbf{Sp}$ has a natural monad structure. Then one defines a *δ-space* to be an object of **Sp** equipped with a W^*-action. (For the details of this definition we refer to [Borg]; note however that in loc. cit. one works over $\mathbb{Z}$ instead of over $\mathbb{Z}_{(p)}$, one uses the big Witt vectors instead of the p-typical Witt vectors, and one uses the terminology of 'Λ-spaces' instead of 'δ-spaces'.) With morphisms of δ-spaces defined in the obvious way we obtain the *category* $\mathbf{Sp}_\delta$ *of δ-spaces*. Then $\mathbf{Aff}_\delta$ can be identified with the full subcategory of $\mathbf{Sp}_\delta$ whose objects are the objects of $\mathbf{Sp}_\delta$ with the property that their image in **Sp** is represented by an object of **Aff**. The category $\mathbf{Sp}_\delta$ can be chosen as a natural framework for δ-geometry as in [Borg]. However, in the present monograph, we found it is more helpful and expedient to present δ-geometry in a semi-simplicial (rather than sheaf theoretic) framework, see Definition 2.1.12 below. This framework is a variant of the one in [Bu05], Ch. 2. In particular, the category $\mathbf{Sp}_\delta$ will play no role in the present monograph.

For $d \in \mathbb{N}$ (so $d \geq 1$), a basic example of filtered δ-ring is given as follows. Let

$$S^r = R[y, y', \ldots, y^{(r)}] := R[y_i^{(l)} | 1 \leq i \leq d, 0 \leq l \leq r], \quad r \geq 0,$$

where

$$y = y^{(0)} = (y_1, \dots, y_d) = (y_1^{(0)}, \dots, y_d^{(0)}),$$
$$y' = y^{(1)} = (y_1', \dots, y_d') = (y_1^{(1)}, \dots, y_d^{(1)}),$$
$$\dots,\ y^{(r)} = (y_1^{(r)}, \dots, y_d^{(r)}),\ \dots$$

are d-tuples of indeterminates. We have inclusions $S^r \to S^{r+1}$ and, for each r, we let $\delta : S^r \to S^{r+1}$ be the unique p-derivation relative to the inclusion $S^r \subset S^{r+1}$ such that we have $\delta y = y', \dots, \delta y^{(r)} = y^{(r+1)}$ (i.e., for integers $1 \leq i \leq d$ and $0 \leq l \leq r$ we have $\delta y_i^{(l)} = y_i^{(l+1)}$). We define the R-algebra of *δ-polynomials*,

$$R[y, y', \dots, y^{(r)}, \dots] := \bigcup_{r=0}^{\infty} S^r; \tag{2.1}$$

this algebra has an induced structure of a filtered δ-ring with filtration $(S^r)_{r \geq 0}$.

Definition 2.1.8 Let $\widehat{\mathbf{FilRing}}_\delta$ be the category whose objects are the filtered δ-rings S with each S^r a p-adically complete, flat R-algebra and with the inclusions $S^r \subset S^{r+1}$ assumed to be R-algebra homomorphisms; the morphisms in the category are the morphisms of filtered δ-rings that induce R-algebra homomorphisms on the rings in the filtrations.

For $S \in \mathrm{Ob}(\widehat{\mathbf{FilRing}}_\delta)$, S is always flat over R but in general it is not p-adically complete (not even p-adically separated); more precisely, S is p-adically complete if and only if there exists a pair $(r_0, n_0) \in \mathbb{Z}_{\geq 0}^2$ such that $p^{n_0} S \subset S^{r_0}$ as one can easily check based on Baire's category theorem.[2] In particular, if S/pS is a finitely generated k-algebra or if S^{r+1}/S^r is torsion free for each integer $r \geq 0$, then S is p-adically complete if and only if there exists $r_0 \in \mathbb{Z}_{\geq 0}$ such that $S = S^{r_0}$. The category $\widehat{\mathbf{FilRing}}_\delta$ has an initial object R with filtration given by $R^r = R$ for all integers $r \geq 0$.

If $X = \mathrm{Spec}(B)$ is an affine smooth scheme over $\mathrm{Spec}(R)$, then we recall from [Bu95a], Sect. 1, that there exists a filtered δ-ring $J(B) \in \mathrm{Ob}(\widehat{\mathbf{FilRing}}_\delta)$, with filtration $(J^r(B))_{r \geq 0}$ starting with $J^0(B) = \widehat{B}$, that satisfies the following universal property: for each $S \in \mathrm{Ob}(\widehat{\mathbf{FilRing}}_\delta)$ and for every R-algebra homomorphism $B \to S^0$ there exists a unique morphism of filtered δ-rings $J(B) \to S$ over R that extends $B \to S^0$. The affine p-adic formal schemes $J^r(X) := \mathrm{Spf}(J^r(B))$ are p-adic completions of affine smooth schemes over $\mathrm{Spec}(R)$. This follows from the fact that if $y = (y_1, \dots, y_d) \subset B^d = \mathcal{O}(X)^d$ define étale coordinates on X (i.e., the R-algebra homomorphism from the ring of polynomials $R[x_1, \dots, x_n]$ into B that maps each indeterminate x_i to y_i is étale), then, by the compatibility of p-derivations with étale maps, we have a natural B-algebra isomorphism

$$\widehat{B[y_1', \dots, y_d', \dots, y_1^{(r)}, \dots, y_d^{(r)}]} \simeq J^r(B) = \mathcal{O}(J^r(X)) \tag{2.2}$$

[2] The following example due to Ofer Gabber shows that in general we cannot take $n_0 = 0$ for $r_0 \gg 1$. If for an integer $r \geq 0$, S^r is the inverse image in the ring $R\{x, y\}$ of restricted formal series in two indeterminates x and y of $k[x, xy, \dots, xy^r]$, then for $S = \cup_{r \geq 0} S^r$ we can take $(r_0, n_0) = (0, 1)$ but n_0 cannot be 0 no matter what r_0 is.

defined by $y_i^{(l)} \to \delta^l y_i$ for all integers $1 \le i \le d$ and $1 \le l \le r$ (see [Bu95a], Prop. 1.4). As explained in [Bu95a], Sect. 1, the construction can be globalized to attach formal schemes $J^r(X)$ to each separated (not necessarily affine) smooth scheme X over Spec(R): if $X = \cup_{i \in I} U_i$ is an affine open cover, then $J^r(X) = \cup J^r(U_i)$ is an open cover with $J^r(U_i)$ glued naturally so that for all $i, j \in I$ we have an identity $J^r(U_i) \cap J^r(U_j) = J^r(U_i \cap U_j)$.

Definition 2.1.9 The p-adic formal scheme $J^r(X)$ is called the r-th *p-jet space* of the separated smooth scheme X over Spec(R).

For a separated smooth scheme X over Spec(R) we may consider the filtered δ-ring $\mathcal{O}^\infty(X) := \bigcup_{r=0}^\infty \mathcal{O}(J^r(X))$ with filtration given by the rings in the union.

By the universal property mentioned above applied to $S = R$, for every R-valued point $P \in X(R)$, $P : \mathrm{Spec}(R) \to X$, there exists a unique sequence $(J^r(P))_{r \ge 0}$ of R-valued points $J^r(P) \in J^r(X)(R)$, $J^r(P) : \mathrm{Spf}(R) \to J^r(X)$, with $J^r(P)$ lifting $J^{r-1}(P)$ for $r \ge 1$, such that $J^0(P)$ is the p-adic completion of P and for every affine open subscheme $U \subset X$ with $P \in U(R)$ the induced sequence of R-algebra homomorphisms $\mathcal{O}(J^r(U)) \to R$ defines a morphism of δ-rings $\mathcal{O}^\infty(U) \to R$. The R-valued points $J^r(P)$ are covariantly functorial in X.

Each element $\varphi \in \mathcal{O}(J^r(X))$ induces a morphism $\varphi : J^r(X) \to \widehat{\mathbb{A}_R^1}$ and a map $\varphi : J^r(X)(R) \to \widehat{\mathbb{A}_R^1}(R) = R$ between sets of Spf(R)-valued points, both still denoted by ϕ. Then we have a well-defined map

$$\varphi_R : X(R) \to R, \quad \varphi_R(P) := \varphi(J^r(P)).$$

The rule $\varphi \mapsto \varphi_R$ defines an R-algebra homomorphism

$$\mathcal{O}(J^r(X)) \to \mathrm{Maps}(X(R), R) \tag{2.3}$$

which is injective (see [Bu05], Prop. 3.19) and thus viewed as an inclusion that allows us to write φ in place of φ_R. Here Maps$(X(R), R)$ denotes the R-algebra of set theoretic maps from $X(R)$ to R.

Definition 2.1.10 A map of sets $X(R) \to R$ is called a *δ-function* of order $\le r$ on $X(R)$ if it belongs to the image of the R-algebra homomorphism of Diagram (2.3).

We say that a δ-function of order $\le r$ has order r if either $r = 0$ or $r \ge 1$ and it does not have order $\le r - 1$. The δ-functions of order 0 are exactly the functions induced by formal functions on $\widehat{X} = J^0(X)$. For $r \ge 1$ the δ-functions of order r are not induced by formal functions on $\widehat{X}$; rather, they are given locally in the Zariski topology by restricted power series in the affine coordinates and their p-derivatives up to order r. If in these restricted power series one formally replaces the p-derivatives of the affine coordinates by polynomials with K-coefficients in the coordinates twisted by the powers of the Frobenius lift ϕ, then in general one gets divergent series (see [Bu05], Rmk. 2.6, for details); this shows that the use of p-derivatives cannot be directly replaced in the theory by that of Frobenius lifts.

One can ask for a geometry based on δ-rings that could be referred to as *δ-geometry*. For our practical purposes we would like a version of δ-geometry that requires a minimum amount of foundational material. Such a formalism was introduced and used in [Bu05]. We shall employ here a slightly different (although related) formalism which is, arguably, expanded and more flexible than the one in [Bu05]. Here are the relevant concepts.

First we recall (see [W], Sect. 8.1) that a semi-simplicial object in a category $\mathbf{C}$ is a collection of objects $X_\bullet = (X_d)_{d\geq 0}$ and morphisms, called *face maps*, $f_{d,0}, \ldots, f_{d,d+1} : X_{d+1} \to X_d$ that satisfy the *semi-simplicial face relations*

$$f_{d,i} \circ f_{d+1,s} = f_{d,s-1} \circ f_{d+1,i} \quad \text{for all integers that satisfy } 0 \leq i < s \leq d+2 \geq 2.$$

A morphism between semi-simplicial objects $(X_d)_{d\geq 0}$ and $(Y_d)_{d\geq 0}$ is a sequence of morphisms $(X_d \to Y_d)_{d\geq 0}$ that commute with the face maps. Assume we are given a functor Γ from the opposite category $\mathbf{C}^{\mathrm{op}}$ to the category $\mathbf{Ab}$ of abelian groups. If we set $B_d := \Gamma(X_d)$ and $u_{d,s} := \Gamma(f_{d,s}) : B_d \to B_{d+1}$ then for $d \geq 0$ we define, as usual, the d-th *cohomology group* of $X_\bullet$ (relative to Γ) by

$$H^d(X_\bullet) := \frac{\mathrm{Ker}(\partial_d)}{\mathrm{Im}(\partial_{d-1})}, \quad \partial_d := \sum_{i=0}^{d+1} (-1)^i u_{d,i} : B_d \to B_{d+1}, \quad \partial_{-1} := 0.$$

Example 2.1.11 Consider now the following example; its only role is to motivate our next Definition 2.1.12 and none of the concepts involved in it will play a role later. Let **Ring** be the category of rings over some ground ring S_0 and let **Aff** be the category of affine schemes over S_0; so we have an equivalence of categories $\mathrm{Spec} : \mathbf{Ring}^{\mathrm{op}} \to \mathbf{Aff}$. (The notation matches the previous uses of **Ring** and **Aff** if $S_0 = \mathbb{Z}_{(p)}$ but S_0 here can be arbitrary.) Consider the category **Spss** whose objects are the semi-simplicial objects in **Aff** and whose morphisms are the morphisms of semi-simplicial objects in **Aff**.[3] We next consider the category **Sch** of Noetherian separated schemes over S_0. Consider a scheme $X \in \mathrm{Ob}(\mathbf{Sch})$ and a *finite affine open cover* $\mathcal{U}$ of X, i.e.,

$$\mathcal{U} = (U_i)_{i\in\{0,\ldots,n\}} \text{ with } n \in \mathbb{N} \cup \{0\}, \quad X = U_0 \cup \cdots \cup U_n, \quad U_i \subset X \text{ affine opens.}$$

(We do not require the U_is to be distinct or non-empty.) One can attach to X and $\mathcal{U}$ an object $X_\bullet = (X_d)_{d\geq 0}$ of **Spss** where X_d are the affine Noetherian schemes

$$X_d := \coprod_{0\leq i_0 < \cdots < i_d \leq n} U_{i_0} \cap \cdots \cap U_{i_d} \quad \text{for } 0 \leq d \leq n+1$$

[3] By analogy with the terminology used in topology one could refer to the objects of this category as *semi-simplicial spaces*; hence the notation.

and $X_d := \emptyset$ for $d > n+1$, with the obvious face maps $f_{d,0}, \ldots, f_{d,d+1}$.[4] Note that the face maps are quasi-finite local isomorphisms. Note also that the natural morphism $X_0 \to X$ is a coequalizer in **Sch** of the pair of morphisms $(f_{d,0}, f_{d,1})$ from X_1 to X_0; so X is uniquely determined, up to isomorphism in **Sch**, by the isomorphism class of $X_\bullet$ in **Spss**.

The cohomologies of X and $X_\bullet$ are related as follows (see [H], Thm. 4.5, p. 222): if $\Gamma : \mathbf{Aff}^{\mathrm{op}} \to \mathbf{Ab}$ is the functor $U \mapsto \Gamma(U) := \mathcal{O}(U) = H^0(U, \mathcal{O}_U)$ and X and $\mathcal{U}$ are as above, then we have natural isomorphisms

$$H^d(X_\bullet) \simeq H^d(X, \mathcal{O}_X),$$

where $H^d(X_\bullet)$ are the cohomology groups of $X_\bullet$ (relative to Γ) and $H^d(X, \mathcal{O}_X)$ are the usual cohomology groups of the scheme X with coefficients in the sheaf $\mathcal{O}_X$.

Next we recall the definition of the (ordered) *nerve* of a cover $\mathcal{U}$ of X as above. It is the semi-simplicial set (i.e., semi-simplicial object in the category of sets) $N_\bullet(\mathcal{U}) = (N_d(\mathcal{U}))_{d\geq 0}$,

$$N_d(\mathcal{U}) := \{(i_0, \ldots, i_d) \mid 0 \leq i_0 < \ldots < i_d \leq n,\ U_{i_0} \cap \ldots \cap U_{i_d} \neq \emptyset\},$$

with the obvious face maps.

Consider the category **Schcov** whose objects are pairs $(X, \mathcal{U})$ where $X \in \mathrm{Ob}(\mathbf{Sch})$ and $\mathcal{U} = (U_i)$ is a finite affine open cover with all U_i connected; a morphism $(X, \mathcal{U}) \to (Y, \mathcal{V})$, where $\mathcal{U} = (U_i)_{i\in\{0,\ldots,n\}}$ and $\mathcal{V} = (V_j)_{j\in\{0,\ldots,m\}}$, is a pair $(\alpha_\bullet, f)$ consisting of a morphism of semi-simplicial sets $\alpha_\bullet : N_\bullet(\mathcal{U}) \to N_\bullet(\mathcal{V})$ and a morphism of schemes $f : X \to Y$ such that $f(U_i) \subset V_{\alpha_0(i)}$ for all $i \in N_0(\mathcal{U})$. We have the forgetful functor $\mathbf{Schcov} \to \mathbf{Sch}$, $(X, \mathcal{U}) \mapsto X$; it induces surjections $\mathrm{Ob}(\mathbf{Schcov}) \to \mathrm{Ob}(\mathbf{Sch})$ and $\mathrm{Mor}(\mathbf{Schcov}) \to \mathrm{Mor}(\mathbf{Sch})$. On the other hand we get a functor $\mathbf{Schcov} \to \mathbf{Spss}$, $(X, \mathcal{U}) \mapsto X_\bullet$, which is easily seen to be fully faithful; so **Schcov** is equivalent to a full subcategory of **Spss**. This suggests taking the semi-simplicial objects $X_\bullet$ (and their morphisms) as a substitute for the pairs $(X, \mathcal{U})$ (and their morphisms) for in so doing 'no information is being lost'.

We would like to take Example 2.1.11 as a blueprint for δ-geometry.

Definition 2.1.12 A *semi-simplicial δ-space* (or simply an *ss-δ-space*) is a semi-simplicial object in the category $\mathbf{Aff}_\delta = \mathbf{Ring}_\delta^{\mathrm{op}}$.

In concrete terms, an ss-δ-space is a sequence $\mathrm{Spec}_\delta(B_\bullet) = (\mathrm{Spec}_\delta(B_d))_{d\geq 0}$ where B_d are δ-rings equipped with δ-ring homomorphisms $u_{d,0}, \ldots, u_{d,d+1} : B_d \to B_{d+1}$ (called *coface maps*) that satisfy the *coface* relations

$$u_{d+1,s} \circ u_{d,i} = u_{d+1,i} \circ u_{d,s-1} \quad \text{for all integers that satisfy } 0 \leq i < s \leq d+2 \geq 2.$$

[4] For each $d \geq 0$, X_d is open and closed in $X'_d := X_0 \times_X \cdots \times_X X_0$ ($d+1$ factors); the family $(X'_d)_{d\geq 0}$, equipped with the natural face maps given by projections, is a semi-simplicial object and the inclusions $X_d \subset X'_d$ are compatible with the face maps. In fact $(X'_d)_{d\geq 0}$ has the richer structure of a *simplicial* object (see [W], Sect. 8.1) which will be ignored here.

Morphisms of ss-δ-spaces are defined as morphisms of semi-simplicial objects in $\mathbf{Aff}_\delta$. In this way we get the *category of ss-δ-spaces* denoted by $\mathbf{Spss}_\delta$. Its study is what we refer to as *δ-geometry*. The category $\mathbf{Spss}_\delta$ has coproducts, fiber products and coequalizers: the coequalizer of a pair of morphisms $\mathrm{Spec}_\delta(B'_\bullet) \rightrightarrows \mathrm{Spec}_\delta(B_\bullet)$ in $\mathbf{Spss}_\delta$ is the ss-δ-space $\mathrm{Spec}_\delta(C_\bullet)$ where for each $d \geq 0$, C_d is the equalizer in $\mathbf{Ring}_\delta$ of the corresponding pair of morphisms $B_d \rightrightarrows B'_d$.

Each δ-ring S defines an ss-δ-space $\mathrm{Spec}_\delta(S_\bullet) = (\mathrm{Spec}_\delta(S_d))_{d\geq 0}$ where $S_d = S$ for all integers $d \geq 0$ and where all face maps are the identity automorphism of $\mathrm{Spec}_\delta(S)$. The functor $\mathbf{Aff}_\delta \to \mathbf{Spss}_\delta$ defined by $\mathrm{Spec}_\delta(S) \mapsto \mathrm{Spec}_\delta(S_\bullet)$ induces an equivalence between $\mathbf{Aff}_\delta$ and a full subcategory of $\mathbf{Spss}_\delta$.

We will usually denote ss-δ-spaces by symbols such as X_δ, Y_δ, etc.

By an ss-δ-space *over* S we mean an ss-δ-space X_δ equipped with a morphism $X_\delta \to \mathrm{Spec}_\delta(S_\bullet)$. We are especially interested in the category of ss-δ-spaces over R whose final object is $\mathrm{Spec}_\delta(R_\bullet)$.

Taking the functor $\Gamma : \mathbf{Aff}_\delta^{\mathrm{op}} = \mathbf{Ring}_\delta \to \mathbf{Ab}$ to be the forgetful functor, if X_δ is a ss-δ-space defined by δ-ring homomorphisms $u_{d,i} : B_d \to B_{d+1}$ as above then we have the d-th cohomology group (relative to Γ), $H^d(X_\delta) := \mathrm{Ker}(\partial_d)/\mathrm{Im}(\partial_{d-1})$, where $\partial_d := \sum_{i=0}^{d+1}(-1)^i u_{d,i} : B_d \to B_{d+1}$ and $\partial_{-1} := 0$. The group $H^0(X_\delta)$ has a structure of a δ-ring which we call the *ring of global functions* on X_δ.

We recall that, if M and N are R-modules, an additive map $f : M \to N$ is called *semilinear* if it satisfies $f(\lambda x) = \phi(\lambda) f(x)$ for each pair $(\lambda, x) \in R \times M$.

If X_δ is an ss-δ-space over R, then its cohomology groups are R-modules equipped with semilinear endomorphisms induced by the Frobenius lift $\phi : R \to R$ and denoted also by ϕ. If $X_\delta \to Y_\delta$ is a morphism of ss-δ-spaces over R, then for each integer $d \geq 0$ we have a naturally induced R-linear map $H^d(Y_\delta) \to H^d(X_\delta)$ compatible with the semilinear endomorphisms ϕ. The map $H^0(Y_\delta) \to H^0(X_\delta)$ is a morphism of δ-rings over R.

If A is a δ-ring we denote by $D := D(A) \subset A$ the multiplicative set of all $a \in A$ such that the image of a in A/pA is a nonzero divisor; then $\phi(D) \subset D$ so the ring $D^{-1}A$ has a unique structure of a δ-ring compatible with that of A. The latter δ-ring is a natural substitute, in our framework, for the total ring of fractions. Note that if B is the zero ring then $D(B) = B$.

Definition 2.1.13 **(a)** An ss-δ-space X_δ is *connected* if the ring $H^0(X_\delta)$ has no nontrivial idempotents.

(b) An ss-δ-space X_δ is *acyclic* if $H^d(X_\delta) = 0$ for all integers $d \geq 1$.

(c) A morphism of ss-δ-spaces over R, $X_\delta \to Y_\delta$, is a *quasi-isomorphism* if the homomorphisms $H^d(Y_\delta) \to H^d(X_\delta)$ are isomorphisms for all integers $d \geq 0$.

(d) An ss-δ-space $\mathrm{Spec}_\delta(B_\bullet)$ over R is *discrete* if for every integer $d \geq 0$ the δ-ring B_d is a finite power of R, i.e., $B_d = R^{\Sigma_d}$, for some finite set Σ_d, with the convention that $R^\emptyset := 0$. (Then $(\Sigma_d)_{d\geq 0}$ has a natural structure of a semi-simplicial set and, conversely, each semi-simplicial set (Σ_d) with Σ_d finite gives rise to a discrete ss-δ-space as above.)

(e) A *δ-point* is an ss-δ-space over R that is connected, discrete, and acyclic.

(f) Given an ss-δ-space X_δ over R, a *δ-point of X_δ* is a morphism over R of the form $P_\delta \to X_\delta$ where P_δ is a δ-point.

(g) For an ss-δ-space $X_\delta = \mathrm{Spec}_\delta(B_\bullet)$ let $D_d := D(B_d) \subset B_d$ for $d \geq 0$ and consider the following condition: for all integers $d \geq 0$ and all $i \in \{0, \dots, d+1\}$ we have $u_{d,i}(D_d) \subset D_{d+1}$. If this condition is satisfied we obtain a naturally associated ss-δ-space $D^{-1}X_\delta := \mathrm{Spec}_\delta(D_\bullet^{-1}B_\bullet)$. The δ-ring $H^0(D^{-1}X_\delta)$ is called the *δ-ring of rational functions*.

(h) An ss-δ-space $X_\delta = \mathrm{Spec}_\delta(B_\bullet)$ is stalk-preserving if for each integer $d \geq 0$ and every $i \in \{0, \dots, d+1\}$, the coface map $u_{d,i} : B_d \to B_{d+1}$ is stalk-preserving.

(i) An ss-δ-space $X_\delta = \mathrm{Spec}_\delta(B_\bullet)$ is quasi-finite if for each integer $d \geq 0$ and every $i \in \{0, \dots, d+1\}$, the coface map $u_{d,i} : B_d \to B_{d+1}$ is defined by a quasi-finite morphism $\mathrm{Spec}(B_{d+1}) \to \mathrm{Spec}(B_d)$.

For each δ-ring S, the ss-δ-space $\mathrm{Spec}_\delta(S_\bullet)$ is acyclic, stalk-preserving, quasi-finite, and moreover we have $H^0(\mathrm{Spec}_\delta(S_\bullet)) = S$. In particular, the ss-$\delta$-space $\mathrm{Spec}_\delta(R_\bullet)$ is a δ-point. Other δ-points will play a role later as well, see Display (2.39). Every δ-point is quasi-finite. For each δ-point P_δ, the canonical morphism $P_\delta \to \mathrm{Spec}_\delta(R_\bullet)$ is a quasi-isomorphism.

We denote by $X_\delta(\delta\text{-pts})$ the set of all δ-points of X_δ. For each morphism $X_\delta \to Y_\delta$ of ss-δ-spaces over R we have an induced map of sets $X_\delta(\delta\text{-pts}) \to Y_\delta(\delta\text{-pts})$.

One can enhance ss-δ-spaces by considering filtrations as follows.

Definition 2.1.14 A *filtered ss-δ-space* is a semi-simplicial object in $\mathbf{FilRing}_\delta^{\mathrm{op}}$.

Filtered ss-δ-spaces form a category, $\mathbf{FilSpss}_\delta$, in the obvious way. This category has coproducts, fiber products, and coequalizers and there exists a forgetful functor $\mathbf{FilSpss}_\delta \to \mathbf{Spss}_\delta$ which forgets the filtrations; so we shall view filtered ss-δ-spaces as ss-δ-spaces with 'additional structure' given by filtrations. The forgetful functor above commutes with coproducts, fiber products, and coequalizers. The ss-δ-space $\mathrm{Spec}(R_\bullet)$ has a natural structure of a filtered ss-δ-space with the constant filtration equal to R in each degree; so we speak about filtered ss-δ-spaces *over R*. If X_δ is a filtered ss-δ-space, then all cohomology groups $H^d(X_\delta)$ inherit an exhaustive increasing filtration by subgroups $H^d(X_\delta)^r$ which, if X_δ is over R, are R-submodules equipped with semilinear homomorphisms $H^d(X_\delta)^r \to H^d(X_\delta)^{r+1}$. If in addition we have $X_\delta = \mathrm{Spec}_\delta(B_\bullet)$, the condition in Definition 2.1.13 (g) is satisfied, and the ring homomorphism $B^r/pB^r \to B/pB$ is injective for each $r \in \mathbb{Z}_{\geq 0}$, then $D^{-1}X_\delta$ has a structure of a filtered ss-δ-space given by the filtration $((D_d^{-1}B_d)^r)_{r\geq 0} = ((D_d^r)^{-1}B_d^r)_{r\geq 0}$ of each $D_d^{-1}B_d$, where $D_d^r := B_d^r \cap D_d$.

A 'first' example of (filtered) ss-δ-space over R is attached to a quasi-compact separated smooth scheme X over $\mathrm{Spec}(R)$ as follows. Given $n \in \mathbb{Z}_{\geq 0}$ and an affine open cover $X = U_0 \cup \cdots \cup U_n$, we consider the filtered ss-δ-space

$$\mathrm{Spec}_\delta\left(\prod_{0\leq i_0<\cdots<i_\bullet\leq n} \mathcal{O}^\infty(U_{i_0} \cap \cdots \cap U_{i_\bullet})\right)$$

equipped with the natural face maps. Here and in what follows, the products of the type above are taken to be equal to the zero ring for $\bullet > n$. However, this is *not* the most natural example appearing in our theory. Rather, the examples that naturally occur come from certain (filtered) *δ-graded rings* and 'Proj$_\delta$' spaces attached to them. To explain this construction we need to introduce a few more concepts, partly following [Bu05], Defs. 2.18 and 2.21.

Let

$$W := \mathbb{Z}[\phi]$$

be the commutative ring of polynomials in an indeterminate ϕ with $\mathbb{Z}$-coefficients. If $w = \sum_{i=0}^{n} a_i \phi^i$ with $n \in \mathbb{Z}_{\geq 0}$ is a weight, $a_i \in \mathbb{Z}$ and $a_n \neq 0$, then we set $\deg(w) := \sum_{i=0}^{n} a_i$ and $\mathrm{ord}(w) := n$. We also set $\deg(0) := 0$ and $\mathrm{ord}(0) := 0$. We endow W with a partial order as follows: we have $w = \sum_{i=0}^{n} a_i \phi^i \geq 0$ if and only if $a_i \geq 0$ for all $i \in \{0, \ldots, n\}$. We set

$$W_+ := \{w \in W | w \geq 0\} \quad \text{and} \quad W(r) := \{w \in W | \mathrm{ord}(w) \leq r\} \text{ for } r \in \mathbb{Z}_{\geq 0}.$$

For a ring endomorphism $\phi : B \to B$ and $b \in B$ we often use the simpler notation $b^\phi := \phi(b)$.

Definition 2.1.15 **(a)** By a *δ-graded ring* we mean a W-graded $\mathbb{Z}_{(p)}$-algebra $B = \bigoplus_{w \in W} B(w)$ equipped with a ring endomorphism $\phi : B \to B$ such that the following three axioms hold.

(GAX1) The ring B is p-adically separated and flat over $\mathbb{Z}_{(p)}$.

(GAX2) For all $w \in W$ we have $\phi(B(w)) \subset B(\phi w)$.

(GAX3) For all $b_1, b_2 \in B(w)$ we have $b_1^p b_2^\phi - b_2^p b_1^\phi \in pB((\phi + p)w)$.

(b) A δ-graded ring as above is *integral* if in addition it satisfies the following two conditions:

(GAX4) The ring B/pB is an integral domain.

(GAX5) Both ϕ and its reduction modulo p are injective.

In the definition above the W-grading of B is with respect to the additive structure of W, i.e., $B(w_1)B(w_2) \subset B(w_1 + w_2)$ for all $w_1, w_2 \in W$.

Note that, by axioms (GAX1) and (GAX4), each integral δ-graded ring is an integral domain.

Some of the main rings in our theory have a structure of δ-graded rings; see the rings $\mathbb{I}, \mathbb{M}, \mathbb{M}'$ introduced after Definition 2.3.5 below, the rings $\mathbb{S}_n$ introduced in Definition 2.3.18 below, and the rings $\mathbb{S}_{X,L}$ introduced in Definition 2.3.19 below. A simpler example is as follows.

Example 2.1.16 Let A be a p-adically separated δ-ring such that both A and A/pA are integral domains. We denote by ϕ the Frobenius lift on A attached to the p-derivation on A. Let $x, x^\phi, x^{\phi^2}, \ldots$ be a sequence of indeterminates and consider the ring of Laurent polynomials

$$A[x^W] := A[x^{\pm 1}, x^{\pm\phi}, x^{\pm\phi^2}, \ldots].$$

So the elements of $A[x^W]$ are formal sums $\Sigma_{w \in W} a_w x^w$ with $a_w \in A$. Multiplication is given by $(ax^w)(bx^v) := abx^{w+v}$ for $a, b \in A$ and $w, v \in W$. Then $B := A[x^W]$ has a natural structure of integral δ-graded ring with homogeneous components $B(w) = Ax^w$ for $w \in W$ and with ring endomorphism $\phi : B \to B$ defined by $\phi(ax^w) := \phi(a)x^{\phi w}$ for $a \in A$ and $w \in W$.

The δ-graded rings are the objects of a category $\mathbf{GrRing}_\delta$ whose morphisms are the morphisms of graded rings that commute with the corresponding ϕs.

Our definition of integral δ-graded rings is closely related to (but is not identical to) the definition of δ-graded rings in [Bu05], Def. 2.18: the latter considers nonunital rings and uses W_+ instead of W.

With the simpler notation above, we write

$$\{b_1, b_2\}_\delta := \frac{b_1^p b_2^\phi - b_2^p b_1^\phi}{p} \in B((\phi + p)w),$$

where division by p is well-defined by axioms (GAX1) and (GAX3).

Note that if B is an integral δ-graded ring then ϕ is a Frobenius lift on B if and only if $B = B(0)$. Thus, in general a δ-graded ring does not have a natural structure of a δ-ring. However, if B is an integral δ-graded ring, then for each $w \in W$ and every $f \in B(w) \setminus pB(w)$, the ring of fractions

$$B_{(f)} := \left\{ \frac{F}{f^v} | F \in B(vw),\ v \in W_+ \right\}$$

has a natural structure of a δ-ring, where its p-derivation δ is defined by the formula

$$\delta\left(\frac{F}{f^v}\right) := \frac{\{f^v, F\}_\delta}{f^{v(\phi+p)}},$$

as one can check easily that the rule $x \mapsto x^p + p\delta(x)$ defines the ring endomorphism of $B_{(f)}$ induced by ϕ. Similarly, the ring of fractions

$$B_{((p))} := \left\{ \frac{F}{G} | F \in B(w),\ G \in B(w) \setminus pB(w),\ w \in W \right\},$$

has a naturally induced structure of a δ-ring; it is a discrete valuation ring with maximal ideal generated by p.

We say that B is a δ-graded ring *over* R if R is a subring of $B(0)$ and ϕ extends the Frobenius lift on R.

Definition 2.1.17 Let B be an integral δ-graded ring over R and let $n \in \mathbb{N}$ and $w \in W$. By a *linear system* in B (of length $n + 1 \geq 2$ and weight $w \in W$) we mean an $n + 1$-tuple

$$\mathbf{f} = (f_0, \ldots, f_n) \in (B(w) \setminus pB(w))^{n+1}.$$

For each such pair $(B, \mathbf{f})$ we define the connected ss-δ-space

$$\mathrm{Proj}_\delta(B, \mathbf{f}) := \mathrm{Spec}_\delta \left(\prod_{0 \le i_0 < \cdots < i_\bullet \le n} B_{\langle f_{i_0} \cdots f_{i_\bullet} \rangle} \right) \tag{2.4}$$

over R with the obvious inclusions as face maps.

The ss-δ-space $X_\delta = \mathrm{Proj}_\delta(B, \mathbf{f})$ clearly satisfies

$$H^0(X_\delta) = \bigcap_{i=0}^{n} B_{\langle f_i \rangle},$$

the intersection being taken inside $B_{((p))}$. Also X_δ above is easily seen to satisfy the condition in Definition 2.1.13 (g) so we can consider the ss-δ-space $D^{-1}X_\delta$ and, in particular, the δ-ring of rational functions $H^0(D^{-1}X_\delta)$ on X_δ. In addition, X_δ is stalk-preserving and quasi-finite because for every index $d \ge 0$, every indices $0 \le i_0 < \ldots < i_{d+1} \le d+1$, and every $j \in \{0, \ldots, d+1\}$, we have an identity $B_{\langle f_{i_0} \cdots f_{i_{d+1}} \rangle} = \Sigma_j^{-1} B_{\langle f_{i_0} \cdots \widehat{f_{i_j}} \cdots f_{i_{d+1}} \rangle}$ where Σ_j is the multiplicative system consisting of the W_+-powers of the fraction

$$\frac{f_{i_j}^{d+1}}{f_{i_0} \cdots \widehat{f_{i_j}} \cdots f_{i_{d+1}}}.$$

Here $\widehat{\ }$ indicates deletion. Note that for all $d \ge 0$ and $i \in \{0, \ldots, d\}$ the homomorphisms $\overline{u_{d,i}} : \overline{B_d} \to \overline{B_{d+1}}$ induced by the coface maps are local isomorphisms.

If a morphism of integral δ-graded rings $u : B \to \tilde{B}$ sends some linear system $\mathbf{f} = (f_0, \ldots, f_n)$ in B into a linear system $\tilde{\mathbf{f}} := (u(f_0), \ldots, u(f_n))$ in $\tilde{B}$ (of the same length and weight), then we call $\tilde{\mathbf{f}}$ the image of $\mathbf{f}$ in $\tilde{B}$ and we have an induced morphism $\mathrm{Proj}_\delta(\tilde{B}, \tilde{\mathbf{f}}) \to \mathrm{Proj}_\delta(B, \mathbf{f})$ in $\mathbf{Spss}_\delta$.

In algebraic geometry the linear systems are vector spaces generated by global sections of a line bundle but our linear systems are tuples and not R-modules, as even the 0-th cohomology groups of the ss-δ-spaces $\mathrm{Proj}_\delta(B, \mathbf{f})$ really depend on $\mathbf{f}$ and not only on the R-submodule $\sum_{i=0}^n Rf_i$ of $B(w)$.

For applications we will need the 'filtered version' of the above concepts.

Definition 2.1.18 A *filtered δ-graded ring* is a δ-graded ring B as in Definition 2.1.15 (a) which is also equipped with a structure of a filtered graded ring, given by a filtration $(B^r)_{r \ge 0}$, for which the following two axioms hold.

(GAX6) For all integers $r \ge 0$ we have $\phi(B^r) \subset B^{r+1}$.

(GAX7) For all integers $r \ge 0$ and all $w \in W$ we have

$$\{B^r(w), B^r(w)\}_\delta := \{\{b_1, b_2\}_\delta | b_1, b_2 \in B^r(w)\} \subset B^{r+1}((\phi + p)w).$$

Note that if (GAX6) holds and $B^r \cap pB^{r+1} = pB^r$ for all integers $r \geq 0$, then (GAX7) follows from (GAX3). Filtered δ-graded rings are the objects of a category $\mathbf{FilGrRing}_\delta$ whose morphisms are the morphisms of δ-graded rings that are also morphisms of filtered rings.

We similarly speak about filtered δ-graded rings over R and about filtered integral δ-graded ring.

If B is a filtered integral δ-graded ring over R, then for each $w \in W$ and every $f \in B(w) \setminus pB(w)$, the δ-ring $B_{\langle f \rangle}$ has a structure of a filtered δ-ring with filtration $((B_{\langle f \rangle})^r)_{r \geq 0}$ given by

$$(B_{\langle f \rangle})^r := \left\{ \frac{F}{f^v} | F \in B^r(vw), v \in W_+ \right\}.$$

Similarly, the δ-ring $B_{((p))}$ has a structure of a filtered δ-ring with the filtration $((B_{((p))})^r)_{r \geq 0}$ given by

$$(B_{((p))})^r := \left\{ \frac{F}{G} | F \in B^r(w), G \in B^r(w) \setminus pB^r(w), w \in W \right\}.$$

For each linear system $\mathbf{f}$ in a filtered integral δ-graded ring B over R, the ss-δ-space $\mathrm{Proj}_\delta(B, \mathbf{f})$ has an induced structure of a filtered ss-δ-space. If a morphism $B \to \tilde{B}$ in $\mathbf{FilGrRing}_\delta$, where B and $\tilde{B}$ are filtered integral δ-graded rings, maps a linear system $\mathbf{f} = (f_0, \ldots, f_n)$ in B into a linear system $\tilde{\mathbf{f}} := (u(f_0), \ldots, u(f_n))$ in $\tilde{B}$, and if $\lambda \in R^\times$, then we have an induced morphism $\mathrm{Proj}_\delta(\tilde{B}, \lambda\tilde{\mathbf{f}}) \to \mathrm{Proj}_\delta(B, \mathbf{f})$ in $\mathbf{FilSpss}_\delta$ where $\lambda\tilde{\mathbf{f}} := (\lambda u(f_0), \ldots, \lambda u(f_n))$.

2.2 Assymptotic Transcendence Degree and Dimension

Throughout this section we use the following convention: a family $(x_i)_{i \in \mathcal{I}}$ of elements of a ring is called *algebraically independent* over a subring if the members of the family are distinct (i.e., $x_i \neq x_j$ for $i, j \in \mathcal{I}, i \neq j$) and the set $\{x_i | i \in \mathcal{I}\}$ is algebraically independent over the subring. We make a similar convention for linearly independent families.

If $B_1 \subset B_2$ is an inclusion of integral domains, let

$$\mathrm{tr.deg}_{B_1}(B_2) \in \mathbb{N} \cup \{0, \infty\}$$

be the supremum of all $d \in \mathbb{N} \cup \{0\}$ such that B_2 (or $\mathrm{Frac}(B_2)$) contains d algebraically independent elements over B_1; here we identify the elements of $\mathbb{N} \cup \{0\}$ with cardinals and we use ∞ to denote an arbitrary infinite cardinal. We have $\mathrm{tr.deg}_{B_1}(B_2) = \mathrm{tr.deg}_{\mathrm{Frac}(B_1)}(\mathrm{Frac}(B_2))$.

In this section we take B_0 to be an integral domain thought of as the 'ground ring'. (In the remaining part of this chapter and in Chap. 3, B_0 will be $R = W(k)$, while in Chap. 4, B_0 will be an algebraically closed field.) All concepts will be relative to B_0 and we will often drop the reference to B_0 if no confusion can arise. Throughout this section B is an integral domain that contains B_0.

Definition 2.2.1 Let B be a filtered ring such that its filtration $(B^r)_{r\geq 0}$ satisfies $B_0 \subset B^0$. We say that B has *asymptotic transcendence degree* $\alpha \in \mathbb{R}$, and we write

$$\text{asy.tr.deg}(B) = \alpha,$$

if $\text{tr.deg}_{B_0}(B^r) < \infty$ for all integers $r \geq 0$ and

$$\lim_{r\to\infty} \frac{\text{tr.deg}_{B_0}(B^r)}{r} = \alpha.$$

We say that B has *asymptotic dimension* $\beta \in \mathbb{R}$, and we write

$$\text{asy.dim}(B) = \beta,$$

if $\dim(B^r \otimes_{B_0} \text{Frac}(B_0)) < \infty$ for all integers $r \geq 0$ and

$$\lim_{r\to\infty} \frac{\dim(B^r \otimes_{B_0} \text{Frac}(B_0))}{r} = \beta.$$

Here dim denotes Krull dimension. If all $\text{Frac}(B_0)$-algebras $B^r \otimes_{B_0} \text{Frac}(B_0)$ are finitely generated, then asy.tr.deg(B) exists if and only asy.dim(B) exists and, in case they exist, they are equal.

A typical situation when asy.dim(B) exists but asy.tr.deg(B) does not arises as follows. Let X be an affine smooth scheme over Spec(R) of relative dimension d and we take

$$B_0 = R, \quad B^r = \mathcal{O}(J^r(X)), \quad B = \mathcal{O}(J^\infty(X)) = \bigcup_{r\geq 0} B^r.$$

Then by Diagram (2.2) we have $\dim(B^r \otimes_R K) = (r+1)d$, so

$$\text{asy.dim}(\mathcal{O}(J^\infty(X))) = d, \tag{2.5}$$

but asy.tr.deg$(\mathcal{O}(J^\infty(X)))$ does not exist if $d \geq 1$ because the transcendence degrees involved in its definition are infinite.

We recall another concept of dimension which comes from difference algebra (see [Le]) and which will play a role in our monograph.

Assume we are given an injective ring endomorphism $\phi : B \to B$ that sends B_0 onto B_0. So ϕ induces a ring endomorphism of Frac(B) that sends Frac(B_0) onto itself. (In Chaps. 2 and 3, where $B_0 = R$, the restriction of ϕ to B_0 will be the unique Frobenius lift on R but ϕ on B will not necessarily be a Frobenius lift. In Sects. 4.2–4.4, where B_0 is an algebraically closed field $\mathbb{K}$, the endomorphism ϕ will be denoted by σ and its restriction to $\mathbb{K}$ will be the identity automorphism $1_{\mathbb{K}}$.) For

$$(w, b) = (\sum_{i=0}^{n} a_i \phi^i, b) \in (W \times B^\times) \cup (W_+ \times B),$$

we extend the simpler notation $b^\phi = \phi(b)$ by setting

$$b^w := b^{a_0}\phi(b)^{a_1} \cdots \phi^n(b)^{a_n} \in B.$$

Definition 2.2.2 Let $\phi : B \to B$ be an endomorphism of an integral domain B that maps a unital subring B_0 of B into itself. Let C be a B_0-subalgebra of B such that $\phi(C) \subset C$ (i.e., a $B_0[\phi]$-subalgebra of B).

(a) A subset $\bowtie$ of $\mathrm{Frac}(B)$ is called *ϕ-algebraically independent* over $\mathrm{Frac}(C)$ if the family $(\phi^j(b))_{(b,j)\in\bowtie\times(\mathbb{N}\cup\{0\})}$ of elements of $\mathrm{Frac}(B)$ is algebraically independent over $\mathrm{Frac}(C)$.

(b) A subset $\mathrm{Frac}(B)$ is called a *ϕ-transcendence basis* of $\mathrm{Frac}(B)$ over $\mathrm{Frac}(C)$ if it is maximal in the set of ϕ-algebraically independent subsets of $\mathrm{Frac}(B)$. As such, the *ϕ-transcendence degree* of B over C is the supremum $\phi\text{-tr.deg}_C(B) \in \mathbb{N} \cup \{0, \infty\}$ of all numbers $d \in \mathbb{N} \cup \{0\}$ such that $\mathrm{Frac}(B)$ has a subset with d elements which is ϕ-algebraically independent over $\mathrm{Frac}(C)$.

(c) A subset $\{b_1, \ldots, b_n\} \subset \mathrm{Frac}(B)$ is said to ϕ-generate the field $\mathrm{Frac}(B)$ over $\mathrm{Frac}(B_0)$ if the set $\{\phi^j(b_i) | (i, j) \in \{1, \ldots, n\} \times (\mathbb{N} \cup \{0\})\}$ generates the field $\mathrm{Frac}(B)$ over $\mathrm{Frac}(B_0)$.

(d) We say $\mathrm{Frac}(B)$ is ϕ-finitely generated over $\mathrm{Frac}(B_0)$ if there exist a finite subset of B that ϕ-generates $\mathrm{Frac}(B)$ over $\mathrm{Frac}(B_0)$.

(e) A subset $\{b_1, \ldots, b_n\} \subset B$ is said to *ϕ-generate* B as a B_0-algebra if the set $\{\phi^j(b_i) | (i, j) \in \{1, \ldots, n\} \times (\mathbb{N} \cup \{0\})\}$ generates B as a B_0-algebra. We also say that B is ϕ-generated by $\{b_1, \ldots, b_n\}$ over B_0.

(f) We say B is ϕ-finitely generated over B_C if there exist a finite subset of B that ϕ-generates B as a B_0-algebra.

We use the simpler notation $\phi\text{-tr.deg}(B) := \phi\text{-tr.deg}_{B_0}(B)$.

We end this section with remarks on interrelations between the various dimension concepts introduced above.

Remark 2.2.3 The following properties hold.

(a) A ϕ-transcendence basis of $\mathrm{Frac}(B)$ over $\mathrm{Frac}(C)$ always exists and every two such ϕ-transcendence bases have the same cardinality (see [Le], Ch. 4, Prop. 4.1.6).

(b) We have the additivity property (see [Le], Ch. 4, Thm. 4.1.9):

$$\phi\text{-tr.deg}(B) = \phi\text{-tr.deg}(C) + \phi\text{-tr.deg}_C(B). \tag{2.6}$$

(c) If $\mathrm{Frac}(B)$ is ϕ-finitely generated over $\mathrm{Frac}(B_0)$, then $\mathrm{Frac}(C)$ is also ϕ-finitely generated over $\mathrm{Frac}(B_0)$ (see [Le], Ch. 4, Thm. 4.4.1).

(d) Each subset of $\mathrm{Frac}(B)$ that ϕ-generates $\mathrm{Frac}(B)$ over $\mathrm{Frac}(C)$ contains a ϕ-transcendence basis (see [Le], Ch. 4, Thm. 4.1.8).

(e) Assume that $\{\beta_1, \dots, \beta_n\} \subset \mathrm{Frac}(C)$ is ϕ-algebraically independent over $\mathrm{Frac}(B_0)$ and that $\{\gamma_1, \dots, \gamma_m\} \subset \mathrm{Frac}(C)$ is a set which ϕ-generates $\mathrm{Frac}(C)$ over $\mathrm{Frac}(B_0)$. Then $m \geq n$ and the set $\{\gamma_1, \dots, \gamma_m\}$ contains a subset with n elements which is ϕ-algebraically independent over $\mathrm{Frac}(B_0)$. To check this statement, we first remark that $\phi\text{-tr.deg}(C) \geq n$. By the property (d) applied to the fields $\mathrm{Frac}(B_0)$ and $\mathrm{Frac}(C)$ we have that $\{\gamma_1, \dots, \gamma_m\}$ contains a ϕ-transcendence basis of $\mathrm{Frac}(C)$ over $\mathrm{Frac}(B_0)$ which, by the property (a), has $\phi\text{-tr.deg}(C)$ elements and the statement follows.

(f) If $\phi\text{-tr.deg}(B) = n \in \mathbb{N}$, then there exists a subset of B with n elements which is ϕ-algebraically independent over $\mathrm{Frac}(B_0)$. To check this statement, let $b_0, \dots, b_n \in B$ be such that $\{\frac{b_1}{b_0}, \dots, \frac{b_n}{b_0}\}$ is a ϕ-transcendence basis of $\mathrm{Frac}(B)$ over $\mathrm{Frac}(B_0)$. Then the statement follows from the property (e) applied to $\{b_0, \dots, b_n\}$ with $C := B_0[\phi^j(b_i) | 0 \leq i \leq n, j \geq 0]$ and $m = n + 1$.

(g) Assume B has a grading by an abelian semigroup in which the homogeneous components are B_0-submodules. Then $\phi\text{-tr.deg}(B)$ is the supremum of all natural numbers d such that B has d homogeneous elements that are ϕ-algebraically independent over $\mathrm{Frac}(B_0)$. To check this statement, let $b_0, \dots, b_n \in B$ be such that $\{\frac{b_1}{b_0}, \dots, \frac{b_n}{b_0}\} \subset \mathrm{Frac}(B)$ is ϕ-algebraically independent over $\mathrm{Frac}(B_0)$ and let $\{\gamma_1, \dots, \gamma_m\}$ be the set of all homogeneous components of $b_0, \dots, b_n$. Taking $C := B_0[\phi^j(\gamma_i) | 0 \leq i \leq m, j \geq 0]$, the statement follows from the property (e).

Remark 2.2.4 We mention the following related concept. Let E be a B_0-module equipped with a semilinear map $\phi : E \to E$, i.e., ϕ on E is additive and for all $\lambda \in B_0$ and all $x \in E$ we have $\phi(\lambda x) = \phi(\lambda)\phi(x)$. By the *$\phi$-rank* of E we mean the supremum, denoted by $\phi\text{-rank}_{B_0}(E) \in \mathbb{N} \cup \{0, \infty\}$, of all $n \in \mathbb{N} \cup \{0\}$ such that there exist elements $x_1, \dots, x_n \in E$ that are *ϕ-linearly independent* over B_0 in the sense that the family $(\phi^j(x_i))_{(i,j) \in \{1,\dots,n\} \times (N \cup \{0\})}$ is B_0-linearly independent.

Remark 2.2.5 Let B be a W-graded ring where $W = \mathbb{Z}[\phi]$. We assume that $B_0 \subset B(0)$ and that we have $\phi(B(w)) \subset B(\phi w)$ for all $w \in W$. We set

$$W_B := \{w \in W | B(w) \neq 0\}.$$

If we view W as a module over the semiring W_+, then W_B is a W_+-submodule of W (recall B is an integral domain and $\phi : B \to B$ is injective). It is easy to check that if $B \otimes_{B_0} \mathrm{Frac}(B_0)$ is ϕ-finitely generated as a $\mathrm{Frac}(B_0)$-algebra, then W_B is a finitely generated W_+-module.

Remark 2.2.6 Assume B is a filtered ring with filtration $(B^r)_{r \geq 0}$ and assume we are given an injective ring endomorphism ϕ with $\phi(B_0) = B_0$ *compatible* with the filtration in the sense that for all integers $r \geq 0$ we have

$$\phi(B^r) \subset B^{r+1}.$$

If B has an asymptotic transcendence degree, then we have an inequality

$$\phi\text{-tr.deg}(B) \leq \text{asy.tr.deg}(B). \tag{2.7}$$

Note that the above inequality is not an equality in general: the right-hand side effectively depends on the filtration whereas the left-hand side does not. If C is B_0-subalgebra of B which is a filtered ring with filtration $(C^r)_{r\geq 0}$ such that we have $C^r \subset B^r$ and $\phi(C^r) \subset C^{r+1}$ for all integers $r \geq 0$, and if C also has an asymptotic transcendence degree, then it is easy to check, using Eq. (2.6) and Inequality (2.7), that we have an implication

$$\text{asy.tr.deg}(C) = \text{asy.tr.deg}(B) \quad \Rightarrow \quad \phi\text{-tr.deg}(C) = \phi\text{-tr.deg}(B). \tag{2.8}$$

Even if either C or B does not have an asymptotic transcendence degree, for an integer $N \geq 0$ we have the implications (1) $\Rightarrow$ (2) $\Rightarrow$ (3) between the following statements.

(1) We have $\text{tr.deg}_{C^r}(B^r) \leq N$ for all integers $r \gg 1$.

(2) We have $\text{tr.deg}_C(B) \leq N$.

(3) We have $\phi\text{-tr.deg}(C) = \phi\text{-tr.deg}(B)$.

Remark 2.2.7 The concepts of ϕ-transcendence degree and asymptotic transcendence degree are related as follows. If B is equipped with an injective ring endomorphism ϕ with $\phi(B_0) = B_0$ such that $\text{Frac}(B)$ is ϕ-finitely generated over $\text{Frac}(B_0)$, then it is easy to see that there exists an increasing exhausting filtration $(B^r)_{r\geq 0}$ by B_0-subalgebras of B such that for all integers $r \geq 0$ we have $\phi(B^r) \subset B^{r+1}$ and the asymptotic transcendence degree exists and is given by

$$\text{asy.tr.deg}(B) = \phi\text{-tr.deg}(B).$$

This will not be used here as the filtrations appearing in the monograph are given (cannot be modified) and Inequality (2.7) is not a priori an equality.

Remark 2.2.8 In this remark we assume that the inclusion $B_0 \subset B$ is a morphism of δ-rings. Let $\phi : B \to B$ be the Frobenius lift attached to $\delta : B \to B$. Then from Remark 2.2.3 (g) we get that ϕ-tr.deg(B) coincides with the supremum of the set of all integers d such that there exist elements $b_1, \ldots, b_d \in B$ with the property that the family

$$(\delta^j(b_i))_{(i,j)\in\{1,\ldots,d\}\times(\mathbb{N}\cup\{0\})}$$

of elements in $\text{Frac}(B)$ is algebraically independent over K. A similar statement holds for the concept of finite ϕ-generation.

Remark 2.2.9 Let $B_0 = R$ and let B be an integral δ-graded ring over R. We assume that $B(0) = R$ and that there exists $w \in W \setminus \{0\}$ such that we have $B(w) \neq pB(w)$. We first state and then argue the following five properties.

(a) We have an equality

$$\phi\text{-tr.deg}(B) = \phi\text{-tr.deg}(B_{((p))}) + 1.$$

(b) The ϕ-transcendence degree ϕ-tr.deg(B) is the supremum of the set of all $n \in \mathbb{N} \cup \{0\}$ such that B contains n elements that are homogeneous of the same weight and ϕ-algebraically independent.

(c) Assume $w \geq 1$ or $w \leq -1$. (For instance, the inequality $w \geq 1$ holds in the example of the 'projective space' introduced below by Display (2.32) and the inequality $w \leq -1$ holds in some of our main examples related to moduli schemes of principally polarized abelian schemes, see Remark 3.6.32 (c) of Sect. 3.6.3 or Family (3.88) below.) Let $f \in B(w) \setminus pB(w)$. Then p is a prime element in the ring $B_{\langle f \rangle}$ and the localization of this ring at the prime ideal (p) satisfies

$$(B_{\langle f \rangle})_{(p)} = B_{((p))}. \tag{2.9}$$

In particular, we have

$$\phi\text{-tr.deg}(B) = \phi\text{-tr.deg}(B_{\langle f \rangle}) + 1.$$

(d) Assume $w \geq 1$ or $w \leq -1$ and assume in addition that B is a filtered δ-graded ring over R. Consider a linear system $\mathbf{f} = (f_0, \ldots, f_n) \in (B(w) \setminus pB(w))^{n+1}$, and let $X_\delta := \text{Proj}_\delta(B_\bullet, \mathbf{f})$. Then, recalling the notation in Definition 2.1.13 (g), defining $N_{n,\bullet} := \binom{n+1}{\bullet + 1}$ we have

$$D_\bullet^{-1} B_\bullet = (B_{((p))})^{N_{n,\bullet}}, \tag{2.10}$$

with the canonical coface maps, so the ss-δ-space $D^{-1}X_\delta$ is acyclic and we have

$$H^0(D^{-1}X_\delta) = B_{((p))}.$$

(e) Assume we are in the situation of part (d) and assume, in addition, that $B^r(v) = 0$ for all integers $r \geq 0$ and all $v \in W$ such that $\text{ord}(v) > r$. (This condition is trivially satisfied in our main examples later in the monograph, see the paragraph after Definition 2.3.5 of Sect. 2.3.1.) Then the canonical filtrations on the two rings in Eq. (2.10) satisfy

$$(D_\bullet^{-1} B_\bullet)^r \subset ((B_{((p))})^r)^{N_{n,\bullet}} \subset (D_\bullet^{-1} B_\bullet)^{r+\text{ord}(w)}. \tag{2.11}$$

In particular we have the following formula for the asymptotic transcendence degree of the δ-ring of rational functions on X_δ:

$$\text{asy.tr.deg}(H^0(D^{-1}X_\delta)) = \text{asy.tr.deg}(B_{((p))}).$$

In what follows we give the arguments for properties (a) to (e).

The inequality '$\geq$' in part (a) follows because each nonzero element in $B(w)$ is ϕ-algebraically independent over $B_{((p))}$. To check the inequality '$\leq$' in part (a) assume ϕ-tr.deg$(B) \geq n+1$ for some integer $n \geq 0$, and consider ϕ-algebraically independent homogeneous elements $b_0, \ldots, b_n \in B$ over R, $b_i \in B(w_i) \setminus pB(w_i)$, see Remark 2.2.3 (g). So the family $(\prod_{i=0}^n b_i^{w_i'})_{(w_0',\ldots,w_n')\in W^{n+1}}$ is linearly independent over R. As $B(0) = R$ we have $w_i \neq 0$ for all $i \in \{0, \ldots, n\}$. Let $v \in W$ be a nonzero common multiple of $w_0, \ldots, w_n$ and for $i \in \{0, \ldots, n\}$ let $v_i \in W$ be such that we have $v = v_i w_i \in W$. As $v_i \neq 0$ for all $i \in \{0, \ldots, n\}$, the family $(\prod_{i=0}^n b_i^{v_i w_i'})_{(w_0',\ldots,w_n')\in W^{n+1}}$ of elements in Frac(B) is linearly independent over R. Thus the family $(b_0^{v_i})_{i\in\{0,\ldots,n\}}$ of elements of Frac(B) is ϕ-algebraically independent over R, hence the family $(b_0^{v_i}/b_0^{v_0})_{i\in\{1,\ldots,n\}}$ is algebraically independent over R. As the reduction of $\phi : B \to B$ modulo p is injective we have $b_i^{v_i}/b_0^{v_0} \in B_{((p))}$ for all $i \in \{1, \ldots, n\}$. Therefore ϕ-tr.deg$(B_{((p))}) \geq n$. From this the inequality '$\leq$' in part (a) follows.

To check part (b), assume again that ϕ-tr.deg$(B) \geq n+1$ and let $b_0, \ldots, b_n$ and the notation be as as in the previous paragraph. For $i \in \{1, \ldots, n\}$, write $b_i^{v_i}/b_0^{v_0} = \beta_i/\gamma_i$, $\beta_i \in B(u_i)$, $\gamma_i \in B(u_i) \setminus pB(u_i)$, $u_i \in W$. We choose u_i such that $u := \sum_{i=1}^n u_i \neq 0$ (as we can multiply β_1 and γ_1 by some element in $B(w) \setminus pB(w)$). Let $\gamma := \prod_{i=1}^n \gamma_i \in B(u)$. Then $\gamma, \beta_1/\gamma_1, \ldots, \beta_n/\gamma_n$ are ϕ-algebraically independent over R (because, as $u \neq 0$, each element in $B(u)$ is ϕ-algebraically independent over $B_{((p))}$). Hence $\gamma, \beta_1\gamma/\gamma_1, \ldots, \beta_n\gamma/\gamma_n \in B(u)$ are ϕ-algebraically independent over R and are homogeneous of the same weight which proves part (b).

To check part (c) note first that, by the axiom (GAX5), p does not divide f^{ϕ^i} for all integers $i \geq 0$. As, by the axiom (GAX4), p is prime in B it follows that p is prime in $B_{\langle f\rangle}$. The inclusion '$\subset$' in Eq. (2.9) is obvious. To check its inclusion '$\supset$', let $F/G \in B_{((p))}$ where $F \in B(v)$ and $G \in B(v) \setminus pB(v)$ for some $v \in W$ with ord$(v) \leq r$. Write $v = v_+ - v_-$ with $v_+, v_- \geq 0$ and ord(v_+), ord$(v_-) \leq r$. Then for $w \geq 1$ we can write

$$\frac{F}{G} = \frac{F^w}{GF^{w-1}} = \frac{(F^w f^{v_-})/f^{v_+}}{(GF^{w-1}f^{v_-})/f^{v_+}} \in (B_{\langle f\rangle})_{(p)} \tag{2.12}$$

while for $w \leq -1$ we can write

$$\frac{F}{G} = \frac{F^{-w}}{GF^{-w-1}} = \frac{(F^{-w} f^{v_+})/f^{v_-}}{(GF^{-w-1}f^{v_+})/f^{v_-}} \in (B_{\langle f\rangle})_{(p)} \tag{2.13}$$

which implies that the inclusion '$\supset$' in Eq. (2.9) holds.

Part (d) follows directly from Eq. (2.9).

To check part (e) note that the first inclusion in Display (2.11) is obvious. To check its second inclusion it is enough to note that if $F/G \in B_{((p))}$ with $F \in B^r(v)$ and $G \in B^r(v) \setminus pB^r(v)$ as in Eq. (2.12) (resp. (2.13)), then F^w and GF^{w-1} (resp. F^{-w} and GF^{-w-1}) have order at most $r + \text{ord}(w)$ while f^{v_+} (resp. f^{v_-}) has order at most ord(w) + ord$(v) \leq$ ord$(w) + r$.

2.3 Concepts of δ-Invariant Theory

We review some basic concepts introduced in [Bu00, BB, Bu03, Bu05] and we introduce some new concepts that are needed in the monograph.

2.3.1 Rings of Abstract δ-Modular Forms

We develop a formalism of δ-modular functions and forms in an abstract (purely categorical) setting of what we will call *moduli data*. In the next subsection we will specialize our study to the case of principally polarized abelian schemes and their polarized isogenies.

Fix a ring S_0 which we view as a 'ground ring' and which, for simplicity, we usually drop from our notation. Recall (see the footnote of Sect. 1.1) that all sets are assumed to belong to a given universe and all categories are small. Denote by **Ring**, **Cat**, **Gr**, and **Set** the category of rings over S_0, the category of categories, the category of groups, and the category of sets. (So in the case of **Ring** the notation matches the one in Sect. 2.1 if we take $S_0 = \mathbb{Z}_{(p)}$; however we will also be interested later in the case $S_0 = R$.) Let $\iota : \mathbf{Gr} \to \mathbf{Cat}$ be the natural functor; so for $H \in \mathrm{Ob}(\mathbf{Gr})$, $\iota(H)$ is a category with one object whose monoid of endomorphisms is H. Let $\mathbb{G}_m : \mathbf{Ring} \to \mathbf{Gr}$ be the multiplicative group functor.

Definition 2.3.1 A *moduli datum* (over S_0) is a quadruple $(\mathfrak{M}, \mathfrak{D}, G, D)$ that consists of:

(MAX1) a functor $\mathfrak{M} : \mathbf{Ring} \to \mathbf{Cat}$;

(MAX2) a morphism $\mathfrak{D} : \mathfrak{M} \to \iota \circ \mathbb{G}_m$ of functors from **Ring** to **Cat**;

(MAX3) a functor $G : \mathbf{Ring} \to \mathbf{Gr}$ that acts on $\mathfrak{M}$ objectwise;

(MAX4) a morphism $D : G \to \mathbb{G}_m$ of functors from **Ring** to **Gr**.

Explicitly, (MAX1)-(MAX4) are given by a rule that functorially attaches to each ring S over S_0 the following:

(i) a category $\mathfrak{M}(S)$ with set of objects $\mathbf{M}(S)$, set of morphisms $\mathbf{M}'(S)$, and source, target, identity, and composition maps

$$\begin{aligned} \pi_1(S), \pi_2(S) : \mathbf{M}'(S) &\to \mathbf{M}(S), \\ \epsilon(S) : \mathbf{M}(S) &\to \mathbf{M}'(S), \\ \mu(S) : \mathbf{M}''(S) &\to \mathbf{M}'(S) \end{aligned}$$

respectively, denoted in what follows simply by $\pi_1, \pi_2, \epsilon, \mu$, where

$$\mathbf{M}''(S) := \{(x_1', x_2') \in \mathbf{M}'(S) \times \mathbf{M}'(S),\ \pi_2(x_1') = \pi_1(x_2')\};$$

(ii) a multiplicative map $\mathfrak{D}(S) : \mathbf{M}'(S) \to S^\times$, denoted in what follows simply by $\mathfrak{D}$;

(iii) a group $G(S)$ acting on the sets $\mathbf{M}(S)$ and $\mathbf{M}'(S)$ such that all the maps π_1, π_2, ϵ are $G(S)$-invariant, and moreover, with $G(S)$ acting naturally on $\mathbf{M}''(S)$, μ is as well $G(S)$-invariant;

(iv) a group homomorphism $D(S) : G(S) \to S^\times$, denoted in what follows simply by D.

We write, as usual, $x'_2 \circ x'_1 := \mu(x'_1, x'_2)$ for $(x'_1, x'_2) \in \mathbf{M}''(S)$. Then multiplicativity in (ii) means that $\mathfrak{D}(x'_2 \circ x'_1) = \mathfrak{D}(x'_2)\mathfrak{D}(x'_1)$.

Remark 2.3.2 Assume $(\mathfrak{M}, \mathfrak{D}, G, D)$ is a moduli datum. Consider the equivalence relation $\sim$ on $\mathbf{M}(S)$ where for $x, y \in \mathbf{M}(S)$ we have $x \sim y$ if and only if there exists $g \in G(S)$ such that the objects y and $g \cdot x$ are isomorphic. We get a functor

$$\mathcal{A} = \mathcal{A}_{\mathfrak{M},G} : \mathbf{Ring} \to \mathbf{Set}$$

that attaches to each $S \in \mathrm{Ob}(\mathbf{Ring})$ the quotient set

$$\mathcal{A}(S) := \mathbf{M}(S)/\sim .$$

Similarly, consider the equivalence relation $\sim'$ on $\mathbf{M}'(S)$ where for $x', y' \in \mathbf{M}'(S)$ we have $x' \sim' y'$ if and only if there exists $g \in G(S)$ and isomorphisms $z'_1, z'_2 \in \mathbf{M}'(S)$ such that $g \cdot x' = z'_1 \circ y' \circ z'_2$. We get a functor

$$\mathcal{A}' = \mathcal{A}'_{\mathfrak{M},G} : \mathbf{Ring} \to \mathbf{Set}$$

that attaches to each $S \in \mathrm{Ob}(\mathbf{Ring})$ the quotient set

$$\mathcal{A}'(S) := \mathbf{M}'(S)/\sim' .$$

We have a natural pair of source and target morphism of functors

$$\mathcal{A}' \overset{\pi_1,\pi_2}{\rightrightarrows} \mathcal{A}. \tag{2.14}$$

Let $\mathcal{A}/\mathcal{A}' : \mathbf{Ring} \to \mathbf{Set}$ be the functor that attaches to each $S \in \mathrm{Ob}(\mathbf{Ring})$ the coequalizer of the two maps $\mathcal{A}'(S) \rightrightarrows \mathcal{A}(S)$ defined by the functors of Diagram (2.14). We have a natural quotient morphism of functors

$$\mathcal{A} \overset{\pi}{\to} \mathcal{A}/\mathcal{A}'. \tag{2.15}$$

Similarly, we may consider the equivalence relation $\sim$ on $\mathbf{M}(S) \times S$ where for $(x, u), (y, v) \in \mathbf{M}(S) \times S$ we have $(x, u) \sim (y, v)$ if and only if there exists $g \in G(S)$ such that the objects y and $g \cdot x$ are isomorphic and we have $v = D(g)u$. We get a functor

$$\mathcal{L} = \mathcal{L}_{\mathfrak{M},G,D} : \mathbf{Ring} \to \mathbf{Set}$$

that attaches to each $S \in \text{Ob}(\mathbf{Ring})$ the quotient set

$$\mathcal{L}(S) := (\mathbf{M}(S) \times S)/\sim .$$

In an entirely similar way we get an equivalence relation $\sim'$ on $\mathbf{M}'(S) \times S$ and a functor

$$\mathcal{L}' = \mathcal{L}'_{\mathfrak{M},G,D} : \mathbf{Ring} \to \mathbf{Set}$$

that attaches to each $S \in \text{Ob}(\mathbf{Ring})$ the quotient set

$$\mathcal{L}'(S) := (\mathbf{M}'(S) \times S)/\sim' .$$

Similar to Diagram (2.14), we have a natural pair of source and target morphism of functors

$$\mathcal{L}' \overset{\pi_1,\pi_2}{\rightrightarrows} \mathcal{L}.$$

If the S_0-schemes are being identified with functors from **Ring** to **Set** then, in the applications we have in mind (e.g., in the situations of Remark 2.3.7 and Sect. 2.3.2 below), the functors $\mathcal{A}$ and $\mathcal{A}'$ are 'very close to being schemes': typically, 'appropriate sheafifications' of them have étale covers that are schemes (as a consequence of results on moduli spaces in 'classical' algebraic geometry, see e.g., [MFK], Ch. 7, Thm. 7.9). Moreover the natural morphisms of functors

$$\mathcal{L} \to \mathcal{A}, \quad \mathcal{L}' \to \mathcal{A}' \tag{2.16}$$

are 'very close to being' line bundles and the pull-backs of $\mathcal{L}$ to $\mathcal{A}'$ via $\mathcal{A}' \overset{\pi_1}{\to} \mathcal{A}$ and $\mathcal{A}' \overset{\pi_2}{\to} \mathcal{A}$ are naturally isomorphic to $\mathcal{L}'$, a situation reminiscent of linearizations of line bundles in geometric invariant theory. However in the applications we have in mind the functor $\mathcal{A}/\mathcal{A}'$ is 'very far from being a scheme'. To remedy this situation one is led to search for a 'refinement' of algebraic geometry, and our δ-geometry does remedy it. This is one of the main motivations for developing δ-geometry and we explain our approach in the rest of this subsection.

Recall the partially ordered ring $W = \mathbb{Z}[\phi]$, the map $\text{ord} : W \to \mathbb{N} \cup \{0\}$, the ring homomorphism $\deg : W \to \mathbb{Z}$, and the cone $W_+ \subset W$ of elements ≥ 0. For $S \in \text{Ob}(\mathbf{FilRing}_\delta)$ and $(w, b) \in (W \times (S^m)^\times) \cup (W_+ \times S^m)$, with $m \in \mathbb{N} \cup \{0\}$ and the writing $w = \sum_{i=0}^n a_i \phi^i$, we set

$$b^w := b^{a_0} \phi(b)^{a_1} \cdots \phi^n(b)^{a_n} \in S^{m+n}.$$

Assume in what follows that the ground ring is $S_0 = R$.

Definition 2.3.3 Let $r \geq 0$ be an integer. A *δ-modular function* of order $\leq r$ for a given moduli datum $(\mathfrak{M}, \mathfrak{D}, G, D)$ is a rule, f, that associates to each object S of $\widehat{\mathbf{FilRing}}_\delta$ and to every $x \in \mathbf{M}(S^0)$ an element $f(x, S) \in S^r$ such that $f(x, S)$ depends only on S and on the isomorphism class of x and such that the association $(x, S) \mapsto f(x, S)$ is functorial in S.

For each integer $r \geq 0$, denote by M^r the set of δ-modular functions of order $\leq r$; it has a structure of a p-adically complete flat R-algebra and there exists a natural injective R-algebra homomorphism $M^r \to M^{r+1}$. We set

$$M := \bigcup_{r \geq 0} M^r.$$

The filtered ring M is an object of $\widehat{\mathbf{FilRing}}_\delta$ with p-derivation defined by

$$(\delta f)(x, S) := \delta(f(x, S))$$

for $f \in M$, $S \in \mathrm{Ob}(\widehat{\mathbf{FilRing}}_\delta)$, $x \in \mathbf{M}(S^0)$. We denote by ϕ the Frobenius lift on M attached to δ.

If in the above part we replace $\mathbf{M}$ by $\mathbf{M}'$ we obtain corresponding rings M'^r and M'. The morphisms π_1, π_2 in the part after Definition 2.3.1 induce morphisms of filtered δ-rings $\pi_1^*, \pi_2^* : M \to M'$, $f \mapsto \pi_i^* f$, $i \in \{1, 2\}$, defined by

$$(\pi_i^* f)(x', S) := f(\pi_i(x'), S).$$

The morphisms π_i^* and their reductions modulo p, $\overline{\pi_i^*} : \overline{M^r} \to \overline{M'^r}$ are injective, as one can see by taking $x' = \epsilon(x)$ in the above equality and using that S^r is flat over R.

Definition 2.3.4 Let $w \in W(r)$. A δ-modular function $f \in M^r$ is called a *δ-modular forms* of *weight* w if for each $g \in G(S^0)$ and every $x \in \mathbf{M}(S^0)$ the following equality holds in S^r:

$$f(g \cdot x, S) = D(g)^{-w} \cdot f(x, S).$$

Here D denotes the map $D(S^0) : G(S^0) \to (S^0)^\times$; see rule (iv) after Definition 2.3.1.

We denote by $M^r(w)$ the R-module of all δ-modular forms M^r of weight w. If in Definition 2.3.4 we replace $\mathbf{M}$ by $\mathbf{M}'$ we obtain R-modules $M'^r(w)$. (The δ-modular forms of weight w can be viewed as 'sections in δ-geometry' of the 'w-power' of the 'line bundle' $\mathcal{L} \to \mathcal{A}$ in Display (2.16).)

Finally, we fix once and for all a rational number $\rho \in \mathbb{Q}$ which we refer to as the *slope* of the theory and consider the ideal of W defined by

$$W^{(\rho)} := \{w \in W | \rho \cdot \deg(w) \in \mathbb{Z}\}.$$

So if $\rho = a/b$ with a, b relatively prime integers with $b \in \mathbb{N}$, then

$$W^{(\rho)} = W^{(\frac{1}{b})} = \{w \in W | b \mid \deg(w)\}.$$

In all our applications we will have $\rho \in \frac{1}{2}\mathbb{Z}$ so $W^{(\rho)}$ will be either W or $W^{(\frac{1}{2})}$ according to ρ being an integer or not.

Definition 2.3.5 Let $w \in W(r) \cap W^{(\rho)}$. We say that a δ-modular form $f \in M^r(w)$ is *covariant* if for each $S \in \mathrm{Ob}(\widehat{\mathbf{FilRing}}_\delta)$ and every $x' \in \mathbf{M}'(S^0)$ the following equality holds in S^r:

$$f(\pi_2(x'), S) = \mathfrak{D}(x')^{\rho \cdot \deg(w)} \cdot f(\pi_1(x'), S).$$

Here $\mathfrak{D}$ denotes the map $\mathfrak{D}(S^0) : \mathbf{M}'(S^0) \to (S^0)^\times$; see rule (ii) after Definition 2.3.1.

For $w \in W(r) \cap W^{(\rho)}$ we denote by $\mathbb{I}^r(w)$ the R-submodule of covariant elements in $M^r(w)$. If $w \in W \setminus (W(r) \cap W^{(\rho)})$ we set $\mathbb{I}^r(w) := 0$.

For $w \in W(r) \cap W^{(\rho)}$ we set $\mathbb{M}^r(w) := M^r(w)$ and $\mathbb{M}'^r(w) := M'^r(w)$; for $w \in W \setminus (W(r) \cap W^{(\rho)})$ we set $\mathbb{M}^r(w) = \mathbb{M}'^r(w) := 0$.

For each integer $r \geq 0$ we consider the W-graded R-algebras

$$\mathbb{I}^r := \bigoplus_{w \in W} \mathbb{I}^r(w), \quad \mathbb{M}^r := \bigoplus_{w \in W} \mathbb{M}^r(w), \quad \mathbb{M}'^r := \bigoplus_{w \in W} \mathbb{M}'^r(w).$$

These rings form filtered graded systems so the R-algebras

$$\mathbb{I} := \bigcup_{r \geq 0} \mathbb{I}^r, \quad \mathbb{M} := \bigcup_{r \geq 0} \mathbb{M}^r, \quad \mathbb{M}' := \bigcup_{r \geq 0} \mathbb{M}'^r$$

have natural structures of filtered graded rings over R with graded pieces of weight w given respectively by

$$\mathbb{I}(w) := \bigcup_{r \geq 0} \mathbb{I}^r(w), \quad \mathbb{M}(w) := \bigcup_{r \geq 0} \mathbb{M}^r(w), \quad \mathbb{M}'(w) := \bigcup_{r \geq 0} \mathbb{M}'^r(w).$$

These graded pieces are 0 for $w \in W \setminus W^{(\rho)}$. The rings $\mathbb{I}, \mathbb{M}, \mathbb{M}'$, equipped with the endomorphisms ϕ induced by the endomorphism ϕ on M, have natural structures of filtered δ-graded rings over R, see Definitions 2.1.15 (a) and 2.1.18. Note that for all $w \in W$ the R-linear map

$$\mathbb{I}(w) \to M \tag{2.17}$$

has a torsion free cokernel. Also if we define the R-algebra automorphism

$$\beta : \mathbb{M}' \to \mathbb{M}' \tag{2.18}$$

by the formula

$$(\beta f)(x', S) := \mathfrak{D}(x')^{\rho \cdot \deg(w)} \cdot f(x', S), \quad f \in \mathbb{M}'(w),$$

then an element $f \in \mathbb{M}$ belongs to $\mathbb{I}$ if and only if we have

$$\beta \pi_1^* f = \pi_2^* f.$$

The natural R-algebra homomorphisms

$$\mathbb{I} \to \mathbb{M} \overset{\beta\pi_1^*, \pi_2^*}{\rightrightarrows} \mathbb{M}' \tag{2.19}$$

are trivially seen to be injective, with injective reductions modulo p. Also, $\mathbb{I}$ is the equalizer in the category of R-algebras (and also in the category of filtered δ-graded rings over R) of the pair of R-algebra homomorphisms $(\beta\pi_1^*, \pi_2^*)$.

Remark 2.3.6 Assume, in this remark, that the filtered δ-graded rings $\mathbb{I}, \mathbb{M}, \mathbb{M}'$ are integral. (This condition is satisfied in our main applications, i.e., for the rings $\mathbb{I}, \mathbb{M}, \mathbb{M}'$ to be considered in Sect. 2.3.2, see Corollary 3.5.2 (d).) For each linear system $\mathbf{f} = (f_0, \dots, f_n)$ in $\mathbb{I}$ we get connected filtered ss-δ-spaces (see Display (2.4))

$$\text{Proj}_\delta(\mathbb{I}, \mathbf{f}) := \text{Spec}_\delta \left(\prod_{0 \le i_0 < \cdots < i_\bullet \le n} \mathbb{I}_{\langle f_{i_0} \cdots f_{i_\bullet} \rangle} \right),$$

$$\text{Proj}_\delta(\mathbb{M}, \mathbf{f}) := \text{Spec}_\delta \left(\prod_{0 \le i_0 < \cdots < i_\bullet \le n} \mathbb{M}_{\langle f_{i_0} \cdots f_{i_\bullet} \rangle} \right),$$

$$\text{Proj}_\delta(\mathbb{M}', \mathbf{f}) := \text{Spec}_\delta \left(\prod_{0 \le i_0 < \cdots < i_\bullet \le n} \mathbb{M}'_{\langle f_{i_0} \cdots f_{i_\bullet} \rangle} \right),$$

and we have naturally induced morphisms of filtered ss-δ-spaces

$$\text{Proj}_\delta(\mathbb{M}', \mathbf{f}) \overset{\pi_1, \pi_2}{\rightrightarrows} \text{Proj}_\delta(\mathbb{M}, \mathbf{f}) \overset{\pi}{\to} \text{Proj}_\delta(\mathbb{I}, \mathbf{f}). \tag{2.20}$$

Here the elements $f_i \in \mathbb{I} \subset \mathbb{M}$ are viewed as elements of $\mathbb{M}'$ by identifying them with the elements $\pi_1^* f_i \in \mathbb{M}'$; the morphism π_2 is then well-defined in view of the display of Definition 2.3.5. Note that the automorphism β of $\mathbb{M}'$ induces the identity on $\text{Proj}_\delta(\mathbb{M}', \mathbf{f})$. It is then trivial to see that Diagram (2.20) is a coequalizer diagram in $\mathbf{Spss}_\delta$.

Assume, in addition, that for every integer $r \ge 0$, weight $w \in W(r)$, and covariant $f \in \mathbb{I}^r(w) \setminus p\mathbb{I}^r(w)$ we have

$$(\mathbb{I}_{\langle f \rangle})^r = (\mathbb{M}_{\langle f \rangle})^r \cap \mathbb{I}_{\langle f \rangle}.$$

(This condition is, again, satisfied in our main applications, i.e., for the rings $\mathbb{I}$, $\mathbb{M}$, $\mathbb{M}'$ to be considered in Sect. 2.3.2, see Corollary 3.5.3 (d).) Then Diagram (2.20) is also a coequalizer in $\mathbf{FilSpss}_\delta$. The pair (π_1, π_2) in Diagram (2.20) may be viewed as an $\mathbf{f}$-realization, in δ-geometry, of the pair of morphisms in Diagram (2.14) while the morphism π in Diagram (2.20) may be viewed as an $\mathbf{f}$-realization of the morphism in Diagram (2.15).

Remark 2.3.7 The δ analogs of the Hecke correspondences on the stack $\mathcal{A}_{g,1,R}$ mentioned in Sects. 1.1–1.4 that will be studied in this monograph fit, as we shall see, into the above formalism. On the other hand our general formalism applies to Hecke correspondences on stacks defined by more general Shimura varieties (see [Bu03, Bu04] and [Bu05], Ch. 8 for the case of Shimura curves) and also to correspondences arising from certain dynamical systems on projective spaces or principally polarized abelian schemes over R (see [Bu05], Ch. 6 for the case of $\mathbb{P}^1_R$ and see [Bu05], Ch. 7 for the case of elliptic curves over R). Here are two illustrations of situations coming from dynamical systems.

(a) Let $1_2 \in \Gamma \subset \mathbf{GL}_2(\mathbb{Z}_{(p)})$ be a submonoid. For each R-algebra S we consider the category $\mathfrak{M}_{\mathbb{P}^1,\Gamma}(S)$ with set of objects

$$\mathbf{M}_{\mathbb{P}^1,\Gamma}(S) := \left\{ v = \begin{pmatrix} v_1 \\ v_2 \end{pmatrix}, \ v_1, v_2 \in S, \ v_1 S + v_2 S = S \right\},$$

set of morphisms

$$\mathbf{M}'_{\mathbb{P}^1,\Gamma}(S) := \Gamma \times \mathbf{M}_{\mathbb{P}^1,\Gamma}(S),$$

source map given by the second projection, target map given by multiplication of matrices with column vectors, composition given by

$$\mu((g, v), (g', gv)) := (g'g, v)$$

and unit defined by $\epsilon(v) := (1_2, v)$. Furthermore, we consider the actions of the group $G(S) = \mathbb{G}_m(S) = S^\times$ on $\mathbf{M}_{\mathbb{P}^1,\Gamma}(S)$ and $\mathbf{M}'_{\mathbb{P}^1,\Gamma}(S)$ given by usual scalar multiplication of elements of S^2 by units of S. We take the homomorphism D to be the identity and define $\mathfrak{D}(g, v) := \det(g)$. We have defined a moduli datum $(\mathfrak{M}_{\mathbb{P}^1,\Gamma}, \mathfrak{D}, \mathbb{G}_m, D)$. Note that if Γ has no invertible elements except the identity, then for every local ring S over R we have $\mathcal{A}_{\mathfrak{M}_{\mathbb{P}^1,\Gamma},\mathbb{G}_m}(S) = \mathbb{P}^1(S)$. Take the slope to be $\rho = \frac{1}{2}$. Then the theory in [Bu05], Ch. 6 provides an analysis of the ring $\mathbb{I} = \mathbb{I}_{\mathbb{P}^1,\Gamma}$ in this case. In particular, if the subgroup of $\mathbf{GL}_2(\mathbb{Z}_{(p)})$ generated by Γ is not solvable, then $\mathbb{I}$ is described in terms of certain explicit δ-modular forms $f^r \in \mathbb{I}^r(-\phi^r - 1)$, $r \geq 1$ (see [Bu05], Prop. 6.17). This example (which, as explained in [Bu05], Ch. 6 is directly related to the classical geometric invariant theory of $\mathbf{SL}_2$ acting on multiple pairs of 2-vectors) will play no role in the present monograph.

(b) Fix an elliptic curve E over R. For each R-algebra S we consider the category $\mathfrak{M}_E(S)$ with set of objects

$$\mathbf{M}_E(S) := \{(P,\omega)|P \in E(S),\ S\omega = H^0(E_S, \Omega_{E_S/S})\},$$

set of morphisms

$$\mathbf{M}'_E(S) := \{(P_1, P_2, \omega, n)|(P_1, \omega), (P_2, \omega) \in \mathbf{M}(S),\ n \in \mathbb{N} \setminus p\mathbb{N},\ P_2 = nP_1\},$$

source and target given by the obvious projections, composition given by

$$\mu((P_1, P_2, \omega, n), (P_2, P_3, \omega, m)) := (P_1, P_3, \omega, nm),$$

and unit defined by $\epsilon(P,\omega) := (P, P, \omega, 1)$. Furthermore we take $G = \mathbb{G}_m$ and we consider the actions of the group $G(S) = \mathbb{G}_m(S) = S^\times$ given by multiplication of ω by elements of $S^\times$. Finally consider the homomorphism D given by $D(\lambda) := \lambda^{-1}$ and the last projection map $\mathfrak{D}(P_1, P_2, \omega, n) := n$. We have defined a moduli datum $(\mathfrak{M}_E, \mathfrak{D}, \mathbb{G}_m, D)$. Note that for every local ring S over R we have an identity $\mathcal{A}_{\mathfrak{M}_E,\mathbb{G}_m}(S) = E(S)$. Take the slope to be $\rho = 1$. Then the theory in [Bu05], Ch. 7 provides an analysis of the ring $\mathbb{I} = \mathbb{I}_E$ in this case. In particular, if $p \neq 2$, E is not a canonical lift, and ψ_ω is the *normalized δ-character* of E over S attached to the 1-form ω (see [Bu95a], Introd. or [Bu05], Thm. 7.22, p. 197, for the case $S = R$, and similarly for a general filtered δ-ring S), then the rule $f(P,\omega) := \psi_\omega(P)$ defines an element $f \in \mathbb{I}^2(\phi^2)$; moreover all the elements in the ring $\mathbb{I}$ can be 'constructed from' this element, see [Bu05], Thm. 7.34. This example plays a key role in a number of contexts [Bu95a, Bu05, BM, BPII] but will play no role in the present monograph.

2.3.2 Rings of Siegel δ-Modular Forms

In what follows we define two moduli data

$$(\mathfrak{M}_g, \mathfrak{D}, \mathbf{GL}_g, \det) \quad \text{and} \quad (\mathfrak{M}_{g,\mathrm{ord}}, \mathfrak{D}_{\mathrm{ord}}, \mathbf{GL}_g, \det)$$

involving principally polarized abelian schemes and Hecke correspondences and we apply to them the constructions of the previous subsection. For background on abelian schemes and Hecke correspondences we refer to the Appendix. The rings $\mathbb{I}, \mathbb{M}, \mathbb{M}'$ that correspond to these moduli data will be called

$$\mathbb{I}_g, \mathbb{M}_g, \mathbb{M}'_g \quad \text{and} \quad \mathbb{I}_{g,\mathrm{ord}}, \mathbb{M}_{g,\mathrm{ord}}, \mathbb{M}'_{g,\mathrm{ord}},$$

respectively; however, if no confusion arises, we will sometimes drop the indices.

Recall that our ground ring is $S_0 = R$. For each R-algebra S, let $\mathbf{M}_g(S)$ denote the set of all triples (A, θ, ω) where A is an abelian scheme of relative dimension g over Spec(S), where $\theta : A \to \check{A}$ is a principal polarization (so $\check{A}$ is the dual of A), and where ω is a *column basis of 1-forms* on A by which we mean a column vector $\omega = (\omega_1 \cdots \omega_g)^{\mathrm{t}}$ whose entries form a basis of the S-module of 1-forms $H^0(A, \Omega^1_{A/\mathrm{Spec}(S)})$.

Let now

$$(A_1, \theta_1, \omega_1), (A_2, \theta_2, \omega_2) \in \mathbf{M}_g(S)$$

and let $u : A_1 \to A_2$ be an isogeny (which is not assumed to be compatible with either the principal polarizations or the column bases of 1-forms). Let

$$u^{\mathrm{t}} := \theta_1^{-1} \circ \check{u} \circ \theta_2 : A_2 \to \check{A}_2 \to \check{A}_1 \to A_1,$$

$$u^*\theta_2 := \check{u} \circ \theta_2 \circ u : A_1 \to A_2 \to \check{A}_2 \to \check{A}_1.$$

We say u is a *polarized isogeny of degree prime to* p *with respect to* θ_1 *and* θ_2 if there exists $d \in \mathbb{N} \setminus p\mathbb{N}$ such that $uu^{\mathrm{t}} = d : A_2 \to A_2$ (equivalently, $u^*\theta_2 = d\theta_1 : A_1 \to \check{A}_1$); for such u we have $\deg(u) = d^g$. If θ_1 and θ_2 are clear from context and p is fixed we will not mention 'of degree prime to p' or 'with respect to θ_1 and θ_2' and simply say that u is a *polarized isogeny*.

We denote by $\mathbf{M}'_g(S)$ the set of all triples (x_1, x_2, u) where

$$x_i := (A_i, \theta_i, \omega_i) \in \mathbf{M}_g(S), \quad i \in \{1, 2\},$$

and $u : A_1 \to A_2$ is a polarized isogeny such that $u^*\omega_2 = \omega_1$. There exist canonical maps of sets

$$\begin{aligned} \pi_1, \pi_2 &: \mathbf{M}'_g(S) \to \mathbf{M}_g(S), \\ \epsilon &: \mathbf{M}_g(S) \to \mathbf{M}'_g(S), \\ \mu &: \mathbf{M}''_g(S) \to \mathbf{M}'_g(S), \end{aligned}$$

where $\pi_i(x_1, x_2, u) := x_i$, $\epsilon(x) := (x, x, \mathrm{id})$, $\mathbf{M}''_g(S)$ is the set of all pairs (y_1, y_2) in $\mathbf{M}'_g(S)^2$ such that $\pi_2(y_1) = \pi_1(y_2)$, and μ is given by the obvious composition of isogenies. We obtain a category $\mathfrak{M}(S)$ whose set of objects is $\mathbf{M}_g(S)$, whose set of morphisms is $\mathbf{M}'_g(S)$, and where $\pi_1, \pi_2, \epsilon, \mu$ are the source, target, identity, and composition maps.

We let the group $\mathbf{GL}_g(S)$ act on $\mathbf{M}_g(S)$ by the rule

$$\lambda \cdot (A, \theta, \omega) := (A, \theta, \lambda\omega)$$

for $\lambda \in \mathbf{GL}_g(S)$ and $(A, \theta, \omega) \in \mathbf{M}_g(S)$ and act on $\mathbf{M}'_g(S)$ by the rule

$$\lambda \cdot (x_1, x_2, u) := (\lambda \cdot x_1, \lambda \cdot x_2, u).$$

We take the group homomorphism $D = \det : \mathbf{GL}_g(S) \to S^\times$, $D(\lambda) = \det(\lambda)$.

Finally we denote by $\mathfrak{D} : \mathbf{M}'_g(S) \to S^\times$ the map defined by letting $\mathfrak{D}(x_1, x_2, u)$ be the image of $d := uu^{\mathrm{t}} \in \mathbb{N} \setminus p\mathbb{N}$ in $S^\times$.

We have defined above a moduli datum $(\mathfrak{M}_g, \mathfrak{D}, \mathbf{GL}_g, \det)$.

Clearly, for every local ring S, the set $\mathcal{A}_{\mathfrak{M}_g,\mathbf{GL}_g}(S)$ attached to this moduli datum (see Remark 2.3.2) identifies with the set $\mathcal{A}_{g,1,R}(S)$ of isomorphism classes of principally polarized abelian schemes (A, θ) over S. Similarly, the set $\mathcal{A}'_{\mathfrak{M}_g,\mathbf{GL}_g}(S)$ (see Remark 2.3.2) identifies with the set $\mathcal{A}'_{g,1,R}(S)$ of isomorphism classes of quintuples $(A_1, \theta_1, A_2, \theta_2, u)$ where $(A_1, \theta_1), (A_2, \theta_2) \in \mathcal{A}_{g,1,R}(S)$ and $u : A_1 \to A_2$ is a polarized isogeny; the set $\mathcal{A}'_{g,1,R}(S)$ may be viewed as the 'union of all prime to p Hecke correspondences' on $\mathcal{A}_{g,1,R}(S)$. We continue to denote by $\mathcal{A}'_{g,1,R}$ the obvious stack inducing the functor $\mathcal{A}'_{g,1,R}$.

In what follows, we take the slope to be $\rho := g/2$; as we shall see, this value is 'forced upon us' by the compatibilities built into our specific situation (see Remark 2.3.10 (c)).

As in Sect. 2.3.1, by adding the index g we can attach to our moduli datum and to our slope $\rho = g/2$ the corresponding R-modules

$$M_g^r(w), M_g^{\prime r}(w), \mathbb{M}_g^r(w), \mathbb{M}_g^{\prime r}(w), \mathbb{I}_g^r(w),$$

and R-algebras

$$M_g^r, M_g, M_g^{\prime r}, M_g', \mathbb{M}_g^r, \mathbb{M}_g^{\prime r}, \mathbb{I}_g^r, \mathbb{M}_g, \mathbb{M}_g', \mathbb{I}_g. \tag{2.21}$$

Definition 2.3.8 The elements of M_g^r are called *Siegel δ-modular functions* of genus g and order $\leq r$. The elements of $M_g^r(w)$ are called *Siegel δ-modular forms* of genus g, order $\leq r$, and weight w. A Siegel δ-modular form $f \in \mathbb{M}_g^r(w)$ is called *Hecke covariant* if it is covariant, i.e., if it belongs to $\mathbb{I}_g^r(w)$.

Remark 2.3.9 **(a)** Corollary 3.5.2 (d) below will prove that the rings $\mathbb{M}_g, \mathbb{M}_g', \mathbb{I}_g$ are filtered integral δ-graded rings. So Remark 2.3.7 applies to these rings.

(b) We later prove that we have $M_g^s(w) \cap M_g^r = 0$ provided $0 \leq r < \text{ord}(w) \leq s$, see Corollary 3.5.3 (a) below. This may be seen as an a posteriori justification for our convention that $\mathbb{M}^r(w) = 0$ for $w \in W \setminus W(r)$.

(c) We later prove that the R-linear maps $\mathbb{M}_g^r \to M_g^r$ are injective, see Corollary 3.5.3 (b) below. On the other hand, for each $w \in W$ the cokernels of the maps $M_g^r(w) \to M_g^r$ and $\mathbb{I}_g^r(w) \to M_g^r$ are trivially seen to be torsion free.

(d) In view of the paragraph after Definition 2.3.5, by taking $\rho = g/2$, from very definitions we get that we have $\mathbb{I}_g^r(w) = \mathbb{M}_g^r(w) = 0$ for $\frac{g \cdot \deg(w)}{2} \notin \mathbb{Z}$.

Remark 2.3.10 **(a)** A Siegel δ-modular function $f \in M_g^r$ is in $M_g^r(w)$ if and only if for every $S \in \text{Ob}(\widehat{\mathbf{FilRing}_\delta})$ and each $\lambda \in \mathbf{GL}_g(S^0)$ the following equality holds in S^r:

$$f(A, \theta, \lambda\omega, S) = \det(\lambda)^{-w} \cdot f(A, \theta, \omega, S). \tag{2.22}$$

Note that if a Siegel δ-modular function $f \in M_g^r$ has a weight then the weight is unique. This follows from the fact that if $w \in W$ is such that $\lambda^w = 1$ for all $\lambda \in R^\times$

then $w = 0$; the latter property follows from the injectivity of the map in Diagram (2.3) applied to $X = \mathrm{Spec}(R[x, x^{-1}])$ and the element $x^w - 1 \in \mathcal{O}(J^{\mathrm{ord}(w)}(X))$.

(b) A Siegel δ-modular form $f \in \mathbb{M}_g^r(w)$ is in $\mathbb{I}_g^r(w)$ if and only if for each $S \in \mathrm{Ob}(\widehat{\mathbf{FilRing}_\delta})$ and every triple $(x_1, x_2, u) \in \mathbf{M}_g'(S^0)$, where $x_1 = (A_1, \theta_1, \omega_1)$, $x_2 = (A_2, \theta_2, \omega_2)$, $uu^{\mathrm{t}} = d \in \mathbb{N} \setminus p\mathbb{N}$, and $u^*\omega_2 = \omega_1$, the following equality holds in S^r:

$$f(A_2, \theta_2, \omega_2, S) = d^{\frac{g \cdot \deg(w)}{2}} \cdot f(A_1, \theta_1, \omega_1, S). \tag{2.23}$$

If $f \in \mathbb{I}_g^r(w)$, then by combining Eqs. (2.22) and (2.23) we get that for every $S \in \mathrm{Ob}(\widehat{\mathbf{FilRing}_\delta})$, every $(A_1, \theta_1, \omega_1), (A_2, \theta_2, \omega_2) \in \mathbf{M}_g(S^0)$, and every polarized isogeny $u : A_1 \to A_2$ with $uu^{\mathrm{t}} = d \in \mathbb{N} \setminus p\mathbb{N}$, the following equality holds in S^r:

$$f(A_2, \theta_2, \omega_2, S) = d^{\frac{g \cdot \deg(w)}{2}} \cdot \det([u])^{-w} \cdot f(A_1, \theta_1, \omega_1, S), \tag{2.24}$$

where $[u] \in \mathbf{GL}_g(S^0)$ is the unique matrix such that $u^*\omega_2 = [u]\omega_1$.

(c) Assume $0 \neq f \in M_g^r(w)$ and $\nu \in \mathbb{Z}$ are such that for each $S \in \mathrm{Ob}(\widehat{\mathbf{FilRing}_\delta})$ and every triple $(x_1, x_2, u) \in \mathbf{M}_g'(S^0)$, where $x_1 = (A_1, \theta_1, \omega_1)$, $x_2 = (A_2, \theta_2, \omega_2)$, $uu^{\mathrm{t}} = d \in \mathbb{N} \setminus p\mathbb{N}$, and $u^*\omega_2 = \omega_1$, we have

$$f(A_2, \theta_2, \omega_2, S) = d^{\nu} \cdot f(A_1, \theta_1, \omega_1, S). \tag{2.25}$$

Then it is easy to check (using Eq. (2.25) for both u and u^{t} plus Eq. (2.22)) that $\frac{g \cdot \deg(w)}{2} \in \mathbb{Z}$ and $\nu = \frac{g \cdot \deg(w)}{2}$.

It turns out that, in order to study the ring $\mathbb{I}_g$, it is crucial to develop a matrix version of some of our definitions. We present this in what follows.

Definition 2.3.11 Let $m \in \mathbb{N}$ and $r \in \mathbb{N} \cup \{0\}$. A *Siegel δ-modular function* of genus g, size m, and order $\leq r$ is a rule f that associates to each pair

$$(S, (A, \theta, \omega)) \in \mathrm{Ob}(\widehat{\mathbf{FilRing}_\delta}) \times \mathbf{M}_g(S^0)$$

an $m \times m$ matrix

$$f(A, \theta, \omega, S) \in \mathbf{Mat}_m(S^r)$$

that depends on S and the isomorphism class of (A, θ, ω) only and that is functorial in S.

In this monograph we will use only the sizes $m = 1$ and $m = g$. The set of Siegel δ-modular functions of genus g, size 1, and order $\leq r$ coincides with M_g^r. We denote by M_{gg}^r the R-algebra of Siegel δ-modular functions of genus g, size g, and order $\leq r$. We have $M_{gg}^r = \mathbf{Mat}_g(M_g^r)$. Similarly to the case of size 1 we set $M_{gg} := \bigcup_{r \geq 0} M_{gg}^r = \mathbf{Mat}_g(M_g)$.

In what follows we introduce the concepts of *weight* and *Hecke covariance* for Siegel δ-modular functions of size g. Let

$$\mathcal{W} := \{\pm\phi^a | a \in \mathbb{Z}_{\geq 0}\} \subset W$$

be the multiplicative submonoid generated by -1 and ϕ. For each $w \in \{\pm\phi^a\} \subset \mathcal{W}$, $m \in \mathbb{N}$, and $S \in \mathrm{Ob}(\widehat{\mathbf{FilRing}_\delta})$, let $\chi_w : \mathbf{GL}_m(S^0) \to \mathbf{GL}_m(S^a)$ be the homomorphism which maps $\lambda \in \mathbf{GL}_m(S^0)$ to

$$\chi_w(\lambda) := \phi^a(\lambda) \text{ if } w = \phi^a,$$

$$\chi_w(\lambda) := ((\phi^a(\lambda))^{\mathrm{t}})^{-1} \text{ if } w = -\phi^a.$$

Definition 2.3.12 Fix a pair $(w', w'') \in \mathcal{W} \times \mathcal{W}$ where w', w'' have order $\leq r$; so $w', w'' \in \{\pm\phi^a | a \in \{0, \dots, r\}\}$. A Siegel δ-modular function $f \in M^r_{gg}$ is called a *Siegel δ-modular form* of weight (w', w'') if for every $S \in \mathrm{Ob}(\widehat{\mathbf{FilRing}_\delta})$, for every $(A, \theta, \omega) \in \mathbf{M}_g(S^0)$ and every $\lambda \in \mathbf{GL}_g(S^0)$ the following equality holds in $\mathbf{Mat}_g(S^r)$:

$$f(A, \theta, \lambda\omega, S) = \chi_{w'}(\lambda) \cdot f(A, \theta, \omega, S) \cdot \chi_{w''}(\lambda^{\mathrm{t}}). \tag{2.26}$$

The R-module of all Siegel δ-modular forms of genus g, size g, order $\leq r$ and weight (w', w'') will be denoted by $M^r_{gg}(w', w'')$. Note that if a nonzero element of M^r_{gg} has a weight (w', w'') then that weight is not a priori unique (it is clearly never unique if $g = 1$); for $g \geq 2$ we do not know if the sum $w' + w''$ is uniquely determined.

Definition 2.3.13 Fix a pair $(w', w'') \in \mathcal{W} \times \mathcal{W}$ where w', w'' have order $\leq r$; so $w', w'' \in \{\pm\phi^a | a \in \{0, \dots, r\}\}$. A Siegel δ-modular form $f \in M^r_{gg}(w', w'')$ is *Hecke covariant* if for each $S \in \mathrm{Ob}(\widehat{\mathbf{FilRing}_\delta})$ and every $(x_1, x_2, u) \in \mathbf{M}'_g(S^0)$, where $x_1 = (A_1, \theta_1, \omega_1)$, $x_2 = (A_2, \theta_2, \omega_2)$, $uu^{\mathrm{t}} = d \in \mathbb{N} \setminus p\mathbb{N} \subset \mathrm{End}(A_2)$, the following equality holds,

$$f(A_2, \theta_2, \omega_2, S) = d^{-\deg(w'+w'')/2} \cdot f(A_1, \theta_1, \omega_1, S). \tag{2.27}$$

As $w', w'' \in \mathcal{W}$ have degree -1 or 1, we have $\deg(w' + w'')/2 \in \{-1, 0, 1\}$. We denote by $\mathbb{I}^r_{gg}(w', w'')$ the R-submodule of $M^r_{gg}(w', w'')$ of Hecke covariant Siegel δ-modular forms.

Lemma 2.3.14 *Let $f \in M^r_{gg}$ and $(w', w'') \in \mathcal{W} \times \mathcal{W}$. Then $f \in \mathbb{I}^r_{gg}(w', w'')$ if and only if for every $S \in Ob(\widehat{\mathbf{FilRing}_\delta})$, every*

$$(A_1, \theta_1, \omega_1), (A_2, \theta_2, \omega_2) \in \boldsymbol{M}_g(S^0),$$

and every polarized isogeny $u : A_1 \to A_2$ with $uu^t = d \in \mathbb{N} \setminus p\mathbb{N}$ we have an identity

$$f(A_2, \theta_2, \omega_2, S) = d^{-deg(w'+w'')/2} \cdot \chi_{w'}([u]) \cdot f(A_1, \theta_1, \omega_1, S) \cdot \chi_{w''}([u]^t), \tag{2.28}$$

where $[u] \in \boldsymbol{GL}_g(S^0)$ *is the unique matrix such that* $u^*\omega_2 = [u]\omega_1$.

Proof. If $f \in \mathbb{I}^r_{gg}(w', w'')$, then using Eqs. (2.26) and (2.27) we compute

$$\begin{aligned} f(A_2, \theta_2, \omega_2, S) &= d^{-\deg(w'+w'')/2} \cdot f(A_1, \theta_1, u^*\omega_2, S) \\ &= d^{-\deg(w'+w'')/2} \cdot f(A_1, \theta_1, [u]\omega_1, S) \\ &= d^{-\deg(w'+w'')/2} \cdot \chi_{w'}([u]) \cdot f(A_1, \theta_1, \omega_1, S) \cdot \chi_{w''}([u]^{\mathrm{t}}) \end{aligned}$$

and so 'only if' part holds.

For the 'if' part, by taking $(A_2, \theta_2) = (A_1, \theta_1)$ and u to be the identity automorphism of A_1, Eq. (2.28) gives that $f \in M^r_{gg}(w', w'')$. Based on this, by taking ω_1 to be $u^*\omega_2$, Eq. (2.28) gives that $f \in \mathbb{I}^r_{gg}(w', w'')$. So the 'if' part holds. □

Remark 2.3.15 Our R-modules $\mathbb{I}^r_{gg}(\phi^a, \phi^b)$ (resp. $\mathbb{I}^r_g(w)$) a priori contain the R-modules $I^r_g(\phi^a, \phi^b) \otimes_{\mathbb{Z}_p} R$ (resp. $I^r_g(w) \otimes_{\mathbb{Z}_p} R$) defined by the $\mathbb{Z}_p$-modules of *isogeny covariant* δ-modular forms $I^r_g(\phi^a, \phi^b)$ (resp. $I^r_g(w)$) introduced in [BB], Sect. 1.3 (resp. Sect. 1.4). We do not know if the inclusion $I^r_g(\phi^a, \phi^b) \otimes_{\mathbb{Z}_p} R \subset \mathbb{I}^r_{gg}(\phi^a, \phi^b)$ (resp. $I^r_g(w) \otimes_{\mathbb{Z}_p} R \subset \mathbb{I}^r_g(w)$) is an equality; but one is tempted to conjecture that this is the case. The $\mathbb{Z}_p$-modules $I^r_g(\phi^a, \phi^b)$ (resp. $I^r_g(w)$) do not play any role in our monograph. We recall, for convenience, that the definition of the spaces $I_g(\phi^a, \phi^b)$ (resp. $I_g(w)$) in [BB] was similar to that of our spaces $\mathbb{I}_{gg}(\phi^a, \phi^b)$ (resp. $\mathbb{I}_g(w)$) here, except that, for the former, one considered '*prolongation sequences* over $\mathbb{Z}_p$' rather than 'filtered δ-rings over R' and, more importantly, a variant of Eq. (2.28) (resp. (2.23)) was required to hold for all isogenies u of degree prime to p, rather than just for polarized isogenies, and this explains the inclusions mentioned. The condition in [BB], Eq. 1.26, p. 1464 (resp. [BB], Eq. 1.37, p. 1468) that defines the module $I^r_g(\phi^a, \phi^b)$ (resp. $I^r_g(w)$) becomes equivalent to Eq. (2.26) (resp. (2.23)) in the case of polarized isogenies.

Remark 2.3.16 If in Definitions 2.3.8–2.3.13 we replace $\mathbf{M}_g(S)$ by the set $\mathbf{M}_{g,\mathrm{ord}}(S)$ of all triples $(A, \theta, \omega) \in \mathbf{M}_g(S)$ with the property that A is ordinary and if we replace $\mathbf{M}'_g(S)$ by the corresponding set $\mathbf{M}'_{g,\mathrm{ord}}(S)$, we obtain a new moduli datum $(\mathfrak{M}_{g,\mathrm{ord}}, \mathfrak{D}_{\mathrm{ord}}, \mathbf{GL}_g, \det)$. Implicitly, the objects $\square_g$ with the index g attached to the datum $(\mathfrak{M}_g, \mathfrak{D}, \mathbf{GL}_g, \det)$ will be replaced by objects $\square_{g,\mathrm{ord}}$ attached to the datum $(\mathfrak{M}_{g,\mathrm{ord}}, \mathfrak{D}_{\mathrm{ord}}, \mathbf{GL}_g, \det)$ and will be referred to as *'ordinary'*. For instance $M^r_g(w)$ will be replaced by $M^r_{g,\mathrm{ord}}(w)$ and the elements of the latter will be referred to as *ordinary Siegel δ-modular forms* of genus g, order $\leq r$, and weight w, etc. All claims in Remarks 2.3.9 and 2.3.10 hold in this new context. Note that here the word 'ordinary' is not an adjective: each Siegel δ-modular form defines an ordinary Siegel δ-modular form, and in fact we have natural R-linear homomorphisms

$$M^r_g(w) \to M^r_{g,\mathrm{ord}}(w), \quad M^r_{gg}(w', w'') \to M^r_{gg,\mathrm{ord}}(w', w'')$$

which turn out to be injective by our Serre–Tate expansion principle (see Proposition 3.5.1 (b) below).

Remark 2.3.17 In this remark we consider the ordinary case, i.e., the case of $\mathbb{I}_{g,\mathrm{ord}}$, etc.; an entirely similar thing holds in the nonordinary case, i.e., the case of $\mathbb{I}_g$, etc.

(a) The R-modules $\mathbb{I}^r_{gg,\mathrm{ord}}(w', w'')$ cannot be naturally assembled into an R-algebra. However, for $w', w'', w''' \in \mathcal{W}$ we have natural maps

$$\mathbb{I}^r_{gg,\mathrm{ord}}(w', w'') \to \mathbb{I}^r_{gg,\mathrm{ord}}(w'', w'), \qquad f \mapsto f^{\mathrm{t}}$$

$$\mathbb{I}^r_{gg,\mathrm{ord}}(w', w'') \to \mathbb{I}^{r+1}_{gg,\mathrm{ord}}(\phi w', \phi w''), \; f \mapsto f^{\phi} := \phi(f),$$

$$\mathbb{I}^r_{gg,\mathrm{ord}}(w', w'') \times \mathbb{I}^r_{gg,\mathrm{ord}}(-w'', w''') \to \mathbb{I}^r_{gg,\mathrm{ord}}(w', w'''), \qquad (f_1, f_2) \mapsto f_1 f_2,$$

where the first map is R-linear, the second one is $\mathbb{Z}_p$-linear, and the third one is R-bilinear. Also, the rule $(A, \theta, \omega, S) \mapsto 1_g$ defines an element of $\mathbb{I}^r_{gg,\mathrm{ord}}(-w', w')$ for each $w' \in \mathcal{W}$. So the set $\mathcal{W}$ can be seen as the set of objects of a category whose set of morphisms from $w' \in \mathcal{W}$ to $w'' \in \mathcal{W}$ identifies with the R-module $\mathbb{I}^r_{gg,\mathrm{ord}}(-w', w'')$ and whose composition law is given by the above R-bilinear maps.

(b) If $f \in \mathbb{I}^r_{gg,\mathrm{ord}}(w', w'')$ has the property that we have

$$f(A, \theta, \omega, S) \in \mathbf{GL}_g(S^r)$$

for all (A, θ, ω, S), then the rule

$$(A, \theta, \omega, S) \mapsto (f(A, \theta, \omega, S))^{-1} \tag{2.29}$$

defines an element

$$f^{-1} \in \mathbb{I}^r_{gg,\mathrm{ord}}(-w'', -w').$$

Similarly, if $f \in \mathbb{I}^r_{g,\mathrm{ord}}(w)$ has the property that we have

$$f(A, \theta, \omega, S) \in (S^r)^{\times}$$

for all (A, θ, ω, S), then the rule of Display (2.29) defines an element

$$f^{-1} \in \mathbb{I}^r_{g,\mathrm{ord}}(-w).$$

The same applies to Siegel δ-modular functions and forms in $M^r_{g,\mathrm{ord}}$ or $M^r_{gg,\mathrm{ord}}$.

(c) For all $w', w'' \in \mathcal{W}$ we have natural maps

$$\det : \mathbb{I}^r_{gg,\mathrm{ord}}(w', w'') \to \mathbb{I}^r_{g,\mathrm{ord}}(-w' - w''), \quad f \mapsto \det(f),$$

which for $g = 1$ are canonical identifications. For $g = 1$ we have additional canonical identifications $M^r_{11,\mathrm{ord}}(w', w'') = M^r_{1,\mathrm{ord}}(-w' - w'')$, etc.

2.3.3 Rings of δ-Homogeneous Polynomials

For an $n+1$-tuple of indeterminates $z := (z_0, \ldots, z_n)$, where $n \geq 1$, the δ-ring of δ-polynomials $R[z, z', z'', \ldots]$ (defined by Display (2.1) but with (y, d) replaced by (z, n)) contains the elements indexed by $i \in \{0, \ldots, n\}$:

$$z_i^{\phi} = z_i^p + pz_i', \quad z_i^{\phi^2} = (z_i^p + pz_i')^p + p((z_i')^p + pz_i''), \text{ etc..}$$

We define $z' := (z_0', \ldots, z_n')$, $z^{\phi} := (z_0^{\phi}, \ldots, z_n^{\phi})$, etc., and we view the R-algebra $R[z, z', z'', \ldots]$ as a K-subalgebra of the K-algebra $K[z, z', z'', \ldots]$ of polynomials in the set of indeterminates $\{z_i, z_i', z_i'', \ldots | i \in \{0, \ldots, n\}\}$. This K-algebra is generated by the set $\{z_i, z_i^{\phi}, z_i^{\phi^2}, \ldots | i \in \{0, \ldots, n\}\}$, i.e.,

$$K[z, z', z'', \ldots] = K[z, z^{\phi}, z^{\phi^2}, \ldots] := K[z_i, z_i^{\phi}, z_i^{\phi^2}, \ldots | i \in \{0, \ldots, n\}]. \quad (2.30)$$

On this K-algebra we consider the W_+-grading defined by letting $z_i^{\phi^j}$ have degree ϕ^j. For $w \in W_+$ we denote by $K[z, z^{\phi}, z^{\phi^2}, \ldots](w)$ the component of degree w in the ring of Eq. (2.30) and consider the intersection

$$\mathbb{S}_n(w) := R[z, z', z'', \ldots] \cap (K[z, z^{\phi}, z^{\phi^2}, \ldots](w))$$

inside the ring of Eq. (2.30).

Definition 2.3.18 The ring of *δ-homogeneous polynomials* in z is the W_+-graded ring

$$\mathbb{S}_n := \bigoplus_{w \in W_+} \mathbb{S}_n(w). \quad (2.31)$$

This ring has a natural structure of an R-algebra and is equipped with a ring endomorphism ϕ, extending the Frobenius lift on R, such that for all $w \in W_+$ we have $\phi(\mathbb{S}_n(w)) \subset \mathbb{S}_n(\phi(w))$. Note however that ϕ is not a Frobenius lift on $\mathbb{S}_n$ and thus this ring does not have a natural structure of a δ-ring; indeed we have $z_i' = \frac{1}{p} z_i^{\phi} - \frac{1}{p} z_i^p \notin \mathbb{S}_n$ because the homogeneous components $\frac{1}{p} z_i^{\phi}$ and $\frac{1}{p} z_i^p$ of z_i' do not belong to $\mathbb{S}_n$. However, if $F_0, \ldots, F_m \in \mathbb{S}_n(w_1)$ and $G \in \mathbb{S}_m(w_2)$, then $G(F_0, \ldots, F_m) \in \mathbb{S}_n(w_1 w_2)$. In particular, as for all $i, j \in \{0, \ldots, n\}$ we have an identity

$$\{z_i, z_j\}_{\delta} = z_i^p z_j' - z_j^p z_i' \in \mathbb{S}_n(\phi + p),$$

it follows that for all $F, G \in \mathbb{S}_n(w)$ we have $\{F, G\}_\delta \in \mathbb{S}_n((\phi + p)w)$. So $\mathbb{S}_n$ has a natural structure of a filtered integral δ-graded ring over R with filtration $(\mathbb{S}_n \cap R[z, z', \ldots, z^{(r)}])_{r \geq 0}$.

It is sometimes useful to shift the filtration and grading on $\mathbb{S}_n$ in a way that it depends on a given integer $r_0 \geq 0$ and on a given $w_0 \in W(r_0)$. More precisely, for a given integer $r_0 \geq 0$ we define a filtration $(\mathbb{S}_n^r)_{r \geq 0}$ by setting $\mathbb{S}_n^r := R$ if $0 \leq r < r_0$ and

$$\mathbb{S}_n^r := \mathbb{S}_n \cap R[z, z', \ldots, z^{(r-r_0)}], \text{ if } r \geq r_0,$$

and for a given $w_0 \in W(r_0)$ we define a new grading on $\mathbb{S}_n$ by letting each $z_i^{\phi^j}$ have weight $\phi^j w_0$. With this new filtration and grading $\mathbb{S}_n$ is, again, a filtered integral δ-graded ring and we refer to the pair (r_0, w_0) as its *shift*. In particular, for each shift (r_0, w_0) (which will not be included in our notation), we view the connected filtered ss-δ-space

$$\mathbb{P}_\delta^n := \mathrm{Proj}_\delta(\mathbb{S}_n, z) = \mathrm{Spec}_\delta\left(\prod_{0 \leq i_0 < \cdots < i_\bullet \leq n} \mathbb{S}_{n,\langle z_{i_0} \cdots z_{i_\bullet}\rangle}\right) \tag{2.32}$$

as an analog in the δ-geometry of the n-th dimensional projective space over R.

We have a K-algebra isomorphism (identification) of W_+-graded algebras

$$\mathbb{S}_n \otimes_R K \simeq K[z, z^\phi, z^{\phi^2}, \ldots].$$

Note that the inclusion $\mathbb{S}_n \subset R[z, z', z'', \ldots]$ is compatible with ϕ acting on the two R-algebras and for all $w \in W$, the R-linear map

$$\mathbb{S}_n(w) \to R[z, z', z'', \ldots]$$

has a torsion free cokernel.

The triple $(\mathbb{S}_n, R[z, z', z'', \ldots], z)$ with shift (r_0, w_0) has the following universal property. Consider an arbitrary triple $(B, C, \mathbf{f})$, where

(i) C is a δ-ring over R, B is a R-subalgebra of C such that $\phi(B) \subset B$ and B is equipped with an integral δ-graded ring structure with respect to the restriction of ϕ to B in such a way that for all $w \in W$, the R-linear map $B(w) \to C$ has a torsion free cokernel;

(ii) $\mathbf{f} = (f_0, \ldots, f_n)$ is a linear system in B of weight $w_0 \in W(r_0)$.

Then there exists a unique morphism of δ-graded rings $\mathbb{S}_n \to B$ such that $z_i \mapsto f_i$ for all $i \in \{0, \ldots, n\}$, where $\mathbb{S}_n$ has shift (r_0, w_0). To check this, note that there exists a unique morphism of δ-rings $R[z, z', z'', \ldots] \to C$ that sends $z_i \mapsto f_i$ for all $i \in \{0, \ldots, n\}$ and this morphism sends $\mathbb{S}_n$ to B.

Let $\mathbb{I}$ stand for either $\mathbb{I}_g$ or $\mathbb{I}_{g,\mathrm{ord}}$ and let M stand for either M_g or $M_{g,\mathrm{ord}}$. Applying the above universal property to $B = \mathbb{I}$ and $C = M$ (see the paragraph of Diagram (2.17)), we get that for every linear system $\mathbf{f} = (f_0, \ldots, f_n)$ in $\mathbb{I}$ whose all entries belong to $\mathbb{I}^{r_0}(w_0)$, we have an induced morphism of filtered δ-graded rings

$$\mathbb{S}_n \to \mathbb{I} \tag{2.33}$$

over R that maps z_i to f_i, where $\mathbb{S}_n$ has shift (r_0, w_0). Then Diagram (2.33) induces a morphism of filtered ss-δ-spaces

$$\mathrm{Proj}_\delta(\mathbb{I}, \mathbf{f}) \to \mathrm{Proj}_\delta(\mathbb{S}_n, z).$$

2.3.4 Rings of Invariants of $\mathbf{SL}_g$-Actions

We now define R-modules and R-algebras that will play key roles in our theory. Let

$$\begin{aligned} T &= T^{(0)} = (T_{ij})_{1\le i,j\le g} = (T_{ij}^{(0)})_{1\le i,j\le g}, \\ T' &= T^{(1)} = (T'_{ij})_{1\le i,j\le g} = (T_{ij}^{(1)})_{1\le i,j\le g}, \\ &\dots, \; T^{(r)} = (T_{ij}^{(r)})_{1\le i,j\le g}, \;\dots \end{aligned}$$

be a sequence of symmetric matrices of indeterminates over R. For all integers $1 \le i \le j \le g$ and $r \ge 0$, $T_{ij}^{(r)} = T_{ji}^{(r)}$ will have degree 1.

For every $s \in \frac{1}{2}\mathbb{Z} = \{\frac{m}{2} | m \in \mathbb{Z}\}$ such that $gs \in \mathbb{Z}$ we denote by $\mathbb{H}_g^r(s)$ the R-module of homogeneous polynomials

$$G \in R[T, \dots, T^{(r)}] := R[T_{ij}^{(l)} | 1 \le i \le j \le g, 0 \le l \le r]$$

of degree gs that satisfy

$$G(\Lambda T \Lambda^{\mathrm{t}}, \dots, \Lambda T^{(r)} \Lambda^{\mathrm{t}}) = \det(\Lambda)^{2s} G(T, \dots, T^{(r)}) \tag{2.34}$$

for all $\Lambda \in \mathbf{GL}_g(R)$. Here and later, for G as above and symmetric matrices $M_0 = (m_{0,ij})_{1\le i,j\le g}, \dots, M_r = (m_{r,ij})_{1\le i,j\le g}$ with entries in an R-algebra we use the notation

$$G(M_1, \dots, M_r) := G(m_{0,11}, m_{0,12}, \dots, m_{0,gg}, \dots, m_{r,11}, m_{r,12}, \dots, m_{r,gg}).$$

A similar notation will be used later (see Sect. 3.5) for G a power series instead of a polynomial.

We also set $\mathbb{H}_g^r(s) := 0$ if $s \in \frac{1}{2}\mathbb{Z}$ and $gs \notin \mathbb{Z}$. Clearly, we have $\mathbb{H}_g^r(s) = 0$ for all $s \in \frac{1}{2}\mathbb{Z}$ with $s < 0$. We then define the R-algebras

$$\mathbb{H}_g^r := \bigoplus_{s\in\mathbb{Z}} \mathbb{H}_g^r(s), \quad \mathbb{H}_{g,\mathrm{tot}}^r := \bigoplus_{s\in\frac{1}{2}\mathbb{Z}} \mathbb{H}_g^r(s),$$

$$\mathbb{H}_g^{r,\mathrm{torus}} := \mathbb{H}_g^r[z_0, \dots, z_{r-1}, \frac{1}{z_0\cdots z_{r-1}}], \quad \mathbb{H}_g^{\mathrm{torus}} := \bigcup_{r\ge 0} \mathbb{H}_g^{r,\mathrm{torus}},$$

$$\mathbb{H}_{g,\mathrm{tot}}^{r,\mathrm{torus}} := \mathbb{H}_{g,\mathrm{tot}}^{r}[z_0, \ldots, z_{r-1}, \frac{1}{z_0 \cdots z_{r-1}}], \quad \mathbb{H}_{g,\mathrm{tot}}^{\mathrm{torus}} := \bigcup_{r \geq 0} \mathbb{H}_{g,\mathrm{tot}}^{r,\mathrm{torus}},$$

where $(z_n)_{n\geq 0}$ is a sequence of indeterminates. We view $\mathbb{H}_{g,\mathrm{tot}}^{\mathrm{torus}}$ as a filtered W-graded ring with filtration given by the subrings in the union that defines it and grading defined by letting each element of $\mathbb{H}_g^r(s)$ have weight $-2s$ and declaring that

$$\text{each } z_i \text{ has weight } \phi^{i+1} - \phi^i.$$

The R-algebras we just introduced are analogs of algebras in the classical geometric invariant theory of multiple quadratic forms we study in Sects. 4.1–4.3.

Similarly, we consider matrices of indeterminates

$$X_1 = (X_{1,ij})_{1\leq i,j\leq g}, \ldots, X_n = (X_{n,ij})_{1\leq i,j\leq g}$$

and we denote by $\mathbb{H}_{g,\mathrm{conj}}^n$ the R-algebra of all polynomials

$$F \in R[X_1, \ldots, X_n] := R[X_{l,ij} | 1 \leq i, j \leq g, 1 \leq l \leq n]$$

such that

$$F(\Lambda X_1 \Lambda^{-1}, \ldots, \Lambda X_n \Lambda^{-1}) = F(X_1, \ldots, X_n)$$

for all $\Lambda \in \mathbf{GL}_g(R)$. The R-algebras $\mathbb{H}_{g,\mathrm{conj}}^n$ are analogs of the algebras that show in the classical geometric invariant theory of multiple endomorphisms and that are studied in Sects. 4.1.1, 4.1.5 and 4.1.6.

Some of our main results will establish links between the algebras $\mathbb{I}_{g,\mathrm{ord}}^r$ and $\mathbb{I}_g^r$ and the algebras $\mathbb{H}_{g,\mathrm{tot}}^{r,\mathrm{torus}}$ and $\mathbb{H}_{g,\mathrm{conj}}^r$ (see Theorems 2.4.7 and 2.4.15 below).

2.3.5 Rings of δ-Power Series

If $f : C \to B$ is a set theoretical map between rings and $n \in \mathbb{N}$, then the rule $(c_{ij})_{1\leq i,j\leq n} \mapsto (f(c_{ij}))_{1\leq i,j\leq n}$ defines a function $f : \mathbf{Mat}_n(C) \to \mathbf{Mat}_n(B)$ denoted also by f.

Our Serre–Tate expansion theory will involve the rings

$$S_{\mathrm{for}}^r := \widehat{R[[T]][T', \ldots, T^{(r)}]} := \widehat{R[[T_{ij} | 1 \leq i \leq j \leq g]][T_{ij}^{(l)} | 1 \leq i \leq j \leq g, 1 \leq l \leq r]}.$$

Their union S_{for} has a natural structure of a filtered δ-ring (with $\delta T_{ij}^{(l)} = T_{ij}^{(l+1)}$ for all integers $1 \leq i \leq j \leq g$ and $l \geq 0$) which is an object of $\widehat{\mathbf{FilRing}}_\delta$. On the other hand the Fourier expansion theory developed in [BB] involved the rings

$$S_{\mathrm{can}}^r := \widehat{R((q))[q', \ldots, q^{(r)}]} := \widehat{R((q))[q'_{ij}, \ldots, q_{ij}^{(r)} | 1 \leq i \leq j \leq g]},$$

where

$$q = q^{(0)} = (q_{ij})_{1 \le i,j \le g} = (q_{ij}^{(0)})_{1 \le i,j \le g},$$
$$q' = q^{(1)} = (q'_{ij})_{1 \le i,j \le g} = (q_{ij}^{(1)})_{1 \le i,j \le g},$$
$$\dots, q^{(r)} = (q_{ij}^{(r)})_{1 \le i,j \le g}, \dots$$

is a sequence of symmetric matrices of indeterminates and

$$R((q)) := R[[q_{ij} | 1 \le i \le j \le g]][q_{ij}^{-1} | 1 \le i \le j \le g].$$

Again, the union S_{can} of these rings is an object of $\widehat{\mathbf{FilRing}}_\delta$, with $\delta q_{ij}^{(l)} = q_{ij}^{(l+1)}$ for all integers $1 \le i \le j \le g$ and $l \ge 0$.

In order to compare later on the Fourier and Serre–Tate expansions of the $f^r_{g,\text{crys}}$s (see Remark 2.4.3 below), we will need to consider the R-subalgebras

$$S^r_{\text{small-can}} := \widehat{R[q_{ij}, q_{ij}^{-1}, q'_{ij}, q''_{ij}, \dots, q_{ij}^{(r)} | 1 \le i \le j \le g]} \subset S^r_{\text{can}}$$

and the R-algebra monomorphisms $S^r_{\text{small-can}} \to S^r_{\text{for}}$, $F \mapsto F_{\text{for}}$ that map $q_{ij}, q'_{ij}, q''_{ij}$, ... to $1 + T_{ij}$, $1 + T'_{ij}$, $1 + T''_{ij}$, ... (respectively). We have an R-algebra monomorphism

$$\mathbf{Mat}_g(S^r_{\text{small-can}}) \to \mathbf{Mat}_g(S^r_{\text{for}})$$

defined by the rule $M = (c_{ij})_{1 \le i,j \le g} \mapsto M_{\text{for}} = ((c_{ij})_{\text{for}})_{1 \le i,j \le g}$. Hence

$$q_{\text{for}} = \mathbb{E}_g + T, \quad q'_{\text{for}} = \delta(\mathbb{E}_g + T), \quad q''_{\text{for}} = \delta^2(\mathbb{E}_g + T), \dots,$$

where we recall that $\mathbb{E}_g$ is the $g \times g$ matrix with all entries equal to 1.

A key role in our monograph will be played by the matrix of series

$$\Psi_q = (\Psi_{q_{ij}})_{1 \le i,j \le g} \in \mathbf{Mat}_g(S^r_{\text{small-can}}) \subset \mathbf{Mat}_g(S^r_{\text{can}}), \tag{2.35}$$

where

$$\Psi_{q_{ij}} := \frac{1}{p} \log \left(1 + p \frac{\delta q_{ij}}{q_{ij}^p} \right) \in S^r_{\text{small-can}},$$

and by its image

$$\Psi := (\Psi_q)_{\text{for}} \in \mathbf{Mat}_g(S^r_{\text{for}}). \tag{2.36}$$

2.3.6 *Rings of δ-Sections of Line Bundles*

Let X be a smooth scheme over $\text{Spec}(R)$ and L a line bundle over X. For simplicity we will assume in what follows that the reduction modulo p, $\overline{X}$, is connected, hence an integral scheme. We denote by

$$\mathbb{L} := \mathbf{Spec}\left(\bigoplus_{i\in\mathbb{Z}} L^{\otimes i}\right)$$

the principal $\mathbb{G}_m$-bundle over X associated to L. The $\mathbb{G}_m$-action on $\mathbb{L}$ induces an $R^\times$-action, denoted by $\star$, on the p-jet spaces $J^r(\mathbb{L})$ and on their rings of global functions $\mathcal{O}(J^r(\mathbb{L}))$. For each $w \in W$ we denote by $\mathbb{S}^r_{X,L}(w)$ the R-module of all $f \in \mathcal{O}(J^r(\mathbb{L}))$ such that we have $\lambda \star f = \lambda^{-w} \cdot f$ for all $\lambda \in R^\times$.

Definition 2.3.19 The ring of *δ-sections* of L is the ring $\mathbb{S}_{X,L}$ defined by

$$\mathbb{S}_{X,L} := \bigcup_{r=0}^{\infty} \mathbb{S}^r_{X,L}, \quad \mathbb{S}^r_{X,L} := \bigoplus_{w\in W(r)} \mathbb{S}^r_{X,L}(w).$$

If $X = \mathrm{Spec}(B)$ is affine, L is trivial over X, and $\{z_L\}$ is a basis of L, then we have isomorphisms

$$\mathcal{O}(J^r(\mathbb{L})) \simeq \overline{\mathcal{O}(J^r(X))[z_L, z_L^{-1}, z_L', \ldots, z_L^{(r)}]} = \overline{J^r(B)[z_L, z_L^{-1}, z_L', \ldots, z_L^{(r)}]},$$

$$\mathbb{S}^r_{X,L}(w) \simeq \mathcal{O}(J^r(X)) \cdot z_L^w = J^r(B) \cdot z_L^w.$$

The R-algebra $\mathbb{S}_{X,L}$ has a structure of a filtered graded ring coming from the filtered graded system $(\mathbb{S}^r_{X,L})_{r\geq 0}$ and with this structure it is a filtered integral δ-graded ring over R. In particular, for each linear system $\mathbf{f} = (f_0, \ldots, f_n)$ in $\mathbb{S}_{X,L}$ we define the filtered ss-δ-space (see Display (2.4))

$$X_{\delta,L,\mathbf{f}} := \mathrm{Proj}_\delta(\mathbb{S}_{X,L}, \mathbf{f}) = \mathrm{Spec}_\delta\left(\prod_{0\leq i_0<\cdots<i_\bullet\leq n} \mathbb{S}_{X,L,\langle f_{i_0}\cdots f_{i_\bullet}\rangle}\right). \tag{2.37}$$

Example 2.3.20 Recall the ring $\mathbb{S}_n$ of Definition 2.3.18 (see Display (2.31)). If we take $X = \mathbb{P}^n_R = \mathrm{Proj}(R[z])$, $z = (z_0, \ldots, z_n)$, and $L = \mathcal{O}(1)$, then one can prove (see [Bu05], Rmk. 6.9, p. 167) that $\mathbb{S}_{\mathbb{P}^n_R,\mathcal{O}(1)} = \mathbb{S}_n$, hence, with notation as in Displays (2.37) and (2.32), we have

$$\mathbb{P}^n_{\delta,\mathcal{O}(1),z} = \mathrm{Proj}_\delta(\mathbb{S}_{\mathbb{P}^n_R,\mathcal{O}(1)}, z) = \mathrm{Proj}_\delta(\mathbb{S}_n, z) = \mathbb{P}^n_\delta.$$

We will abbreviate $X_{\delta,L,\mathbf{f}}$ in Display (2.37) by X_δ if L and $\mathbf{f}$ are clear from the context. In view of the above example this is consistent with the notation in Display (2.32) in case $X = \mathbb{P}^n_R$, $L = \mathcal{O}(1)$, $\mathbf{f} = z$.

Going back to the case of a general X and $\mathbf{f}$, let $P \in X(R)$ be an R-valued point of X. Then we will construct an n-tuple

$$(f_0(P), \ldots, f_n(P)) \in R^{n+1}, \tag{2.38}$$

well-defined up to multiplication by an element of $R^\times$, as follows. We can assume X is affine and L is trivial on X. Fix a basis $\{z_L\}$ of L, write $f_i = \varphi_i \cdot z_L^w$, with $\varphi_i \in \mathcal{O}(J^r(X))$, view φ_i as a morphism $\varphi_i : J^r(X) \to \widehat{\mathbb{A}^1_R}$, consider the point $J^r(P) \in J^r(X)(R)$, and set $f_i(P) := \varphi_i(J^r(P)) \in \widehat{\mathbb{A}^1_R}(R) = R$. In other words, $P \mapsto f_i(P)$ is the δ-function defined by φ_i (see Definition 2.1.10).

Definition 2.3.21 Let $P \in X(R)$. We say that P is:

(a) f-*semistable* if there exists $i \in \{0, \ldots, n\}$ such that $f_i(P) \in R^\times$.

(b) f-*unstable* if $f_0(P) = \cdots = f_n(P) = 0$.

The definition makes sense even though the $n + 1$-tuple of Display (2.38) is only defined up to multiplication by a scalar in $R^\times$. We denote by $X(R)^{ss}_{\mathbf{f}}$ and $X(R)^{u}_{\mathbf{f}}$ the sets of **f**-semistable and **f**-unstable (respectively) points in $X(R)$. We have

$$X(R)^{ss}_{\mathbf{f}} \sqcup X(R)^{u}_{\mathbf{f}} \subset X(R)$$

and in general the inclusion is strict.

Example: for $X = \mathbb{P}^n_R$ and $\mathbf{f} = z = (z_0, \ldots, z_n)$, we have $\mathbb{P}^n_R(R)^{ss}_z = \mathbb{P}^n_R(R)$.

For general X and **f** we have a canonical evaluation map

$$\mathcal{E}_{X,\mathbf{f}} : X(R)^{ss}_{\mathbf{f}} \to X_\delta(\delta\text{-pts})$$

defined as follows. For $P \in X(R)^{ss}_{\mathbf{f}}$, we consider the nonempty set of indices

$$I := \{i \in \{0, \ldots, n\} | f_i(P) \in R^\times\},$$

and for each increasing sequence $0 \leq i_0 < \cdots < i_d \leq n$ let $R_{i_0 \ldots i_d}$ be R if $\{i_0, \ldots, i_d\} \subset I$ and be 0 if $\{i_0, \ldots, i_d\} \not\subset I$. Then consider the discrete ss-δ-space

$$P_\delta := \text{Spec}\left(\prod_{0 \leq i_0 < \cdots < i_\bullet \leq n} R_{i_0 \ldots i_\bullet}\right), \tag{2.39}$$

with obvious face maps; one can easily check that P_δ is connected and acyclic, hence a δ-point. Next we consider the R-algebra homomorphisms

$$\mathbb{S}_{X,L,\langle f_{i_0} \cdots f_{i_\bullet} \rangle} \to R_{i_0 \ldots i_\bullet}$$

defined when $\{i_0, \ldots, i_\bullet\} \subset I$ as the evaluation at P. The induced morphism $P_\delta \to X_\delta$ is $\mathcal{E}_{X,\mathbf{f}}(P)$. Note that there exists a canonical diagram

$$\begin{array}{ccccc} X(R)^{ss}_{\mathbf{f}} & \xrightarrow{\Phi_{\mathbf{f}}} & \mathbb{P}^n_R(R)^{ss}_z & = & \mathbb{P}^n_R(R) \\ \downarrow{\scriptstyle \mathcal{E}_{X,\mathbf{f}}} & & \downarrow{\scriptstyle \mathcal{E}_{\mathbb{P}^n_R, z}} & & \\ X_\delta(\delta\text{-pts}) & \longrightarrow & \mathbb{P}^n_\delta(\delta\text{-pts}), & & \end{array}$$

where

$$\Phi_{\mathbf{f}}(P) := (f_0(P) : \cdots : f_n(P)).$$

As with δ-functions (see Definition 2.1.10) the maps $\Phi_{\mathbf{f}}$ are not induced by morphisms of formal schemes from open formal subschemes of $\widehat{X}$ to $\widehat{\mathbb{P}^n_R}$. Rather, they are induced by morphisms of formal schemes from open formal subschemes of $J^r(X)$ to $\widehat{\mathbb{P}^n_R}$ and hence they are given, locally in the Zariski topology of X, by restricted power series in local affine coordinates $x_1, \ldots, x_d$ and their p-derivatives $\delta x_1, \ldots, \delta x_d, \ldots, \delta^r x_1, \ldots, \delta^r x_d$; such maps are referred to in [Bu95a], Introd. and [Bu05], Sect. 2.1 as *δ-maps*.

Example 2.3.22 We recall that for an integer $N \geq 3$ prime to p the moduli scheme $\mathcal{X} := \mathcal{A}_{g,1,N,R}$ of principally polarized abelian schemes of relative dimension g equipped with simplectic similitude level-N structure is smooth over Spec(R) (see Appendix). We denote by $\mathcal{X}_{\text{ord}} := \mathcal{A}_{g,1,N,R,\text{ord}}$ the open subscheme of $\mathcal{X}$ which is the union of the generic fiber $\mathcal{X}_K$ and of the ordinary locus $\mathcal{X}_{\text{ord},k}$ of the special fiber $\mathcal{X}_k$.

Let $X \subset \mathcal{X}$ be an open subscheme with $\overline{X}$ connected and let $\pi : A_X \to X$ the universal abelian scheme with principal polarization θ_X. Recall the *Hodge line bundle* over X,

$$L := \pi_*(\wedge^g \Omega_{A_X/X}).$$

With X and L as above we set

$$\mathbb{M}^r_X = \bigoplus_{w \in W(r)} \mathbb{M}^r_X(w) := \mathbb{S}^r_{X,L} = \bigoplus_{w \in W(r)} \mathbb{S}^r_{X,L}(w), \quad \mathbb{M}_X := \mathbb{S}_{X,L}. \tag{2.40}$$

We define a canonical morphism of filtered δ-graded rings

$$\mathbb{M}_g \to \mathbb{M}_{\mathcal{X}}. \tag{2.41}$$

If X is affine and there exists a column basis $\omega_X = (\omega_{X,1}, \ldots, \omega_{X,g})^{\mathrm{t}}$ of 1-forms on A_X, then for $f \in M^r_g(w)$ we consider the element

$$f_X := f(A_X, \theta_X, \omega_X, \mathcal{O}^\infty(X)) \in \mathcal{O}(J^r(X)) \simeq \mathcal{O}(J^r(X)) \cdot z^w_L, \tag{2.42}$$

where we identify z_L with the canonical global section of L defined by the wedge product $\omega_{X,1} \wedge \ldots \wedge \omega_{X,g}$. The rule $f \mapsto f_X$ defines R-linear maps

$$M^r_g(w) \to \mathcal{O}(J^r(X)) \simeq \mathcal{O}(J^r(X)) \cdot z^w_L \tag{2.43}$$

that depend on ω_X. For X arbitrary, the above maps induce an R-algebra homomorphism in Diagram (2.41) which is intrinsic, i.e., is independent of the choices made to get such local column bases of 1-forms. By Corollary 3.5.2 (d) below, the homomorphism in Diagram (2.41) is injective with injective reduction modulo p.

If $X \subset \mathcal{X}_{\text{ord}}$ then we similarly define a canonical morphism of filtered δ-graded rings $\mathbb{M}_{g,\text{ord}} \to \mathbb{M}_X$ which is injective with injective reduction modulo p. For instance, as above, if X is affine, then the rule $f \mapsto f_X$ defines R-linear maps

$$M^r_{g,\text{ord}}(w) \to \mathcal{O}(J^r(X)) \simeq \mathcal{O}(J^r(X)) \cdot z_L^w \tag{2.44}$$

that depend on ω_X.

2.3.7 Analogies with Geometric Invariant Theory

Before proceeding to explain our main results (see Sect. 2.4) we give here a quick summary of the various ss-δ-spaces introduced so far and we point out some analogies with geometric invariant theory as in [MFK].

Let $(\mathbb{I}, \mathbb{M}, \mathbb{M}')$ be one of the two triples

$$(\mathbb{I}_g, \mathbb{M}_g, \mathbb{M}'_g) \quad \text{or} \quad (\mathbb{I}_{g,\text{ord}}, \mathbb{M}_{g,\text{ord}}, \mathbb{M}'_{g,\text{ord}}), \tag{2.45}$$

see Display (2.21) and its ordinary analog, and recall that $\mathbb{I}$, $\mathbb{M}$, $\mathbb{M}'$ are filtered integral δ-graded rings. Recall also the filtered integral δ-graded rings $\mathbb{S}_n$ and $\mathbb{M}_X$, where $X \subset \mathcal{X} = \mathcal{A}_{g,1,N,R}$ or $X \subset \mathcal{X}_{\text{ord}}$, $N \geq 3$ an integer prime to p, $\overline{X}$ connected, see Definition 2.3.18 and Display (2.40). We have homomorphisms of filtered δ-graded rings

$$\mathbb{S}_n \to \mathbb{I} \to \mathbb{M} \to \mathbb{M}_X \quad \text{and} \quad \mathbb{M} \rightrightarrows \mathbb{M}',$$

where the homomorphism $\mathbb{S}_n \to \mathbb{I}$ is defined by a linear system $\mathbf{f}$ of length $n+1$ in $\mathbb{I}$ (see Diagram (2.33)) and the other homomorphisms (see Diagrams (2.19) and (2.41)) are the canonical ones which, as stated, are injective, with injective reduction modulo p. The main geometric objects of our theory are then the filtered ss-δ-spaces:

$$\begin{aligned} X_\delta &:= \text{Proj}_\delta(\mathbb{M}_X, \mathbf{f}), \\ Y'_\delta &:= \text{Proj}_\delta(\mathbb{M}', \mathbf{f}), \\ Y_\delta &:= \text{Proj}_\delta(\mathbb{M}, \mathbf{f}), \\ Z_\delta &:= \text{Proj}_\delta(\mathbb{I}, \mathbf{f}), \\ \mathbb{P}^n_\delta &:= \text{Proj}_\delta(\mathbb{S}_n, z), \end{aligned} \tag{2.46}$$

where we continue to denote by $\mathbf{f}$ the images of $\mathbf{f}$ in the rings $\mathbb{M}$, $\mathbb{M}'$, $\mathbb{M}_X$. There exist naturally induced morphisms of filtered ss-δ-spaces

$$X_\delta \to Y_\delta \to Z_\delta \to \mathbb{P}^n_\delta \quad \text{and} \quad Y'_\delta \rightrightarrows Y_\delta. \tag{2.47}$$

We also have an induced map of sets

$$\Phi_{\mathbf{f}} : X(R)^{ss}_{\mathbf{f}} \to \mathbb{P}^n_R(R) \tag{2.48}$$

compatible with the map $X_\delta(\delta\text{-pts}) \to \mathbb{P}^n_\delta(\delta\text{-pts})$.

The morphism

$$Y_\delta \to Z_{\hat{\delta}} \tag{2.49}$$

is a categorical quotient (coequalizer) in $\mathbf{FilSpss}_\delta$ of the pair of morphisms

$$Y'_\delta \rightrightarrows Y_\delta ; \tag{2.50}$$

see Remark 2.3.6.

By analogy with geometric invariant theory (see below), the pair of morphisms in Diagram (2.50) can be viewed as an analog in δ-geometry of the pair of morphisms of functors from **Ring** to **Set**

$$\mathcal{A}'_{g,1,R} \overset{\pi_1,\pi_2}{\rightrightarrows} \mathcal{A}_{g,1,R} \tag{2.51}$$

from the union $\mathcal{A}'_{g,1,R}$ of all prime to p Hecke correspondences on $\mathcal{A}_{g,1,R}$. (A similar picture holds for the ordinary locus.) Note that as $N \geq 3$ is prime to p the coequalizer in the category of schemes over Spec(R) of the correspondence

$$\mathcal{A}'_{g,1,N,R} \overset{\pi_1,\pi_2}{\rightrightarrows} \mathcal{A}_{g,1,N,R} \tag{2.52}$$

obtained by taking the union of all prime to p Hecke correspondences on $\mathcal{A}_{g,1,N,R}$ is trivial, i.e., is isomorphic to Spec(R). This is so as, for an algebraically closed field κ, the generic orbits of the Hecke correspondences are Zariski dense in $\mathcal{A}_{g,1,N,\kappa}$ (see [Cha], Thm. 2, for κ of characteristic p). Moreover, Diagram (2.49) can be viewed as an analog in δ-geometry of the coequalizer

$$\mathcal{A}_{g,1,R} \to \mathcal{A}_{g,1,R}/\mathcal{A}'_{g,1,R} \tag{2.53}$$

of Diagram (2.51) in the category of functors from **Ring** to **Set**, or of the similar map in the ordinary case. Similarly, we can view the morphism

$$X_\delta \to Z_{\hat{\delta}}$$

as an analog of the coequalizer of Diagram (2.52) in the category of functors from **Ring** to **Set**.

The various properties of the ring $\mathbb{I}$ to be proved in Chap. 3 (such as ϕ-transcendence degree, asymptotic transcendence degree, ϕ-finite generation, etc.) correspond to 'geometric properties' of the morphisms in Display (2.47) and of the maps in Diagram (2.48). We will not pursue this in a systematic way but some of these geometric properties will be explained in Sect. 2.4. Here is an example of such a property. Let us say that a subset $C \subset X(R)$ is a *polarized isogeny class* if there exists $P \in C$ such that C consists of all points $P' \in X(R)$ with the property that there exists a polarized isogeny between the principally polarized abelian schemes over Spec(R) that correspond to P and P'. Directly from the definition of Hecke covariance we get that the maps in Diagram (2.48) are constant on polarized isogeny classes.

Here are some comments on the analogy between the situation described above and the situation in classical geometric invariant theory as in [MFK]. Given an action $\rho : X' := G \times X \to X$ of an algebraic group G on a variety X over some algebraically closed field, one says that a morphism $\pi : X \to Y$ of varieties is a categorical quotient for this action if it is a coequalizer, in the category of varieties, of the two morphisms

$$\rho, \pi_2 : X' \rightrightarrows X \tag{2.54}$$

where π_2 is the second projection. In our setting, Diagram (2.54) is replaced by Hecke correspondences on the moduli stack $\mathcal{A}_{g,1,R}$ over Spec(R), see Diagram (2.51). The fact that Diagram (2.49) is a coequalizer in $\mathbf{FilSpss}_\delta$ of the pair of morphisms in Diagram (2.50), suggests, following the language of the geometric invariant theory, to regard Z_δ as a 'quotient', in δ-geometry, of an analog of $\mathcal{A}_{g,1,N,R}$ or of its ordinary variant by the Hecke correspondences. As we shall see, Z_δ is a highly nontrivial object, and this is one of the main motivations here (and in [Bu00, BB, Bu03, Bu05]) for enriching classical algebraic geometry with the δ-geometry.

In the same vein, we view the maps $\Phi_{\mathbf{f}}$ in Diagram (2.48) as analogs, in the δ-geometry of Hecke correspondences, of maps to projective spaces defined by G-invariant sections of G-linearized bundles over a variety with G-action in classical geometric invariant theory, see [MFK], Ch. 1, Sect. 3. The role of G-linearization is played, in our formalism, by the consideration of the isomorphism β in Diagram (2.18). Our $\mathbf{f}$-semistable points are then an analog of the G-semistable points in [MFK], Ch. 1, Sect. 4.

One can ask for a δ-geometric analog of the stable points in [MFK], Ch. 1, Sect. 4. Such a concept could be of an arithmetic nature as follows. Denoting $R_{\text{alg}} := \{x \in R | x \text{ algebraic over } \mathbb{Q}\}$, $\mathcal{X}$ comes via base change from a moduli scheme $\mathcal{X}_{\text{alg}}$ over Spec(R_{alg}) and we take X such that it comes via base change from an open subscheme X_{alg} of $\mathcal{X}_{\text{alg}}$. Then our tentative definition is as follows. A point $P \in X_{\text{alg}}(R_{\text{alg}}) \subset X(R)$ is called $\mathbf{f}$-*stable* if it is $\mathbf{f}$-semistable and the intersection $\Phi_{\mathbf{f}}^{-1}(\Phi_{\mathbf{f}}(P)) \cap X_{\text{alg}}(R_{\text{alg}})$ taken inside $X(R)$ is a finite union of polarized isogeny classes. We will not further investigate this concept.

2.4 Main Results

2.4.1 Serre–Tate Expansions and the Basic δ-Modular Forms

Our approach is based on the following *Serre–Tate expansion principle* which is analogous to the Fourier expansion principle in [BB], Thm. 3.1. For the next statement fix an ordinary elliptic curve E_0 over $k = R/pR$ and a generator e_0 of (the $\mathbb{Z}_p$-module defined by the Tate module) $T_p(E_0)$ and set $A_0 := E_0^g$. We endow A_0 with the standard diagonal principal polarization θ_0 and with the diagonal column basis e of $T_p(A_0)$ defined by e by which we mean a column vector $e = (e_{0,1} \ \cdots \ e_{0,g})^{\mathrm{t}}$ where each $e_{0,i}$ with $i \in \{1, \ldots, g\}$ is e_0 but viewed as a generator of the Tate module of the i-th copy $E_{0,i}$ of E_0 in A_0.

Theorem 2.4.1 *The following properties hold.*

(a) *For every $w', w'' \in \mathcal{W}$ there exist natural injective R-linear maps*

$$\mathcal{E}_{A_0,\theta_0,e} : \mathbb{I}^r_{gg}(w', w'') \to \boldsymbol{Mat}_g(S^r_{\text{for}}), \quad \mathcal{E}_{A_0,\theta_0,e} : \mathbb{I}^r_{gg,ord}(w', w'') \to \boldsymbol{Mat}_g(S^r_{\text{for}}),$$

whose reductions modulo p are also injective.

(b) *For each $w \in W$ there exist natural injective R-linear map*

$$\mathcal{E}_{A_0,\theta_0,e} : \mathbb{I}^r_g(w) \to S^r_{\text{for}}, \quad \mathcal{E}_{A_0,\theta_0,e} : \mathbb{I}^r_{g,ord}(w) \to S^r_{\text{for}},$$

whose reductions modulo p are also injective.

All these injective R-linear maps will be compatible with the R-linear maps $\mathbb{I}^r_{gg}(w', w'') \to \mathbb{I}^r_{gg,\text{ord}}(w', w'')$ and $\mathbb{I}^r_g(w) \to \mathbb{I}^r_{g,\text{ord}}(w)$ (which are therefore injective with injective reduction modulo p) and will be compatible in a natural way with multiplication. We refer to the R-linear maps in Theorem 2.4.1 as *Serre–Tate expansion maps*. On our way of proving Theorem 2.4.1 we will prove that the R-algebra homomorphism of Diagram (2.41) is injective with injective reductions modulo p (see Proposition 3.5.1 (b)).

Recall the matrix Ψ in Display (2.36). We will prove the following existence result (see Theorem 3.6.16 (a)).

Theorem 2.4.2 *There exists a generator e_0 of the Tate module of E_0 with the following property. For every integer $r \geq 1$ there exists a unique Hecke covariant Siegel δ-modular form*

$$f^r := f^r_{g,crys} \in \mathbb{I}^r_{gg}(\phi^r, 1)$$

whose Serre–Tate expansion equals

$$\mathcal{E}_{A_0,\theta_0,e}(f^r_{g,crys}) = \Psi^{\phi^{r-1}} + p\Psi^{\phi^{r-2}} + \cdots + p^{r-1}\Psi.$$

Remark 2.4.3 We will refer to the forms f^r as the *basic* Hecke covariant Siegel δ-modular forms of weight $(\phi^r, 1)$. These δ-modular forms were introduced in [BB], Sect. 4.1 and their (crystalline) construction will be recalled in Sect. 3.4. By construction, if $p > 2$ (resp. if $p = 2$) these δ-modular forms vanish on all quadruples (A, θ, ω, R) in which A is an abelian scheme over Spec(R) which is a canonical lift (resp. quasi-canonical lift) of its reduction A_k modulo p. For $r = 1$ the converse holds: if $p > 2$ (resp. if $p = 2$) and if f^1 vanishes at some quadruple (A, θ, ω, R) with A ordinary, then A is an abelian scheme over Spec(R) which is a canonical lift (resp. quasi-canonical lift) of A_k.

Recall from [BB], Sect. 2.3 that for $w \in W$ and $w', w'' \in \mathcal{W}$ we have natural *Fourier expansion maps*

$$\mathcal{E}_\infty : \mathbb{I}^r_{gg}(w', w'') \to \mathbf{Mat}_g(S^r_{\text{can}}), \quad \mathcal{E}_\infty : \mathbb{I}^r_g(w) \to S^r_{\text{can}}.$$

These maps will not play any role in the present monograph. Note however that by [BB], Eq. (4.4) the Fourier expansion of $f^r_{g,\text{crys}}$ is given by

$$\mathcal{E}_\infty(f^r_{g,\text{crys}}) = \Psi_q^{\phi^{r-1}} + p\Psi_q^{\phi^{r-2}} + \cdots + p^{r-1}\Psi_q. \tag{2.55}$$

Combining Eq. (2.55) and Theorem 2.4.2 we get that the Fourier and Serre–Tate expansions of the basic δ-modular forms $f^r_{g,\text{crys}}$ coincide up to a natural identification:

$$(\mathcal{E}_\infty(f^r_{g,\text{crys}}))_{\text{for}} = \mathcal{E}_{A_0,\theta_0,e}(f^r_{g,\text{crys}}).$$

We have the following rank 1 result (see Corollaries 3.6.21 and 3.6.22 below).

Theorem 2.4.4 *The following properties hold.*

(a) *The R-modules $\mathbb{I}^1_{gg}(\phi, 1)$ and $\mathbb{I}^1_{gg,ord}(\phi, 1)$ coincide and are free of rank* 1 *generated by $f^1_{g,crys}$.*

(b) *For all integers $n \geq 1$, the R-modules $\mathbb{I}^1_g(-n\phi - n)$ and $\mathbb{I}^1_{g,ord}(-n\phi - n)$ coincide and are free of rank* 1 *generated by* $(\det(f^1_{g,crys}))^n$.

In addition, generalizing a construction for $g = 1$ in [Bar] and [Bu05], in Sect. 3.6.3 we will prove the following theorem (see Theorem 3.6.17 and Corollaries 3.6.31 (e) and 3.6.34).

Theorem 2.4.5 *The following properties hold.*

(a) *The R-module $\mathbb{I}^1_{gg,ord}(-\phi, 1)$ is free of rank* 1 *and has a generator $f^\partial_{g,crys}$ whose Serre–Tate expansion is the identity matrix:*

$$\mathcal{E}_{A_0,\theta_0,e}(f^\partial_{g,crys}) = 1_g.$$

(b) *The form* $\det(f^\partial_{g,crys})$ *is invertible in the ring $\mathbb{I}^1_{g,ord}$.*

(c) *If $w \in W(r)$ is such that $deg(w) = 0$ (equivalently, if $w \in (\phi - 1)W$), then the R-module $\mathbb{I}^r_{g,ord}(w)$ is free of rank 1 generated by*

$$\det(f^{\partial}_{g,crys})^{\frac{w}{\phi-1}}.$$

(d) *We have $\mathbb{I}^1_g(\phi - 1) = \mathbb{I}^1_g(1 - \phi) = 0$.*

Remark 2.4.6 We shall refer to $f^{\partial}_{g,\mathrm{crys}}$ as the *basic* ordinary Siegel δ-modular form of weight $(-\phi, 1)$: the free R-modules $\mathbb{I}^r_{gg,\mathrm{ord}}(w', w'')$ for $w', w'' \in \mathcal{W}$ of order $\leq r$ can be fully described using $f^{\partial}_{g,\mathrm{crys}}$ (see Theorem 3.6.17 below). These modules have rank r if $w', w'' > 0$, have rank 1 if $w'w'' < 0$, and vanish (have rank 0) if $w' < 0$, $w'' < 0$. However, the concrete (explicit) description of the submodules $\mathbb{I}^r_{gg}(w', w'') \subset \mathbb{I}^r_{gg,\mathrm{ord}}(w', w'')$ for all $w', w'' \in \mathcal{W}$ is an open problem.

2.4.2 *Structure of the Ring* $\mathbb{I}_{g,ord}$

We next present a series of applications of our Serre–Tate expansion theory to the structure of the rings $\mathbb{I}_{g,\mathrm{ord}}$ (see Theorem 3.6.20 (b) below).

Theorem 2.4.7 *There exist natural monomorphisms of graded R-algebras*

$$\diamondsuit : \mathbb{H}^{r-1,torus}_{g,tot} \to \mathbb{I}^r_{g,ord}, \quad \diamondsuit : \mathbb{H}^{torus}_{g,tot} \to \mathbb{I}_{g,ord},$$

whose cokernels are torsion.

Remark 2.4.8 By Theorem 2.4.7 we have induced K-algebra isomorphisms

$$\diamondsuit : \mathbb{H}^{r-1,\mathrm{torus}}_{g,\mathrm{tot}} \otimes_R K \to \mathbb{I}^r_{g,\mathrm{ord}} \otimes_R K, \quad \diamondsuit : \mathbb{H}^{\mathrm{torus}}_{g,\mathrm{tot}} \otimes_R K \to \mathbb{I}_{g,\mathrm{ord}} \otimes_R K;$$

so the study of the rings $\mathbb{I}^r_{g,\mathrm{ord}} \otimes_R K$ gets reduced to the study of the rings $\mathbb{H}^r_g \otimes_R K$, which is a topic in the geometric invariant theory of multiple quadratic forms. This invariant theory has classical roots and will be revisited, and further developed, in Sects. 4.1–4.3. In particular, as we shall see, the K-algebras $\mathbb{H}^r_{g,\mathrm{tot}} \otimes_R K$ are finitely generated, Cohen–Macauley, and unique factorization domains, thus Gorenstein, and hence so are the K-algebras $\mathbb{I}^r_{g,\mathrm{ord}} \otimes_R K$. Also we will see that (see Corollary 3.6.31 (b) and (c)):

- if g is odd, then we have $\mathbb{H}^r_{g,\mathrm{tot}} = \mathbb{H}^r_g$ hence $\mathbb{H}_{g,\mathrm{tot}} = \mathbb{H}_g$ which implies $\mathbb{I}^r_{g,\mathrm{ord}}(w) = 0$ for all integers $r \geq 1$ and $w \in W(r)$ with $\deg(w)$ odd;
- if g is even, then $\mathbb{H}_{g,\mathrm{tot}} \neq \mathbb{H}_g$ and hence the R-algebras $\mathbb{I}^r_{g,\mathrm{ord}}(w)$ do not vanish for all integers $r \geq 1$ and $w \in W(r)$ with $\deg(w)$ odd.

Section 4.2 provides, in particular, a detailed study of the structure of the algebra $\mathbb{H}^r_{g,\mathrm{tot}} \otimes_R K$ tensored with an algebraic closure of K. The results there easily translate into results about the algebra $\mathbb{I}^r_{g,\mathrm{ord}} \otimes_R K$; we leave this translation to the reader. In particular we will introduce in Sect. 4.2.6 some special 'Θ and Υ invariants' and we will compare the algebras of all invariants with the algebra generated by the Θ invariants (see, for instance, Theorems 4.2.8 and 4.2.13 and Corollary 4.2.23). We will deduce information on the Hilbert series of the rings of invariants (see Corollaries 4.2.14 and 4.2.15), the smallest number of homogeneous generators (see Proposition 4.2.10 (a)), and transcendence bases (see Theorem 4.2.17). For $g = 2$, Sect. 4.3 gives finer results on Hilbert series (see Proposition 4.3.4 and Corollary 4.3.8), the smallest number of homogeneous generators (see Proposition 4.3.3), and relations (see Corollary 4.3.6).

On the other hand, the R-algebra homomorphism $\diamondsuit : \mathbb{H}^{\mathrm{torus}}_{g,\mathrm{tot}} \to \mathbb{I}_{g,\mathrm{ord}}$ is not an isomorphism. It would be interesting to compute the cokernel of the R-linear map $\diamondsuit$. A related open problem is to compute the cokernel (which is, again, torsion) of the R-linear map given by the inclusion

$$R[z, z^{\phi}, z^{\phi^2}, \dots] \to \mathbb{S}_n.$$

Proposition 3.6.24 (b) below will provide a lower bound for the torsion of the cokernel of $\diamondsuit$ in the following sense. If for an R-module M we denote by $M[p]$ the k-linear space of all elements of M annihilated by p, then for $s \in \frac{1}{2}\mathbb{Z}_{\geq 0}$ we will show:

$$\dim_k \left(\frac{\mathbb{I}^{r+1}_{g,\mathrm{ord}}(-2(p+1)s)}{\diamondsuit(\mathbb{H}^r_g((p+1)s))}[p] \right) \geq 2 \dim_K(\mathbb{H}^{r-1}_g(s)_K) - 3.$$

It would be interesting to find the precise value of the left-hand side of the above inequality.

Remark 2.4.9 By combining the proofs of Theorems 2.4.2, 2.4.5, and 2.4.7, the Serre–Tate expansions of all elements of $\mathbb{I}_{g,\mathrm{ord}}$ will be made explicit in Remark 3.6.32 (a).

Remark 2.4.10 Let $X \subset \mathcal{X}_{\mathrm{ord}} = \mathcal{A}_{g,1,N,R,\mathrm{ord}}$ be an open subscheme with $\overline{X}$ connected. We check that for each polarized isogeny class $C \subset X(R)$ there exist 'many' Hecke covariant Siegel δ-modular forms that vanish on C. First, by Theorem 2.4.7, there exist (many) integers $r \geq 1$ and (many) weights $w \in W(r)$ such that the (finitely generated) R-module $\mathbb{I}^r_{g,\mathrm{ord}}(w)$ has 'large' rank. For a point $P \in C$ consider the principally polarized abelian scheme (A, θ) over $\mathrm{Spec}(R)$ that corresponds to it and the submodule $\mathbb{I}^{r,[P]}_{g,\mathrm{ord}}(w)$ of all $f \in \mathbb{I}^r_{g,\mathrm{ord}}(w)$ that vanish at P in the sense that $f(A, \theta, \omega, R) = 0$ for one (equivalently every) ω. Note that, by Hecke covariance, each $f \in \mathbb{I}^{r,[P]}_{g,\mathrm{ord}}(w)$ vanishes on the whole of C. We claim that the factor module $\mathbb{I}^r_{g,\mathrm{ord}}(w)/\mathbb{I}^{r,[P]}_{g,\mathrm{ord}}(w)$ is either 0 or (noncanonically) isomorphic to R. Indeed, if this factor module is nonzero, let $0 \neq f_0 \in \mathbb{I}^r_{g,\mathrm{ord}}(w)$ be such that the element $f_0(A, \theta, \omega, R) \in R$ has minimum p-adic valuation and we conclude by noting that $\mathbb{I}^{r,[P]}_{g,\mathrm{ord}}(w)$ is the kernel of the surjective R-linear homomorphism

$$\mathbb{I}^r_{g,\mathrm{ord}}(w) \to R, \quad f \mapsto \frac{f(A,\theta,\omega,R)}{f_0(A,\theta,\omega,R)}.$$

Remark 2.4.11 Let $X \subset \mathcal{X}_{\mathrm{ord}}$ be, as before, an open subscheme with $\overline{X}$ connected. By the proof of Theorem 2.4.7 it will also follow that for each $w \in W$, every $f \in \mathbb{I}_{g,\mathrm{ord}}(w)$ vanishes on the set of ordinary CM_R-points of X, i.e., of R-valued points of X that correspond to ordinary abelian schemes over $\mathrm{Spec}(R)$ that have *complex multiplication* (see Remark 3.6.32 (b)). If $p > 2$, then an ordinary CM_R-point is the same as a CL_R point, i.e., an R-valued point of $\mathcal{X}_{\mathrm{ord}}$ that corresponds to an ordinary abelian scheme over $\mathrm{Spec}(R)$ which is the canonical lift of its reduction modulo p (see [V], Thm. 1.6.3). Similarly, if $p = 2$, then an ordinary CM_R-point is the same as a QCL_R point, i.e., an R-valued point of $\mathcal{X}_{\mathrm{ord}}$ that correspond to an ordinary abelian scheme over $\mathrm{Spec}(R)$ which is a *quasi-canonical lift* of its reduction modulo 2, which means that it defines the same filtered F-crystal over k as the canonical lift (see [V], Thm. 1.6.3). So for every linear system $\mathbf{f}$ in $\mathbb{I}_{g,\mathrm{ord}}$, the set of $\mathbf{f}$-unstable points $X(R)^{\mathrm{u}}_{\mathbf{f}}$ in $X(R)$ contains the set of ordinary CM_R-points of X. It would also be interesting to decide when, for suitable **f**s, the locus of **f**-stable points of $X(R)$ is nonempty and, when so, to compute it.

Combining Theorem 2.4.7 with results from Sects. 4.1 and 4.2 we obtain the following results (see Theorem 3.6.29 (c), (e) and (g)).

Theorem 2.4.12 *The following properties hold.*

(a) *We have equalities:*

$$\text{asy.tr.deg}(\mathbb{I}_{g,ord}) = \phi\text{-tr.deg}(\mathbb{I}_{g,ord}) = \frac{g(g+1)}{2} + 1.$$

(b) *The field* $\mathit{Frac}(\mathbb{I}_{g,ord})$ *is* ϕ*-generated over* K *by finitely many elements in* $\mathbb{I}^4_{g,ord}$.

Here and from now on, ϕ-tr.deg := ϕ-tr.deg$_R$.

Remark 2.4.13 As the K-algebras $\mathbb{H}^{r-1,\mathrm{torus}}_g \otimes_R K$ are finitely generated by classical geometric invariant theory, we also get that

$$\text{asy.dim}(\mathbb{I}_{g,\mathrm{ord}}) = \frac{g(g+1)}{2} + 1.$$

On the other hand we do not know whether $\mathbb{I}_{g,\mathrm{ord}} \otimes_R K$ is ϕ-finitely generated as a K-algebra.

Remark 2.4.14 We recall the R-algebra monomorphism

$$\mathbb{I}_{g,\mathrm{ord}} \to \mathbb{M}_X \tag{2.56}$$

with injective reduction modulo p, where $X \subset \mathcal{X}_{\mathrm{ord}}$ is an arbitrary open subscheme with $\overline{X}$ connected (see Diagram (2.41)). If X is affine, then by viewing $\mathbb{M}_X$ as filtered by the subrings $\mathbb{M}^r_X$, we have (see Eq. (2.5))

$$\text{asy.dim}(\mathbb{M}_X) = \frac{g(g+1)}{2} + 1.$$

Diagram (2.49) is induced by Diagram (2.56) and we recall that we view Diagram (2.49) as an avatar of Diagram (2.53). Then the equality of the asymptotic dimensions of the source and target in Diagram (2.56) says that the 'quotient map' in Diagram (2.53) is a map between geometric objects 'of the same dimension' in 'δ-geometry'.

Note that we have K-algebra monomorphisms

$$\mathbb{I}^r_{g,\mathrm{ord}} \otimes_R K \to \mathbb{M}^r_X \otimes_R K \tag{2.57}$$

induced by Diagram (2.56). By Theorem 3.6.29 (b) below and Remark 2.4.13 the K-algebra $\mathbb{I}^r_{g,\mathrm{ord}} \otimes_R K$ is finitely generated and has dimension

$$\dim(\mathbb{I}^r_{g,\mathrm{ord}} \otimes_R K) = r\frac{g(g+1)}{2} - g^2 + 1 + r \tag{2.58}$$

while, for X affine, the (noetherian but not finitely generated) K-algebra $\mathbb{M}^r_X \otimes_R K$ has dimension

$$\dim(\mathbb{M}^r_X \otimes_R K) = (r+1)\frac{g(g+1)}{2} + r. \tag{2.59}$$

The difference between the two dimensions is therefore

$$\dim(\mathbb{M}^r_X \otimes_R K) - \dim(\mathbb{I}^r_{g,\mathrm{ord}} \otimes_R K) = g^2 + \frac{g(g+1)}{2} - 1$$

which is independent of r; note that this difference is less than (rather than equal to) the dimension $\dim(\mathbf{GSp}_{2g,K}) = 2g^2 + g + 1$ of the group of symplectic similitudes of rank $g+1$ over K, as one might expect from the analogy with the differential algebraic picture [Bu95b], Introd. and the complex analytic picture [FC], Ch. VII, Sect. 3, p. 255. It would be interesting to study the fibers of the morphism of spectra induced by Diagram (2.57).

We are especially interested in the structure of the rings $\mathbb{M}^r_X/\mathbb{I}^{r,[P]}_{g,\mathrm{ord}}\mathbb{M}^r_X$ and $\mathbb{M}_X/\mathbb{I}^{[P]}_{g,\mathrm{ord}}\mathbb{M}_X$ for $P \in X(R)$, where $\mathbb{I}^{r,[P]}_{g,\mathrm{ord}} \subset \mathbb{I}^r_{g,\mathrm{ord}}$ and $\mathbb{I}^{[P]}_{g,\mathrm{ord}} \subset \mathbb{I}_{g,\mathrm{ord}}$ are the prime ideals (see Remark 2.4.10)

$$\mathbb{I}^{r,[P]}_{g,\mathrm{ord}} := \bigoplus_{w \in W} \mathbb{I}^{r,[P]}_{g,\mathrm{ord}}(w), \quad \mathbb{I}^{[P]}_{g,\mathrm{ord}} := \bigcup_{r \geq 1} \mathbb{I}^{r,[P]}_{g,\mathrm{ord}}.$$

The ring $\mathbb{M}_X/\mathbb{I}^{[P]}_{g,\mathrm{ord}}\mathbb{M}_X$ can be viewed as an avatar, in δ-geometry, of the polarized isogeny class of P. It has a natural structure of a filtered W-graded ring equipped with an endomorphism induced by ϕ. Note that, by Remark 2.4.10, the ring $\mathbb{I}_{g,\mathrm{ord}}/\mathbb{I}^{[P]}_{g,\mathrm{ord}}$ has a natural structure of a filtered W-graded integral domain all of whose homogeneous components are either 0 or (noncanonically) isomorphic to R and is equipped with an endomorphism induced by ϕ. One would like to know, for instance, if or when

$\mathbb{M}_X/\mathbb{I}^{[P]}_{g,\text{ord}}\mathbb{M}_X$ is an integral domain and one would like to compute its asymptotic dimension. A related problem is to understand the fibers of the maps $\Phi_{\mathfrak{f}}$ in Diagram (2.48). In the same vein one would like to understand the image of the maps $\Phi_{\mathfrak{f}}$, e.g., the ideal of all δ-homogeneous polynomials in $\mathbb{S}_n$ that vanish on this image. Another related open problem is to compute the transcendence degrees $\text{tr.deg}_k(\mathbb{I}^r_{g,\text{ord}} \otimes_R k)$.

In Chap. 3 we will also obtain estimates/asymptotics for the (finite) ranks of the R-modules $\mathbb{I}^r_{g,\text{ord}}(w)$ (see Theorem 3.6.20 (a)). Also, for $r \leq 2$ we will obtain W-graded K-algebra isomorphisms

$$\mathbb{I}^1_{g,\text{ord}} \otimes_R K \simeq K[\Theta_0, z_0, z_0^{-1}],$$

$$\mathbb{I}^2_{g,\text{ord}} \otimes_R K \simeq K[\Theta_0, \ldots, \Theta_g, z_0, z_1, \frac{1}{z_0 z_1}],$$

where $\Theta_0, \ldots, \Theta_g, z_0, z_1$ are indeterminates of weights $-2, \ldots, -2, \phi - 1, \phi^2 - \phi$ (respectively); these isomorphisms will be entirely explicit (see Corollary 3.6.33). The structure of the rings $\mathbb{I}^r_{g,\text{ord}}$ in case $r \geq 3$ will turn out to be rather complicated. However the rings $\mathbb{I}^r_{g,\text{ord}} \otimes_R K$ are isomorphic to the rings $\mathbb{H}^{r-1,\text{torus}}_{g,\text{tot}} \otimes_R K$ (see Theorem 2.4.7) and the rings $\mathbb{H}^{r-1}_{g,\text{tot}} \otimes_R K$ will be studied in detail in Sects. 4.1–4.3.

Recalling the notation in Remark 2.2.4, we get, for g odd, an equality of W_+-modules

$$W_{\mathbb{I}_{g,\text{ord}}} = \{w \in W | \deg(w) \in 2\mathbb{Z},\ w \leq 0\}.$$

The structure of $W_{\mathbb{I}_{g,\text{ord}}}$ for g even is more complicated. However in view of the isomorphism $\mathbb{I}_{g,\text{ord}} \otimes_R K \simeq \mathbb{H}^{\text{torus}}_{g,\text{tot}} \otimes_R K$ (see Theorem 2.4.7) the problem of understanding the structure of $W_{\mathbb{I}_{g,\text{ord}}}$ trivially reduces to the problem of determining which homogeneous components of $\mathbb{H}^{\text{torus}}_{g,\text{tot}}$ (with respect to a natural gradation) are nonzero. This latter problem reduces to a similar problem for the ring $\mathbb{H}_{g,\text{tot}}$ which, in its turn, is partially addressed (although not completely solved) in Chap. 4; see Lemma 4.2.3 (a) and (b) and Sect. 4.2.6.

2.4.3 *Structure of the Ring* $\mathbb{I}_g$

Next we turn our attention to $\mathbb{I}_g$. We will prove that $\mathbb{I}_g$ has an asymptotic transcendence degree that is equal to that of $\mathbb{I}_{g,\text{ord}}$ so the two rings are 'close' in the sense of asymptotic transcendence degrees. Indeed, using the geometric invariant theory of multiple endomorphisms (see Sects. 4.1.1, 4.1.5 and 4.1.6), we will prove (see Theorem 3.7.1) the following theorem.

Theorem 2.4.15 *For each integer $r \geq 4$ there exists a natural map of W-graded R-algebras*

$$\square : \mathbb{H}^{r-2}_{g,conj}[x_1, \ldots, x_r, \xi_1, \ldots, \xi_r] \to \mathbb{I}^r_g$$

whose image satisfies

$$\text{tr.deg}_R(Im(\square)) \geq \text{tr.deg}_R(\mathbb{I}^r_{g,ord}) - \frac{g(g+1)}{2} - 2.$$

Here x_i and ξ_i are indeterminates of degree $-\phi^{i-1} - \phi^i$ and $-1 - \phi^i$ (respectively) and the maps $\square$ are compatible as r varies.

Remark 2.4.16 Theorem 2.4.15 and Remark 2.2.6 imply the inequality

$$\text{tr.deg}_{\mathbb{I}_g}(\mathbb{I}_{g,\text{ord}}) \leq \frac{g(g+1)}{2} + 2. \tag{2.60}$$

It would be interesting to compute the exact value of the transcendence degree $\text{tr.deg}_{\mathbb{I}_g}(\mathbb{I}_{g,\text{ord}})$. We emphasize that we do not know even if the field extension $\text{Frac}(\mathbb{I}_g) \subset \text{Frac}(\mathbb{I}_{g,\text{ord}})$ is not an equality; this question is open even for $g = 1$. On the other hand, by Theorem 2.4.15 and Eqs. (2.58) and (2.59) we get

$$\dim(\mathbb{M}^r_X \otimes_R K) - \text{tr.deg}_R(\mathbb{I}^r_g) \leq 2g^2 + g + 1 = \dim(\mathbf{GSp}_{2g,K}) \tag{2.61}$$

with X as in Remark 2.4.14. The value of $\text{tr.deg}_R(\mathbb{I}^r_g)$ is not known even for $g = 1$. The analogies with the differential algebraic picture [Bu95b], Introd. and with the complex analytic picture [FC], Ch. VII, Sect. 3, p. 255, suggest that Inequality (2.61) might be an equality. A related open problem is to compute the transcendence degrees $\text{tr.deg}_k(\mathbb{I}^r_g \otimes_R k)$; again, this is not known even for $g = 1$.

Using Inequality (2.60) and Theorem 2.4.12 (a) we will deduce (see Corollary 3.7.2 (c)) the following result which solves, in particular, a 'main open problem' in [BB] (see [BB], paragraph after Thm. 1.13) and which can be easily used to reobtain for $g \in \{1, 2\}$ the results [BB], Thms. 1.13 and 1.14.

Theorem 2.4.17 *We have*

$$\text{asy.tr.deg}(\mathbb{I}_g) = \phi\text{-tr.deg}(\mathbb{I}_g) = \frac{g(g+1)}{2} + 1.$$

Remark 2.4.18 As $\text{Frac}(\mathbb{I}_{g,\text{ord}})$ is ϕ-finitely generated over K so is its subfield $\text{Frac}(\mathbb{I}_g)$ by Remark 2.2.3 (c). Also, as $\mathbb{I}^r_g(w) \subset \mathbb{I}^r_{g,\text{ord}}(w)$, it follows that for all linear systems $\mathbf{f}$ of weight w in $\mathbb{I}^r_{g,\text{ord}}$ the set of $\mathbf{f}$-unstable points of $\mathcal{X}(R) = \mathcal{A}_{g,1,N,R}(R)$ contains the set of all ordinary CM_R-points. Here are some questions that we leave open. Are the K-algebras $\mathbb{I}^r_g \otimes_R K$ finitely generated? Is $\mathbb{I}_g \otimes_R K$ a ϕ-finitely generated K-algebra? What are the transcendence degrees of $\mathbb{I}^r_g \otimes_R K$ over K? What is the structure of the rings $\mathbb{M}_X/\mathbb{I}^{[P]}_g\mathbb{M}_X$ where X is open in $\mathcal{X}$, with $\overline{X}$ connected, $P \in X(R)$, and $\mathbb{I}^{[P]}_g$ is the prime ideal of $\mathbb{I}_g$ defined as in Remark 2.4.14 by replacing $\mathbb{I}_{g,\text{ord}}$ with $\mathbb{I}_g$?

Remark 2.4.19 It would be interesting to compute the W_+-module $W_{\mathbb{I}_g}$, see Remark 2.2.4. As $\det(f^r) \in \mathbb{I}_g^r(-1-\phi^r)$ we have $-1-\phi^r \in W_{\mathbb{I}_g}$ for all integers $r \geq 1$. So $W_{\mathbb{I}_g}$ contains the $\mathbb{Z}$-linear span W_{cyc} of the elements of the form $-\phi^a - \phi^b$ with $0 \leq a < b$ integers. Note that W_{cyc} is a W_+-submodule of $W_{\mathbb{I}_g}$. All examples of homogeneous elements in $\mathbb{I}_g$ appearing in our monograph have weights in W_{cyc}. We do not know if or when $W_{\mathbb{I}_g} = W_{\text{cyc}}$. On the other hand we claim that the W_+-module W_{cyc} is not finitely generated. Indeed, if $w_1, \ldots, w_n$ generate the W_+-module W_{cyc}, then for every integer $r \geq 1$ we can write $-1-\phi^r = \sum_{i=1}^n v_{r,i} w_i$ with $v_{r,1}, \ldots, v_{r,n} \in W_+$. Taking degrees and using the fact that $\deg(w_i) \in -2\mathbb{N}$ we get that $-1-\phi^r = w_{i(r)}$ for some $i(r) \in \{1, \ldots, n\}$ which for $r \gg 1$ represents a contradiction, proving our claim. As a consequence of the claim and of Remark 2.2.4, if for some $g \in \mathbb{N}$ we have $W_{\mathbb{I}_g} = W_{\text{cyc}}$, then $\mathbb{I}_g \otimes_R K$ is not ϕ-finitely generated as a K-algebra.

For the next result, for a polarized abelian scheme (A, θ) over R, we denote by $\text{End}(A,\theta)^{(p)}$ the subring of $\text{End}(A)$ generated by all polarized isogenies $A \to A$. The theorem below shows that the locus in $\mathcal{A}_{g,1,N,R}(R)$ of all (A,θ)s for which $\text{End}(A,\theta)^{(p)}$ is noncommutative is contained in the vanishing locus of a nonzero Hecke covariant Siegel δ-modular form (see Theorem 3.7.8 below for a more precise result).

Theorem 2.4.20 *There exists $w \in W$ with the property that $\mathbb{I}_g(w)$ contains a nonzero δ-modular form $\mathcal{D}$ such that for all $(A, \theta, \omega) \in \boldsymbol{M}_g(R)$ with $\mathcal{D}(A, \theta, \omega) \neq 0$, the ring $\text{End}(A,\theta)^{(p)}$ is commutative.*

Let us say that the polarized isogeny class of a principally polarized abelian scheme (A, θ) over R is *decomposable* if there exist nonzero principally polarized abelian schemes (A_1, θ_1) and (A_2, θ_2) and a polarized isogeny $A \to A_1 \times A_2$, where $A_1 \times A_2$ is equipped with the polarization $\theta_1 \times \theta_2$. We say that the polarized isogeny class of (A, θ) is *indecomposable* if it is not decomposable. The theorem below shows that the locus in $\mathcal{A}_{g,1,N,R}(R)$ of all (A,θ)s whose polarized isogeny class are decomposable is contained in the vanishing locus of a nonzero Hecke covariant Siegel δ-modular form (see Theorem 3.7.9 below for a more precise result).

Theorem 2.4.21 *There exists $w \in W$ with the property that $\mathbb{I}_g(w)$ contains a nonzero δ-modular form $\mathcal{G}$ such that for all $(A, \theta, \omega) \in \boldsymbol{M}_g(R)$ with $\mathcal{G}(A, \theta, \omega) \neq 0$, the polarized isogeny class of (A, θ) is indecomposable.*

Theorems 2.4.20 and 2.4.21 are arithmetic analogs of results in differential algebraic geometry, see [Bu95b], Prop. 6.8 and Cor. 5.11.

The following 'p-adic approximation result' (see Theorem 3.8.13 (c) below) shows that $\mathbb{I}_g$ is 'close to' $\mathbb{I}_{g,\text{ord}}$ in yet another way.

Theorem 2.4.22 *Let $\mathbb{D}_g$ be the multiplicative set of all homogeneous elements of $\mathbb{I}_g \backslash p\mathbb{I}_g$ whose weights have even degree. Then the natural homomorphism*

$$\widehat{\mathbb{D}_g^{-1}\mathbb{I}_g} \to \widehat{\mathbb{D}_g^{-1}\mathbb{I}_{g,ord}}$$

is an isomorphism.

The above theorem implies that $\mathbb{I}_g$ is 'p-adically sufficiently close to' $\mathbb{I}_{g,\mathrm{ord}}$ in the sense that each element in $\mathbb{I}_{g,\mathrm{ord}}$ can be p-adically approximated by fractions with denominators homogeneous elements of $\mathbb{I}_g \setminus p\mathbb{I}_g$.

2.4.4 Flows and Reciprocity

Our next application involves the notion of δ-flow introduced in [Bu17], Ch. 3, Sect. 3.6, Def. 3.122. Let X be an affine smooth scheme over $\mathrm{Spec}(R)$ and let $J^1(X)$ be its first p-jet space. We recall that an ideal $\mathfrak{a}$ in $\mathcal{O}(J^1(X))$ is called a *δ-flow* on X if the composite projection

$$\mathrm{Spf}(\mathcal{O}(J^1(X))/\mathfrak{a}) \subset J^1(X) \to J^0(X) = \widehat{X}$$

is an isomorphism. Our terminology here is suggested by the analogies between p-derivations and derivations (vector fields) that are in the background of [Bu95a, Bu05]. Note that by the universal property of $J^1(X)$ every δ-flow on X induces a Frobenius lift on $\widehat{X}$, i.e., an endomorphism of $\widehat{X}$ over $\mathbb{Z}_p$ whose reduction modulo p is the p-power Frobenius. We will prove the following result (see Theorem 3.9.12).

Theorem 2.4.23 *We consider an affine open subscheme X of the moduli scheme $\mathcal{X} = \mathcal{A}_{g,1,N,R}$ over $\mathrm{Spec}(R)$, where $N \geq 3$ is an integer prime to p, such that $\overline{X}$ is contained in the ordinary locus and there exists a column basis ω_X of 1-forms on A_X. Let $\mathfrak{a}_X$ be the ideal of $\mathcal{O}(J^1(X))$ locally generated by the entries of $f^1_{g,crys}(A_X, \theta_X, \omega_X)$ where (A_X, θ_X) is the universal principally polarized abelian scheme over X. Then $\mathfrak{a}_X$ is a δ-flow on X.*

Remark 2.4.24 It would be interesting to check whether the Frobenius lift on the p-adic completion $\widehat{\mathcal{A}_{g,1,N,R,\mathrm{ord}}}$ induced by the δ-flow in Theorem 2.4.23 coincides with the Frobenius lift considered in [BG], p. 3 that arises from the functor that attaches to every ordinary abelian scheme its quotient by the canonical subgroup. If this were the case, then we would get the interesting consequence that the Frobenius lift in [BG], although not extendable to the whole of $\widehat{\mathcal{A}_{g,1,N,R}}$, 'comes from a global object (f^1_{crys}) defined on the whole of $\mathcal{A}_{g,1,N,R}$'.

Using Theorem 2.4.23 and the arithmetic analog in [Bu97], Cor. 1.7 of Manin's Theorem of the Kernel [Man], Sect. 5, Thm. 2 we will prove the following result (see Theorem 3.10.2 below) which extends the 'Reciprocity Theorem for CL_R points' in [BPII], Thm. 3.5, to the case of arbitrary genus $g \geq 1$ and all primes p. We refer to the part before Theorem 3.10.2 for more details.

Theorem 2.4.25 *Assume we are given morphisms of schemes over R*

$$\Pi : Y \to X, \quad \Phi : Y \to A$$

with $X \subset \mathcal{X}_{ord}$ affine, Y affine, Π étale, and A an abelian scheme of relative dimension d. Assume Y, X, A, Π, Φ descend to corresponding objects Y', X', A', Π', Φ' defined over a finite normal extension R' of $\mathbb{Z}_p$ contained in R and assume A' is nondegenerate in the sense of Definition 3.10.1 below. Then there exists a morphism of p-adic formal schemes

$$\Phi^+ : \widehat{Y} \to \widehat{\mathbb{G}_a^d}$$

that satisfies the following property. For all $P_1, \dots, P_n \in \Pi^{-1}(CL_R) \subset Y(R)$ and all $m_1, \dots, m_n \in \mathbb{Z}$ we have

$$\sum_{i=1}^{n} m_i \Phi(P_i) \in A(R)_{\text{tors}} \iff \sum_{i=1}^{n} m_i \Phi^+(P_i) = 0 \in R^d.$$

Remark 2.4.26 As a consequence of Theorem 2.4.25 we will have (see Corollary 3.10.3) that if $\Pi^{-1}(\text{CL}_R) \not\subset \Phi^{-1}(A(R)_{\text{tors}})$, then the image of

$$\Pi^{-1}(\text{CL}_R) \cap \Phi^{-1}(A(R)_{\text{tors}})$$

via the reduction modulo p map $Y(R) \to Y(k)$ is not Zariski dense in $\overline{Y}$.

This statement is a higher dimensional analog of the following result in [BPII], Thm. 1.3. Suppose that $X \stackrel{\Pi}{\longleftarrow} Y \stackrel{\Phi}{\longrightarrow} E$ is a correspondence (with Y a curve) between a modular (or Shimura) curve X and an elliptic curve E over some number field and assume that p is a sufficiently large 'good' prime in the sense of [BPII], Def. 3.3. Then there exists a constant c that depends on p such that for each subgroup $\Gamma \leq E(R)$ with $r := \text{rank}(\Gamma) < \infty$, the set $\Phi(\Pi^{-1}(\text{CL}_R)) \cap \Gamma$ is finite of cardinality at most cp^r. This latter result was deduced in [BPII] from a one-dimensional analog of Theorem 2.4.25 (see [BPII], Thm. 3.5).

A global version of the local result in [BPII], Thm. 1.3 stated above was proved in [BPI], Thm. 1.1 as follows. Suppose that $X \stackrel{\Pi}{\longleftarrow} Y \stackrel{\Phi}{\longrightarrow} E$ is a correspondence (with Y a curve) between a modular curve X and an elliptic curve E over $\overline{\mathbb{Q}}$ and let $\Gamma \leq E(\overline{\mathbb{Q}})$ be a finite rank subgroup. Then the set $\Phi(\Pi^{-1}(\text{CM}))$ is finite. Here $\text{CM} \subset X(\overline{\mathbb{Q}})$ is the set of CM-points.

This latter global result generalized a result in [NS], Thm. 1.5 according to which there exist only finitely many torsion Heegner points on each elliptic curve over $\mathbb{Q}$.

The papers [BPI, BPII] also contained results in which the sets CL_R and CM were replaced by isogeny classes.

For a proof of the conjecture [BPI], Conj. 1.7 on isogeny classes we refer to [Bal], Thm. 1.4. For extensions of some of the results in [BPI, Bal] and for relations with conjectures of Andreé–Oort–Pink–Zilber we refer to [PT].

2.4.5 Cohomological Computations

It would be interesting to explore the cohomology modules of the 'main' ss-δ-spaces in Display (2.46) that appears in our theory. We will leave this problem open but, with notation as in Display (2.46), we will present some easy computations that illustrate the flavor of the problem.

First, consider the cohomology groups $H^d(\mathbb{P}^1_\delta)$. We will check (see Proposition 3.11.1) that the following equalities hold:

$$H^0(\mathbb{P}^1_\delta) = R, \quad \phi\text{-rank}_R(H^1(\mathbb{P}^1_\delta)) = \infty.$$

This is in stark contrast with the situation in usual algebraic geometry where $H^1(\mathbb{P}^1_S, \mathcal{O}) = 0$ for the projective line $\mathbb{P}^1_S$ over (the spectrum of) a ring S.

Similarly, let us consider the case $g = 1$ and the basic δ-modular forms

$$f^r := f^r_{1,\text{crys}} \in I^r_{11}(\phi^r, 1) = I^r_1(-1 - \phi^r)$$

and $f^\partial_{1,\text{crys}} \in \mathbb{I}^1_{11,\text{ord}}(-\phi, 1) = \mathbb{I}^1_{1,\text{ord}}(\phi - 1)$ referred to above and abbreviated as f^∂. Consider also the forms

$$f^{\langle r\rangle} := (f^1)^{\phi^{r-1}} \cdot (f^\partial)^{\frac{\phi^r+\phi^{r-1}-2}{\phi-1}} \in \mathbb{I}_{1,\text{ord}}(-2); \tag{2.62}$$

these δ-modular forms (and their generalizations for arbitrary g) will play a key role in the monograph. For each integer $n \geq 2$, consider the ss-δ-space

$$Z_{\delta,n} := \text{Proj}_\delta(\mathbb{I}_{1,\text{ord}}, (f^{\langle 1\rangle}, \ldots, f^{\langle n\rangle})).$$

We will then check (see Proposition 3.11.2) that the following equalities hold:

$$H^0(Z_{\delta,n}) = \mathbb{I}_{1,\text{ord},\langle f^{\langle n\rangle}\rangle}, \quad H^d(Z_{\delta,n}) = 0 \text{ for } d \geq 1.$$

On the other hand, if we consider a matrix $\lambda = (\lambda_{ij})_{1\leq i,j\leq 2} \in \mathbf{GL}_2(R)$ and we set $(f_1^\lambda, f_2^\lambda) := (f^{\langle 1\rangle}, f^{\langle 2\rangle})\lambda^t$ and

$$Z^\lambda_{\delta,2} := \text{Proj}_\delta(\mathbb{I}_{1,\text{ord}}, (f_1^\lambda, f_2^\lambda)),$$

then we will show (see Theorem 3.11.3) that if each $\lambda_{ij} \neq 0$, then the following equalities hold:

$$H^0(Z^\lambda_{\delta,2}) = R, \quad \phi\text{-rank}_R(H^1(Z^\lambda_{\delta,2})) = \infty.$$

Note that, if $\lambda = 1_2$, then we have $Z^\lambda_{\delta,2} = Z_{\delta,2}$. So the cohomology of $Z^\lambda_{\delta,2}$ effectively depends on λ. This is, again, in stark contrast with the situation in usual algebraic geometry where the Čech cohomology groups of an affine open cover of a separated scheme do not depend on the cover. Indeed, for each λ the pair $(f_1^\lambda, f_2^\lambda)$ can be intuitively viewed as defining an 'affine open cover' of a 'separated' ss-δ-space that does not depend on λ (as the R-module spanned by $f_1^{\lambda}, f_2^{\lambda}$ does not depend on λ).

Finally, for arbitrary g, we will prove (see Proposition 3.11.5) that for a 'generic choice' of a linear system $\mathbf{f}$ of weight -2 in $\mathbb{I}_{g,\text{ord}}$ we have $H^0(\text{Proj}_{\delta}(\mathbb{I}_{g,\text{ord}}, \mathbf{f})) = R$. For a weaker version of this statement for $\mathbb{I}_g$ see Proposition 3.11.6 below.

Chapter 3
Proofs of the Main Results

All the main results are proved in this chapter.

Section 3.1 records some basic symmetry properties of the bilinear maps over $\mathbb{Z}_p$ appearing in the Serre–Tate deformation theory of abelian schemes.

Section 3.2 introduces and studies some multiplicative monoids contained in endomorphism algebras of abelian varieties over k; these monoids, and some related unitary groups, will play a key role in our arguments.

Section 3.3 translates the conditions of lifting from k to R suitable isogenies between abelian schemes into the language of matrices and relates them to the previously introduced unitary groups.

Section 3.4 reviews a construction of Hecke covariant Siegel δ-modular forms $f^r_{g,\mathrm{crys}}$ introduced in [BB], Sect. 4.1; it also introduces and studies the ordinary Hecke covariant Siegel δ-modular forms $f^\partial_{g,\mathrm{crys}}$ which generalize the one $f^\partial_{1,\mathrm{crys}}$ introduced and studied in [Bar] and [Bu05], Ch. 8. Out of these ordinary δ-modular forms we construct some new ordinary Hecke covariant Siegel δ-modular forms $f^{\langle r\rangle}$ of weight -2 that play a key role later.

In Sect. 3.5 we introduce one of our main tools, the Serre–Tate expansion of a Siegel δ-modular form, and we prove a *Serre–Tate expansion principle* (see Proposition 3.5.1 (b)).

In Sect. 3.6 we use the Serre–Tate expansions to construct embeddings (comparison maps) $\heartsuit$ of the R-modules $\mathbb{I}^r_{g,\mathrm{ord}}(w)$ of Hecke covariant Siegel δ-modular forms into R-modules $\mathbb{H}^r_g(-\deg(w)/2)$ of $\mathbf{SL}_g$-invariant functions on certain R-modules of multiple quadratic forms (see Corollary 3.6.10). On the other hand we use the forms $f^{\langle r\rangle}$ to construct embeddings (comparison maps) $\diamondsuit$ of the R-modules $\mathbb{H}^{r-1}_g(-\deg(w)/2)$ into R-modules $\mathbb{I}^r_{g,\mathrm{ord}}(-\deg(w))$. A key Lemma 3.6.12 on 'elimination of logarithms' will help us use the embeddings $\heartsuit$ and $\diamondsuit$ to provide a complete description, up to torsion, of the R-algebra $\mathbb{I}_{g,\mathrm{ord}}$ in terms of the R-algebra of invariants $\mathbb{H}_g$ (see Theorem 3.6.29(a)).

In Sect. 3.7 we study the R-algebra $\mathbb{I}_g$ of Hecke covariant Siegel δ-modular forms. The main idea is to use the forms $f^r_{g,\mathrm{crys}}$ to find embeddings (comparison maps)

A. Buium and A. Vasiu, *Delta-Invariant Theory for Hecke Correspondences*,
Lecture Notes in Mathematics 2396, https://doi.org/10.1007/978-3-032-20571-1_3

□ of the R-algebras $\mathbb{H}^r_{g,\mathrm{conj}}$ of $\mathbf{SL}_g$-invariant functions on R-modules of multiple endomorphisms into the ring $\mathbb{I}^r_g$ (see Theorem 3.7.1).

Section 3.8 contains the proof of our main p-adic approximation result Theorem 2.4.22.

In Sect. 3.9 we prove Theorem 2.4.23 on the δ-flow structure defined by $f^1_{g,\mathrm{crys}}$.

In Sect. 3.10 we prove the 'reciprocity' Theorem 2.4.25.

Section 3.11 contains a few cohomological computations.

3.1 Symmetry

Let A_0 be an ordinary abelian variety over $\mathrm{Spec}(k)$ of dimension g. Let $\mathcal{M}_{A_0} = \mathrm{Spf}(R_g)$ be the formal deformation space of A_0; it is well-known that we can identify

$$R_g = R[[T_{ij}|1 \leq i, j \leq g]],$$

where T_{ij}s are algebraically independent indeterminates over R.[1]

For an abelian scheme A over $\mathrm{Spec}(R)$ that lifts A_0, so $A \times_{\mathrm{Spec}(R)} \mathrm{Spec}(k) = A_0$, let

$$q_A : T_p(A_0) \times T_p(\check{A}_0) \to 1 + \mathfrak{m}_R = 1 + pR$$

be the Serre–Tate bilinear map attached to A, where $1 + \mathfrak{m}_R$ is a group under multiplication. Let $e = (e_1 \cdots e_g)^{\mathrm{t}}$ be a column basis of the Tate module $T_p(A_0)$ of A_0 by which we mean that its entries form a $\mathbb{Z}_p$-basis of $T_p(A_0)$.

We assume that there exists a principal polarization on A defined by an isomorphism $\theta : A \to \check{A}$. Let $\theta_0 : A_0 \to \check{A}_0$ be the isomorphism obtained from θ by reduction modulo p. We consider the bilinear map

$$q_{A,\theta} := q_A \circ [T_p(1_{A_0}) \times T_p(\theta_0)] : T_p(A_0) \times T_p(A_0) \to 1 + \mathfrak{m}_R$$

induced by q_A (where 1_{A_0} is the identity automorphism of A_0) and the isomorphism

$$T_p(\theta_0) : T_p(A_0) \to T_p(\check{A}_0).$$

Let $q_{\mathrm{univ},A_0} : T_p(A_0) \times T_p(\check{A}_0) \to 1 + \mathfrak{m}_{R_g}$ be the Serre–Tate bilinear map attached to the formal deformation of A_0 over $\mathcal{M}_{A_0} = \mathrm{Spf}(R_g)$. We similarly get a universal Serre–Tate bilinear map

$$q_{\mathrm{univ},\theta_0} := q_{\mathrm{univ},A_0} \circ [T_p(1_{A_0}) \times T_p(\theta_0)] : T_p(A_0) \times T_p(A_0) \to 1 + \mathfrak{m}_{R_g}.$$

---

[1] For instance, see [Ka81], Thm. 2.1 (a) and (b) for the Serre–Tate deformation theory in the case of ordinary abelian varieties.

Note that q_A and $q_{A,\theta}$ are composites of q_{univ,A_0} and $q_{\mathrm{univ},\theta_0}$ (respectively) with a uniquely determined epimorphism $1+\mathfrak{m}_{R_g} \to 1+\mathfrak{m}_R$ of multiplicative groups induced by an R-algebra retraction $R_g \to R$.

Let I_{θ_0} be the ideal of R_g such that $\mathrm{Spf}(R_g/I_{\theta_0})$ is the formal deformation space $\mathcal{M}_{A_0,\theta_0}$ of (A_0, θ_0).

Proposition 3.1.1 *The bilinear map $q_{A,\theta}$ is symmetric.*

Proof For $(\alpha, \beta) \in T_p(A_0) \times T_p(\check{A}_0)$ we have $q_{\mathrm{univ},\check{A}_0}(\beta, \alpha) = q_{\mathrm{univ},A_0}(\alpha, \beta)$. This follows from Cartier duality and duality of abelian schemes, which in particular allow us to identify $A_0 = \check{\check{A}}_0$ and $\mathrm{Spf}(R_g) = \mathcal{M}_{A_0} = \mathcal{M}_{\check{\check{A}}_0}$. More precisely, if $\mathfrak{R}$ is a local artinian R-algebra of residue field k, then a short exact sequence

$$0 \to \boldsymbol{\mu}_{p^\infty,\mathfrak{R}} \to D_{\mathfrak{R}} \to (\mathbb{Q}_p/\mathbb{Z}_p)_{\mathfrak{R}} \to 0$$

of p-divisible groups over $\mathrm{Spec}(\mathfrak{R})$ is automatically self-dual with respect to the Cartier duality and this implies the identity $q_{\mathrm{univ},\check{A}_0}(\beta, \alpha) = q_{\mathrm{univ},A_0}(\alpha, \beta)$.

The isomorphism $\theta_0 : A_0 \to \check{A}_0$ induces an isomorphism

$$\iota_{\theta_0} : \mathrm{Spf}(R_g) = \mathcal{M}_{A_0} \simeq \mathrm{Spf}(R_g) = \mathcal{M}_{\check{A}_0},$$

and the ideal I_{θ_0} of R_g is the smallest one so that ι_{θ_0} modulo I_{θ_0} is the identity automorphism of R_g/I_{θ_0}. In other words, the formal deformation space $\mathcal{M}_{A_0,\theta_0}$ of (A_0, θ_0) is the locus where the two maps

$$q_{\mathrm{univ},A_0},\, q_{\mathrm{univ},\check{A}_0} \circ (T_p(\theta_0) \times T_p(\check{\theta}_0)^{-1}) : T_p(A_0) \times T_p(\check{A}_0) \to 1 + \mathfrak{m}_{R_g}$$

coincide (cf. also [Ka81], Thm. 2.1 4)).

From the very definition of a polarization we have the well-known symmetry property $\theta_0 = \check{\theta}_0$. From this and the identity $q_{\mathrm{univ},\check{A}_0}(\beta, \alpha) = q_{\mathrm{univ},A_0}(\alpha, \beta)$, by composing the last two maps with the isomorphism

$$T_p(1_{A_0}) \times T_p(\theta_0) : T_p(A_0) \times T_p(A_0) \to T_p(A_0) \times T_p(\check{A}_0)$$

we get that the resulting two maps from $T_p(A_0) \times T_p(A_0)$ to $1 + \mathfrak{m}_{R_g}$ differ from each other via the composite with the switching involution of $T_p(A_0) \times T_p(A_0)$ defined by (x, y) goes to (y, x). This implies that

$$I_{\theta_0} = (q_{\mathrm{univ},\theta_0}(e_i, e_j) - q_{\mathrm{univ},\theta_0}(e_j, e_i) | 1 \le i, j \le g). \tag{3.1}$$

Thus the symmetry of reductions of $q_{\mathrm{univ},\theta_0}$ via quotients of R_g holds precisely for R_g/I_{θ_0} (i.e., for $\mathcal{M}_{A_0,\theta_0}$) and hence it holds for the bilinear map $q_{A,\theta}$ which is defined via the R-valued point of $\mathcal{M}_{A_0,\theta_0}$ that corresponds to (A, θ). □

The entries of $q_{A,\theta}$ satisfy certain 'linear relations'. More precisely, the formal closed subscheme $\mathcal{M}_{A_0,\theta_0} = \mathrm{Spf}(R_g/I_{\theta_0})$ of $\mathcal{M}_{A_0} = \mathrm{Spf}(R_g)$ (i.e., the formal completion of the moduli stack $\mathcal{A}_{g,1,R}$ at the point corresponding to (A_0, θ_0)) is a formal subtorus of $\mathcal{M}_{A_0} = \mathrm{Spf}(R_g)$ and one gets $g(g-1)/2$ 'linear relations' for the entries of $q_{A,\theta}$ which can be read out from θ_0 as follows.

First, $\check{e} = (\check{e}_1 \cdots \check{e}_g)^{\mathrm{t}} := (T_p(\theta_0)(e_1) \cdots T_p(\theta_0)(e_g))^{\mathrm{t}}$ is a column basis of $T_p(\check{A}_0)$ and we get the T_{ij}s via the rule $T_{ij} = q_{ij} - 1 = q_{\mathrm{univ},\theta_0}(e_i, e_j) - 1$. We have $q_{\mathrm{univ},\theta_0}(e_i, e_j) - q_{\mathrm{univ},\theta_0}(e_j, e_i) = T_{ij} - T_{ji}$ for all $1 \leq i < j \leq g$, and, in view of Equation (3.1), we get directly the following consequence.

Corollary 3.1.2 *We have* $\mathcal{M}_{A_0,\theta_0} = Spf(S^0_{for})$, *with* $S^0_{for} = R[[T_{ij} | [1 \leq i \leq j \leq g]]$ *identified canonically with* $R_g/(T_{ij} - T_{ji} | 1 \leq i < j \leq g)$.

If f is an arbitrary column basis of $T_p(\check{A}_0)$, then the 'linear relations' would involve the matrix representation of $T_p(\theta_0)$ with respect to e and f.

3.2 Unitary Groups

For the principally polarized abelian variety (A_0, θ_0) over k we consider the monoid

$$\mathcal{U}_0 := \mathcal{U}_{A_0,\theta_0} := \{u_0 \in \mathrm{End}(A_0) | u_0 u_0^\dagger \in \mathbb{N} \backslash p\mathbb{N}\},$$

where $u_0 \to u_0^\dagger := \theta_0^{-1} \circ \check{u}_0 \circ \theta_0$ is the rule that defines the Rosati involution of $\mathrm{End}(A_0)$ relative to the principal polarization θ_0. With $e = (e_1 \cdots e_g)^{\mathrm{t}}$ as above, let

$$U_0 = U_{A_0,\theta_0,e} \subset \mathbf{GL}_g(\mathbb{Z}_p)$$

be the image of $\mathcal{U}_0$ under the injective algebra homomorphism

$$\mathrm{End}(A_0) \to \mathrm{End}(T_p(A_0)) \simeq \mathbf{Mat}_g(\mathbb{Z}_p), \quad v_0 \mapsto M_{v_0}, \tag{3.2}$$

where $\mathrm{End}(A_0) \to \mathrm{End}(T_p(A_0))$ maps $v_0 \in \mathrm{End}(A_0)$ to $T_p(v_0)$ and where the isomorphism in Display (3.2) is defined by e, so we have $T_p(v_0)(e) = M_{v_0} e$. Then U_0 is equipped with an involution $M \mapsto M^\dagger$ induced by the Rosati involution, so $M_{v_0^\dagger} = M^\dagger_{v_0}$. For each $M = (m_{ij})_{1 \leq i,j \leq g} \in U_0$ we write $M^\dagger = (m^\dagger_{ij})_{1 \leq i,j \leq g}$. The involution on U_0 does not extend, in general, to an involution on $\mathbf{Mat}_g(\mathbb{Z}_p)$, and is a priori unrelated to the transposition on $\mathbf{Mat}_g(\mathbb{Z}_p)$. The subgroup of $\mathbf{GL}_g(\mathbb{Z}_p)$ generated by U_0 is:

$$\mathbb{Z}^\times_{(p)} \cdot U_0 := \{\lambda M | \lambda \in \mathbb{Z}^\times_{(p)} \text{ and } M \in U_0\} \subset \mathbf{GL}_g(\mathbb{Z}_p).$$

Example 3.2.1 Consider the special case

$$\begin{aligned} A_0 &= E_0^g, \\ \theta_0 &= \text{standard diagonal principal polarization}, \\ e &= (e_1 \cdots e_g)^{\mathrm{t}} = \text{a diagonal basis}; \end{aligned} \tag{3.3}$$

so $A_0 = E_0^g$ is the g-fold product of an ordinary elliptic curve E_0 over $\mathrm{Spec}(k)$, θ_0 is obtained from (i.e., is the g-fold product of) the Abel–Jacobi map of E_0, we fix a basis e_0 of $T_p(E_0)$, and each e_i equals e_0 but viewed as an element of the i-th direct summand $T_p(E_0)$ of $T_p(A_0) = T_p(E_0)^g$. Let $\mathcal{O} := \mathrm{End}(E_0)$ and consider the canonical embedding

$$\mathcal{O} \to \mathrm{End}(T_p(E_0)) = \mathbb{Z}_p.$$

Recall that $\mathcal{O}$ is an order in an imaginary quadratic field $\mathcal{K}$ and that the Rosati involution on $\mathcal{O}$ is the complex conjugation $\lambda \mapsto \overline{\lambda}$.

Let $M = (m_{ij})_{1 \le i,j \le g} \in \mathbf{Mat}_g(\mathcal{O}) \cap \mathbf{GL}_g(\mathbb{Z}_p)$. Let

$$u_0 : A_0 = E_0^g \to A_0 = E_0^g$$

be the isogeny defined naturally by the matrix M. We have $u_0(P) = MP$ for all $P \in A_0(k) = E_0(k)^g$, viewed as a column vector with entries in $E_0(k)$. Note that $T_p(u_0)(e) = Me$, hence $M_{u_0} = M$ and $u_0^{\dagger} P = M^{\dagger} P$ for all $P \in A_0(k)$.

Lemma 3.2.2 *If (A_0, θ_0, e) is as in Display (3.3), then for all $M \in U_0$ we have $M^{\dagger} = \overline{M}^t$.*

Proof The case $g = 1$ is well-known. Let now $i, j \in \{1, \ldots, g\}$. Let $E_{0,i}$ be E_0 but viewed as the i-th factor of $A_0 = E_0^g$. The ij entry m_{ij} of M defines an endomorphism $m_{ij} : E_{0,j} \to E_{0,i}$ inducing $\check{m}_{ij} : \check{E}_{0,i} \to \check{E}_{0,j}$ and thus, via the standard principal polarization on E_0, inducing an endomorphism $E_{0,i} \to E_{0,j}$ that defines the ji entry $m_{ji}^{\dagger}$ of $M^{\dagger}$. Thus we have

$$m_{ji}^{\dagger} = \check{m}_{ij} = \overline{m_{ij}},$$

where the last equality follows from the case $g = 1$ by dropping the indexes i and j in the domains and codomains of m_{ij} and $\check{m}_{ij}$. □

An element $\lambda \in \mathcal{U}_0$ is called a *scalar* if M_λ is a scalar matrix. If λ is a scalar, then so is $\lambda^{\dagger}$. In this case we will view $\lambda, \lambda^{\dagger}$ as elements of $\mathbb{Z}_p^{\times}$; so also $\lambda/\lambda^{\dagger} \in \mathbb{Z}_p^{\times}$. Consider the following conditions, thought as unitary axioms, on (A_0, θ_0).

(UAX1) There exists a scalar $\lambda \in \mathcal{U}_0$ with $\lambda/\lambda^{\dagger}$ not a root of unity.

(UAX2) The group $\mathbb{Z}_p^{\times} \cdot U_0$ is Zariski dense in $\mathbf{GL}_{g,\mathbb{Z}_p}$, i.e., each polynomial in g^2 indeterminates with coefficients in $\mathbb{Z}_p$ which vanishes on $\mathbb{Z}_p^{\times} \cdot U_0$ is the zero polynomial.

Lemma 3.2.3 *Each (A_0, θ_0) as in Display (3.3) satisfies condition (UAX1).*

Proof To check this we can assume $g = 1$. Based on Lemma 3.2.2 it suffices to show that there exists a nonzero $\lambda \in \mathcal{K}$ such that $\lambda/\overline{\lambda}$ is not a root of unity. But this is well-known (for instance, see [Bu05], Lem. 8.21). □

Proposition 3.2.4 *Each (A_0, θ_0) as in Display (3.3) satisfies condition (UAX2).*

We first need the following general lemma.

Lemma 3.2.5 *Let $\mathcal{G}$ be a connected reductive group scheme over Spec($\mathbb{Z}_{(p)}$). Then $\mathcal{G}(\mathbb{Z}_{(p)})$ is Zariski dense in $\mathcal{G}_{\mathbb{Q}}$ and is dense in $\mathcal{G}(\mathbb{R})$.*

Proof suggested by Offer Gabber. Let κ be a field equipped with a reasonable field topology: nondiscrete, Hausdorff, and the ring operations and inversions are continuous. This defines a topology on $X(\kappa)$ for every κ-scheme locally of finite type, compatible with morphisms, products, open and closed embeddings.

Let $\mathcal{B}$ be a connected smooth affine group over Spec(κ) which is unirational (e.g., this holds if κ is perfect, see [Bore], Ch. V, Thm. 18.2).

We check that each nonempty open $\mathcal{V}$ of $\mathcal{B}(\kappa)$ (in the mentioned topology) is Zariski dense in $\mathcal{B}$. The unirationality means that there exists a nonempty open $\mathcal{Y}$ in an affine space $\mathbb{A}^n_\kappa$ and a dominant morphism $f : \mathcal{Y} \to \mathcal{B}$. By translations on the source and target we can assume that the origin 0 of $\mathbb{A}^n_\kappa$ is in $\mathcal{Y}$ and is mapped by f into the identity of $\mathcal{B}$. Then the Zariski density of $\mathcal{V}$ in $\mathcal{B}$ follows from the observation that nonempty opens $U \subset \kappa^n = \mathbb{A}^n_\kappa(\kappa)$ are Zariski dense in $\mathbb{A}^n_\kappa$. Indeed we may assume U is a product of n nonempty balls in $\mathbb{A}^1_\kappa(\kappa)$; then the observation follows by induction on n and the fact that every infinite set, in particular every nonempty ball, in $\mathbb{A}^1_\kappa(\kappa)$ is Zariski dense.

Lemma follows by taking $\kappa = \mathbb{Q}$ and $\mathcal{B} = \mathcal{G}_{\mathbb{Q}}$ and using the fact that nonempty opens of $\mathbb{Q}^n = \mathbb{A}^n_\kappa(\kappa)$ in the p-adic topology are dense in $\mathbb{R}^n$ (in its metric topology), by the approximation theorem for independent valuations of $\mathbb{Q}$. □

Proof of Proposition 3.2.4. Consider an embedding $\mathcal{K} = \text{Frac}(\mathcal{O}) \subset \mathbb{C}$ and note that the group $\mathbb{Z}^\times_{(p)} \cdot U_0$ is contained in the subgroup $\mathbf{GL}_g(\mathcal{K})$ of $\mathbf{GL}_g(\mathbb{Q}_p)$. By considering a $\mathcal{K}$-embedding of $\mathbb{Q}_p$ into $\mathbb{C}$ we get that to prove Proposition 3.2.4 it suffices to show that $\mathbb{Z}^\times_{(p)} \cdot U_0$ is Zariski dense in $\mathbf{GL}_g(\mathbb{C})$.

Let now $\mathbf{U}_g$ be the reductive (unitary) group scheme over $\mathbb{Z}_{(p)}$ whose functor of valued points attaches to each $\mathbb{Z}_{(p)}$-algebra B the group

$$\mathbf{U}_g(B) := \{M \in \mathbf{GL}_g(B \otimes_{\mathbb{Z}} \mathcal{O}) | M\overline{M}^{\mathrm{t}} = 1_g\},$$

where the upper bar conjugation acts on the second factor of $B \otimes_{\mathbb{Z}} \mathcal{O}$. From Lemma 3.2.5 we get that $\mathbf{U}_g(\mathbb{Z}_{(p)})$ is dense in $\mathbf{U}_g(\mathbb{R})$ in the metric topology. Also from Lemma 3.2.3 we get that $\mathbf{U}_g(\mathbb{Z}_{(p)}) \subset \mathbb{Z}^\times_{(p)} \cdot U_0$. We conclude that Proposition 3.2.4 holds by the classical fact that $\mathbf{U}_g(\mathbb{R})$ is dense in $\mathbf{GL}_g(\mathbb{C})$ in the Zariski topology.

We include a proof of this classical fact. Let Γ_g be the Zariski closure of $\mathbf{U}_g(\mathbb{R})$ in $\mathbf{GL}_{g,\mathbb{C}}$. Then Γ_g is a complex algebraic group and to check that $\Gamma_g = \mathbf{GL}_{g,\mathbb{C}}$ it is enough to check that its Lie algebra Lie(Γ_g) equals Lie$(\mathbf{GL}_{g,\mathbb{C}})$ (see [Bore], Ch. I, Sect. 7.1). But Lie(Γ_g) contains the $\mathbb{C}$-span of the Lie algebra Lie$(\mathbf{U}_g(\mathbb{R}))$ over $\mathbb{R}$ and thus it contains all the antisymmetric matrices of size $g \times g$ and all matrices of the form $\sqrt{-1}$ times a symmetric real matrix of size $g \times g$. Hence Lie(Γ_g) contains all the real matrices and thus all the complex matrices of size $g \times g$. We conclude that $\Gamma_g = \mathbf{GL}_{g,\mathbb{C}}$. □

Remark 3.2.6 **(a)** The set of points in the moduli stack $\mathcal{A}_{g,1,k}$ over Spec(k) that correspond to ordinary principally polarized abelian varieties over Spec(k) for which conditions (UAX1) and (UAX2) hold is Zariski dense. This follows from Lemma 3.2.3, Proposition 3.2.4 and Larsen's example in [Cha], p. 443.

(b) If $\lambda \in \mathcal{U}_0 \backslash \mathbb{Q}$ is a scalar, then $\mathcal{F} := \mathbb{Q}[\lambda]$ is a totally imaginary quadratic extension of $\mathbb{Q}$ and the Mumford–Tate group of the generic fiber over $K = \text{Frac}(R)$ of the canonical lift of A_0 is naturally identified with the 2 dimensional torus which is the Weil restriction $\text{Res}_{\mathcal{F}/\mathbb{Q}}\mathbb{G}_{m,\mathcal{F}}$. Using this, it is easy to see that if the condition (UAX1) holds, then A_0 is isogeneous to E_0^g for some elliptic curve E_0 over Spec(k) and the Rosati involution defined by the principal polarization θ_0 of A_0 induces the complex conjugation on $\mathcal{F}$.

(c) Similarly, if the condition (UAX2) holds, then A_0 is isogeneous to E_0^g for some elliptic curve E_0 over Spec(k). One can check this by first remarking that A_0 is isogeneous to a power of a simple ordinary abelian variety A_{00} over Spec(k) which has a model over a finite field whose Frobenius endomorphism generates a totally imaginary extension of its subfield fixed by the Rosati involution defined by θ_0 and thus it is an imaginary quadratic extension of $\mathbb{Q}$. As A_{00} is ordinary, the Honda–Serre–Tate theory implies that A_{00} is an elliptic curve over Spec(k).

3.3 Lifting Condition

Assume $Q_1, Q_2 \in \mathbf{Mat}_g(R)$ are symmetric. Let (A_1, θ_1), (A_2, θ_2) be the principally polarized abelian schemes over Spec(R) that lift A_0 and such that we have $(q_{A_1,\theta_1}(e_i, e_j))_{1 \le i,j \le g} = Q_1$ and $(q_{A_2,\theta_2}(e_i, e_j))_{1 \le i,j \le g} = Q_2$. By the functoriality part of Serre–Tate theory (see [Ka81], Thm. 2.1 4)) the condition that the isogeny $u_0 : A_0 \to A_0$ lifts to a homomorphism $u : A_1 \to A_2$ is that for all $1 \le i, j \le g$ we have

$$q_{A_1,\theta_0}(e_i, T_p(u_0^\dagger)(e_j)) = q_{A_2,\theta_0}(T_p(u_0)(e_i), e_j). \tag{3.4}$$

For all $c \in \mathbb{Z}_p$ and $m \in 1 + \mathfrak{m}_R$ we let $c \star m = m \star c := m^c \in 1 + \mathfrak{m}_R$; hence $\star$ denotes multiplication in the (left and right) $\mathbb{Z}_p$-module structure of the multiplicative abelian group $1 + \mathfrak{m}_R$. More generally, we want to consider left and right multiplication of matrices with entries in the $\mathbb{Z}_p$-module $1 + \mathfrak{m}_R$ by matrices with entries in $\mathbb{Z}_p$. It is useful to introduce a special notation for this as follows. Let $\mathbf{Mat}_g(1 + \mathfrak{m}_R)$

be the subset of $\mathbf{Mat}_g(R)$ formed by matrices whose all entries are in $1+\mathfrak{m}_R$. If $M \in \mathbf{Mat}_g(\mathbb{Z}_p)$ and $X \in \mathbf{Mat}_g(1+\mathfrak{m}_R)$ we denote by $M \star X \in \mathbf{Mat}_g(1+\mathfrak{m}_R)$ the matrix with ij entries given by

$$(m_{i1} \star x_{1j}) \cdots (m_{ig} \star x_{gj}) = \prod_{l=1}^{g} x_{lj}^{m_{il}}$$

and we denote by $X \star M \in \mathbf{Mat}_g(1+\mathfrak{m}_R)$ the matrix with ij entries given by

$$(x_{i1} \star m_{1j}) \cdots (x_{ig} \star m_{gj}) = \prod_{l=1}^{g} x_{il}^{m_{lj}}.$$

These operations satisfy the following four 'associativity' properties:

$$M_1 \star (X \star M_2) = (M_1 \star X) \star M_2,$$

$$(M_1 \cdot M_2) \star X = M_1 \star (M_2 \star X),$$

$$X \star (M_1 \cdot M_2) = (X \star M_1) \star M_2,$$

$$(X \star M)^{\mathrm{t}} = M^{\mathrm{t}} \star X^{\mathrm{t}}.$$

Here (and later) $M_1 \cdot M_2 = M_1 M_2$ denotes the usual product of the $g \times g$ matrices M_1 and M_2.

As an example, for $g = 2$, we have

$$M \star X = \begin{pmatrix} x_{11}^{m_{11}} x_{21}^{m_{12}} & x_{12}^{m_{11}} x_{22}^{m_{12}} \\ x_{11}^{m_{21}} x_{21}^{m_{22}} & x_{12}^{m_{21}} x_{22}^{m_{22}} \end{pmatrix} \quad \text{and} \quad X \star M = \begin{pmatrix} x_{11}^{m_{11}} x_{12}^{m_{21}} & x_{11}^{m_{12}} x_{12}^{m_{22}} \\ x_{21}^{m_{11}} x_{22}^{m_{21}} & x_{21}^{m_{12}} x_{22}^{m_{22}} \end{pmatrix}.$$

Returning to Eq. (3.4), set $M := M_{u_0}$ (see Display (3.2)). Then Eq. (3.4) is equivalent to the equality

$$Q_2 = M^{-1} \star Q_1 \star (M^{\dagger})^{\mathrm{t}}. \tag{3.5}$$

Lemma 3.3.1 *For $M \in U_0$ with $MM^{\dagger} = d \cdot 1_g$, $d \in \mathbb{N}\backslash p\mathbb{N}$, let $u_0 : A_0 \to A_0$ be the endomorphism corresponding to M; so $M = M_{u_0}$. Assume (A_1, θ_1) is a principally polarized abelian scheme over R that lifts (A_0, θ_0) and let Q_1 be the symmetric matrix with coefficients in R that corresponds to A_1. Then the matrix $Q_2 := M^{-1} \star Q_1 \star (M^{\dagger})^t$ is symmetric. Moreover, if (A_2, θ_2) is the principally polarized abelian scheme over Spec(R) which lifts (A_0, θ_0) and corresponds to Q_2, then there exists an isogeny $u : A_1 \to A_2$ that lifts the isogeny $u_0 : A_0 \to A_0$ and such that $uu^t = d$.*

Proof The symmetry of Q_2 follows from the fact that $Q_1 \star (d \cdot 1_g)$ is symmetric and we have

$$Q_2 = M^{-1} \star Q_1 \star (M^\dagger)^t = M^{-1} \star (Q_1 \star (d \cdot 1_g)) \star (M^{-1})^t.$$

As Eq. (3.5) holds for A_1, A_2 and $M = M_{u_0}$ we get that there exists an isogeny $u : A_1 \to A_2$ that lifts the isogeny $u_0 : A_0 \to A_0$. As the reduction modulo p map $\mathrm{End}(A_1) \to \mathrm{End}(A_0)$ is injective, $u_0 u_0^\dagger = d$ implies that $uu^t = d$. □

Let dt/t be the invariant 1-form on the formal multiplicative group $\mathbb{G}_{m,R}^{\mathrm{for}}$ of the group scheme $\mathbb{G}_{m,R} = \mathrm{Spec}(R[t, t^{-1}])$ over $\mathrm{Spec}(R)$.

For $c \in \{1, 2\}$, as A_c is ordinary one has that the formal group A_c^{for} is isomorphic to $(\mathbb{G}_{m,R}^{\mathrm{for}})^g$ over R; see [Maz], Lem. 4.27. Let $\zeta_{ci} \in \mathrm{Hom}(A_c^{\mathrm{for}}, \mathbb{G}_{m,R}^{\mathrm{for}})$ be the image via the isomorphism

$$T_p(\check{A}_0) \simeq \mathrm{Hom}(A_c^{\mathrm{for}}, \mathbb{G}_{m,R}^{\mathrm{for}}) \tag{3.6}$$

of the $i1$ entry of the column basis $T_p(\theta_0)(e)$ of $T_p(\check{A}_0)$. Defining

$$\omega_{ci} := \zeta_{ci}^*(dt/t),$$

we get an R-basis $\omega_c = \{\omega_{c1}, \ldots, \omega_{cg}\}$ of global 1-forms on A_c^{for} which we identify with 1-forms on A_c. We say that the ω_c is *attached* to the basis e.

Lemma 3.3.2 *We assume that $M \in U_0$ and $Q_2 = M^{-1} \star Q_1 \star (M^\dagger)^t$ and we consider the isogeny $u : A_1 \to A_2$ provided by Lemma 3.3.1. If $[u] \in \boldsymbol{GL}_g(R)$ is such that $u^*(\omega_2) = [u]\omega_1$, then we have $[u] = M^\dagger$.*

Proof By the functoriality of the isomorphism of Diagram (3.6) we get that the following diagram is commutative:

$$\begin{array}{ccc} T_p(\check{A}_0) & \simeq & \mathrm{Hom}(A_1^{\mathrm{for}}, \mathbb{G}_{m,R}^{\mathrm{for}}) \\ T_p(\check{u}_0) \uparrow & & \uparrow (u^{\mathrm{for}})^* \\ T_p(\check{A}_0) & \simeq & \mathrm{Hom}(A_2^{\mathrm{for}}, \mathbb{G}_{m,R}^{\mathrm{for}}). \end{array} \tag{3.7}$$

Here $(u^{\mathrm{for}})^*(\zeta) = \zeta \circ u^{\mathrm{for}}$ for all $\zeta \in \mathrm{Hom}(A_2^{\mathrm{for}}, \mathbb{G}_{m,R}^{\mathrm{for}})$. Recalling the identities $M^\dagger = M_{u_0}^\dagger = M_{u_0^\dagger}$ and using Diagram (3.7) we get that for all $i \in \{1, \ldots, g\}$ we have an equality of formal homomorphisms

$$\zeta_{2i} \circ u^{\mathrm{for}} = \sum_{j=1}^{g} m_{ij}^\dagger \zeta_{1j},$$

hence

$$(u^{\mathrm{for}})^*(\zeta_{2i}^*(dt/t)) = \sum_{j=1}^{g} m_{ij}^\dagger \zeta_{1j}^*(dt/t).$$

Thus we have $[u] = M^\dagger$ as one can see at the level of formal Lie groups. □

3.4 Ordinary Crystalline δ-Modular Forms

We first recall from [BB], Sect. 4.1, the construction for each $r \in \mathbb{N}$ of a Hecke covariant Siegel δ-modular forms $f^r_{g,\text{crys}} \in \mathbb{I}^r_{gg}(\phi^r, 1)$ which we refer to as the *basic* Siegel δ-modular form of weight $(\phi^r, 1)$. Then, generalizing a construction for $g = 1$ in [Bar], Constr. 3.2, p. 248 (cf. also [Bu05], Sect. 8.4.3), we introduce an ordinary Hecke covariant Siegel δ-modular form $f^\partial_{g,\text{crys}} \in \mathbb{I}^1_{gg,\text{ord}}(-\phi, 1)$ which we refer to as the *basic* Siegel δ-modular form of weight $(-\phi, 1)$.

Let S and $\mathbf{M}_g(S)$ be as in Sect. 2.3.2, let $(A, \theta, \omega) \in \mathbf{M}_g(S)$ and consider the de Rham S-modules $H^1_{\text{dR}}(A/S)$ and $H^1_{\text{dR}}(\check{A}/S)$. We recall from [FC], Ch. III, Sect. 9, paragraph before Prop. 9.1, that $H^1_{\text{dR}}(A/S)$ has a canonical submodule $F^1 H^1_{\text{dR}}(A/S)$ canonically isomorphic to (and will be identified with) $H^0(A, \Omega_{A/S})$ and such that

$$H^1_{\text{dR}}(A/S)/F^1 H^1_{\text{dR}}(A/S) \simeq H^1(A, \mathcal{O}) \simeq H^0(\check{A}, \Omega^*_{\check{A}/S}).$$

From loc. cit. we also get that there exists a canonical perfect pairing

$$\langle\ ,\ \rangle_A : H^1_{\text{dR}}(\check{A}/S) \times H^1_{\text{dR}}(A/S) \to S \tag{3.8}$$

which induces the canonical pairing between $F^1 H^1_{\text{dR}}(A/S) = H^0(A, \Omega_{A/S})$ and $H^1_{\text{dR}}(\check{A}/S)/F^1 H^1_{\text{dR}}(\check{A}/S) = H^0(A, \Omega^*_{A/S})$. In view of this compatibility, for all $(\omega, \omega') \in H^0(A, \Omega_{A/S}) \times H^0(\check{A}, \Omega_{\check{A}/S})$ we have $\langle \omega', \omega \rangle_A = 0$.

We claim that the pairing $\langle\ ,\ \rangle_A$ is compatible with polarized isogenies in the sense that if $u : A_1 \to A_2$ is a polarized isogeny of principally polarized abelian schemes, then for all $\check{\alpha}_1 \in H^1_{\text{dR}}(\check{A}_1/S)$ and $\alpha_2 \in H^1_{\text{dR}}(A_2/S)$ the following formula holds:

$$\langle \check{\alpha}_1, u^* \alpha_2 \rangle_{A_1} = \langle \check{u}^* \check{\alpha}_1, \alpha_2 \rangle_{A_2}. \tag{3.9}$$

To see this, working in the étale topology of $\text{Spec}(S)$, we can assume that there exists $N \geq 3$ relatively prime to p and such that the two principally polarized abelian schemes over S are endowed with symplectic similitudes level N-structures. Consider the union $\mathcal{A}'_{g,1,N,R}$ of all prime to p *Hecke correspondences* on $\mathcal{A}_{g,1,N,R}$ parameterizing polarized isogenies that are compatible with the symplectic similitude level-N structures (see Appendix). Then $\mathcal{A}'_{g,1,N,R}$ is a union of affine open subschemes that are spectra of integral domains which are smooth over R and hence of characteristic 0. So it is enough to check the claim for each integral domain S of characteristic 0. Again, by compatibility with base change and direct limits, we can assume S is a finitely generated $\mathbb{Z}$-algebra.

As every finitely generated $\mathbb{Z}$-algebra that is an integral domain of characteristic 0 can be embedded into the complex field, we can assume $S = \mathbb{C}$.

As $S = \mathbb{C}$, we use the analytic description of our objects as follows. If V is a finite dimensional complex vector space and $\Lambda \subset V$ is a lattice, then the first Betti homology group $H_1(V/\Lambda, \mathbb{Z})$ is naturally identified with Λ. Assume now A is a complex abelian variety. Then by [GH], Ch. II, Sect. 6, pp. 327 and 328 we have canonical identifications of complex manifolds

$$A(\mathbb{C}) = H^0(A, \Omega)^*/H_1(A(\mathbb{C}), \mathbb{Z}), \quad \check{A} = H^1(A, \mathcal{O})/H_1(A(\mathbb{C}), \mathbb{Z})^*.$$

So the natural pairing

$$H_1(A(\mathbb{C}), \mathbb{Z})^* \times H_1(A(\mathbb{C}), \mathbb{Z}) \to \mathbb{Z} \tag{3.10}$$

induces a pairing on the Betti cohomology complex vector spaces:

$$H^1(\check{A}(\mathbb{C}), \mathbb{C}) \times H^1(A(\mathbb{C}), \mathbb{C}) \to \mathbb{C}.$$

The latter can be identified with the pairing in Diagram (3.8) under the natural identifications of the Betti and de Rham cohomologies complex vector spaces. Indeed, by [FC], Ch. III, Sect. 9, paragraph before Prop. 9.1, the (first) Chern class $c(\mathcal{L}) \in H^2_{\mathrm{dR}}(A \times \check{A}/\mathbb{C})$ of the Poincaré line bundle $\mathcal{L}$ on $A \times \check{A}$ lies in $H^1_{\mathrm{dR}}(A/\mathbb{C}) \otimes_{\mathbb{C}} H^1_{\mathrm{dR}}(\check{A}/\mathbb{C}) \subset H^2_{\mathrm{dR}}(A \times \check{A}/\mathbb{C})$ and it induces the perfect pairing of Diagram (3.8); on the other hand, using the analytic description of the Poincaré line bundle $\mathcal{L}$ in [GH], Ch. II, Sect. 6, p. 328 one checks directly that the pairing between $H^1_{\mathrm{dR}}(A/\mathbb{C}) \simeq H^1(A(\mathbb{C}), \mathbb{C})$ and $H^1_{\mathrm{dR}}(\check{A}/\mathbb{C}) \simeq H^1(\check{A}(\mathbb{C}), \mathbb{C})$ induced by $c(\mathcal{L})$ coincides with the paring in Diagram (3.10). Finally note that the pairing in Diagram (3.10) is compatible (in the obvious sense) with isogenies hence so is the pairing in Diagram (3.8).

We consider the pullback isomorphism $\theta^* : H^1_{\mathrm{dR}}(\check{A}/S) \to H^1_{\mathrm{dR}}(A/S)$ induced by $\theta : A \to \check{A}$ and define a bilinear map

$$\langle\ ,\ \rangle_\theta : H^1_{\mathrm{dR}}(A/S) \times H^1_{\mathrm{dR}}(A/S) \to S$$

by the rule

$$\langle \alpha, \beta \rangle_\theta := \langle (\theta^*)^{-1}(\alpha), \beta \rangle_A,$$

where $\alpha, \beta \in H^1_{\mathrm{dR}}(A/S)$. By what was said above about the pairing $\langle\ ,\ \rangle_A$ we get that the bilinear map $\langle\ ,\ \rangle_\theta$ is a perfect pairing and has $H^0(A, \Omega^1)$ as an isotropic S-submodule. Clearly the formation of this paring commutes with base change $S \to S'$. Moreover, we have the following compatibility.

Lemma 3.4.1 *Let $u : A_1 \to A_2$ be a polarized isogeny between principally polarized abelian schemes over Spec(S) with principal polarizations θ_1 and θ_2 (respectively). Let*

$$u^* : H^1_{dR}(A_2/S) \to H^1_{dR}(A_1/S), \quad u^{t*} : H^1_{dR}(A_1/S) \to H^1_{dR}(A_2/S)$$

be the induced isomorphisms. Then for all $\alpha \in H^1_{dR}(A_1/S)$ and $\beta \in H^1_{dR}(A_2/S)$ we have

$$\langle \alpha, u^*(\beta)\rangle_{\theta_1} = \langle u^{t*}(\alpha), \beta\rangle_{\theta_2}.$$

Proof Using Eq. (3.9) we compute

$$\langle \alpha, u^*(\beta)\rangle_{\theta_1} = \langle (\theta_1^*)^{-1}(\alpha), u^*(\beta)\rangle_{A_1} = \langle \check{u}^* \circ (\theta_1^*)^{-1}(\alpha), \beta\rangle_{A_2}$$
$$= \langle (\theta_2^*)^{-1} \circ u^{t*}(\alpha), \beta\rangle_{A_2} = \langle u^{t*}(\alpha), \beta\rangle_{\theta_2}.$$

□

Let now $S \in \mathrm{Ob}(\widehat{\mathbf{FilRing}_\delta})$ and $(A, \theta, \omega) \in \mathbf{M}_g(S^0)$. Then for all integers $i \geq 0$, the crystalline theory provides ϕ-linear maps (see [Bu00] or [Bu05], Sect. 8.1.9):

$$\Phi : H^1_{\mathrm{dR}}(A/S^0) \otimes_{S^0} S^i \to H^1_{\mathrm{dR}}(A/S^0) \otimes_{S^0} S^{i+1}.$$

Thus we have $\Phi(\lambda x) = \phi(\lambda)\Phi(x)$ for each pair $(\lambda, x) \in S^i \times [H^1_{\mathrm{dR}}(A/S^0) \otimes_{S^0} S^i]$.

Following [BB], Sect. 4.1, for each integer $r \geq 1$ we have the r-th iterates $\Phi^r : H^1_{\mathrm{dR}}(A/S^0) \otimes_{S^0} S^i \to H^1_{\mathrm{dR}}(A/S^0) \otimes_{S^0} S^{i+r}$ of Φ and we define a Siegel δ-modular function $f^r_{g,\mathrm{crys}}$ of genus g, size g, and order $\leq r$ by the formula

$$f^r_{g,\mathrm{crys}}(A, \theta, \omega, S) := p^{-1}\langle \Phi^r \omega, \omega^{\mathrm{t}}\rangle_\theta \in \mathbf{Mat}_g(S^r).$$

Writing $\omega = (\omega_1 \cdots \omega_g)^{\mathrm{t}}$ as a column vector of 1-forms, $\langle \Phi^r \omega, \omega^{\mathrm{t}}\rangle_\theta$ is the matrix whose each ij entry is $\langle \Phi^r(\omega_i), \omega_j\rangle_\theta \in S^r$.

Lemma 3.4.2 *The rule $f^r_{g,crys}$ defines a Hecke covariant Siegel δ-modular function of genus g, size g, order $\leq r$, and weight $(\phi^r, 1)$, i.e.,*

$$f^r_{g,crys} \in \mathbb{I}^n_{gg}(\phi^r, 1).$$

Proof This was proved in [BB], Sect. 4.1. Alternatively one can prove this by a direct computation using Lemma 3.4.1 (see the proof of Lemma 3.4.4 below for a similar computation). □

Remark 3.4.3 By construction, the form $f^1_{g,\mathrm{crys}}$ vanishes at some (A, θ, ω, R) if and only if the filtered F-crystal over Spec(k) of the abelian scheme A over Spec(R) is a direct sum of $2g$ filtered F-crystals over Spec(k) of rank 1. Thus (A, θ) is either the canonical lift of an ordinary principally polarized abelian variety (A_k, θ_k) over Spec(k) or, if $p = 2$, one of the $2^{\frac{g(g+1)}{2}}$ quasi-canonical lifts of the ordinary principally polarized abelian variety (A_k, θ_k) over Spec(k) (i.e., corresponds to a Spf(R)-valued point of order 2 of the formal torus of dimension $\frac{g(g+1)}{2}$ over Spf(R) of formal deformations of (A_k, θ_k)). See [Me], App. and [V], Thm. 1.6.3 and Prop. 9.5.1.

Assume now that $S \in \mathrm{Ob}(\widehat{\mathbf{FilRing}_\delta})$ and $(A, \theta, \omega) \in \mathbf{M}_{g,\mathrm{ord}}(S^0)$. Recall that, as A is ordinary, we can consider the *unit root subspace* U of $H^1_{\mathrm{dR}}(A/S^0)$, see [Ka73b], Thm. 4.1. Strictly speaking, loc. cit. is stated for the case when the reductions of S^0

modulo powers of p are smooth over the corresponding reductions of R but working in the étale topology, (A, θ) is the pullback of a principally polarized abelian scheme over the p-adic completion of a smooth R-algebra to which loc. cit. applies. The unit root subspace U is a projective S^0-module which is a complement of $H^0(A, \Omega_{A/S^0})$ in $H^1_{\mathrm{dR}}(A/S^0)$. Assume it is free on S^0 and pick an arbitrary column vector $\eta = (\eta_1 \cdots \eta_g)^{\mathrm{t}}$ whose entries form an S^0-basis of U.

Consider the $g \times g$ matrix $\phi(\langle \eta, \omega^{\mathrm{t}} \rangle_\theta)$ whose ij entries are $\phi(\langle \eta_i, \omega_j \rangle_\theta) \in S^1$ and the $g \times g$ matrix $\langle \Phi\eta, \omega^{\mathrm{t}} \rangle_\theta$ whose ij entries are $\langle \Phi\eta_i, \omega_j \rangle_\theta \in S^1$.

Then we consider the product of invertible matrices

$$f^{\partial}_{g,\mathrm{crys}}(A, \theta, \omega, S) := (\phi(\langle \eta, \omega^{\mathrm{t}} \rangle_\theta))^{-1} \cdot \langle \Phi\eta, \omega^{\mathrm{t}} \rangle_\theta \in \mathbf{GL}_g(S^1).$$

Lemma 3.4.4 *The rule $f^{\partial}_{g,crys}$ defines an ordinary Hecke covariant Siegel δ-modular form of weight $(-\phi, 1)$, size g, and order 1, i.e.,*

$$f^{\partial}_{g,crys} \in \mathbb{I}^1_{gg,ord}(-\phi, 1).$$

In particular, its inverse is also an ordinary Siegel δ-modular form

$$f_{\partial,g,crys} := (f^{\partial}_{g,crys})^{-1} \in \mathbb{I}^1_{gg,ord}(-1, \phi)$$

and their determinants are ordinary Siegel δ-modular forms

$$\det(f^{\partial}_{g,crys}) \in \mathbb{I}^1_{g,ord}(\phi - 1),$$

$$\det(f_{\partial,g,crys}) = (\det(f^{\partial}_{g,crys}))^{-1} \in \mathbb{I}^1_{g,ord}(1 - \phi).$$

Proof The fact that $f^{\partial}_{g,\mathrm{crys}}$ has weight $(-\phi, 1)$ follows from the following computation:

$$\begin{aligned} f^{\partial}_{g,\mathrm{crys}}(A, \theta, \lambda\omega, S) &= (\phi(\langle \eta, \omega^{\mathrm{t}}\lambda^{\mathrm{t}} \rangle_\theta))^{-1} \cdot \langle \Phi\eta, \omega^{\mathrm{t}}\lambda^{\mathrm{t}} \rangle_\theta \\ &= ((\phi(\lambda))^{\mathrm{t}})^{-1} \cdot f^{\partial}_{g,\mathrm{crys}}(A, \theta, \omega, S) \cdot \lambda^{\mathrm{t}}. \end{aligned}$$

To check the Hecke covariance consider an object $S \in \mathrm{Ob}(\widehat{\mathbf{FilRing}_\delta})$ and triples $(A_1, \theta_1, \omega_1), (A_2, \theta_2, \omega_2) \in \mathbf{M}_{g,\mathrm{ord}}(S^0)$, let $u : A_1 \to A_2$ be a polarized isogeny with $u^*\omega_2 = \omega_1, uu^{\mathrm{t}} = d \in \mathbb{N}\backslash p\mathbb{N}$, and we assume there exists a column vector η_1 whose entries form a basis of the unit root subspace of A_1. Then, using Lemma 3.4.1 we have:

$$\begin{aligned} f^{\partial}_{g,\mathrm{crys}}(A_1, \theta_1, \omega_1, S) &= (\phi(\langle \eta_1, \omega_1^{\mathrm{t}} \rangle_{\theta_1}))^{-1} \cdot \langle \Phi\eta_1, \omega_1^{\mathrm{t}} \rangle_{\theta_1} \\ &= (\phi(\langle \eta_1, u^*\omega_2^{\mathrm{t}} \rangle_{\theta_1}))^{-1} \cdot \langle \Phi\eta_1, u^*\omega_2^{\mathrm{t}} \rangle_{\theta_1} \\ &= (\phi(\langle u^{\mathrm{t}*}\eta_1, \omega_2^{\mathrm{t}} \rangle_{\theta_2}))^{-1} \cdot \langle u^{\mathrm{t}*}\Phi\eta_1, \omega_2^{\mathrm{t}} \rangle_{\theta_2} \\ &= (\phi(\langle u^{\mathrm{t}*}\eta_1, \omega_2^{\mathrm{t}} \rangle_{\theta_2}))^{-1} \cdot \langle \Phi u^{\mathrm{t}*}\eta_1, \omega_2^{\mathrm{t}} \rangle_{\theta_2} \\ &= f^{\partial}_{g,\mathrm{crys}}(A_2, \theta_2, \omega_2, S). \end{aligned}$$

From this identity and Eq. (2.27), applied to $\deg(-\phi + 1) = 0$, we get that $f^{\partial}_{g,\text{crys}} \in \mathbb{I}^1_{gg,\text{ord}}(-\phi, 1)$. □

Let us abbreviate $f^{\partial} := f^{\partial}_{g,\text{crys}}$ and for an integer $b \geq 1$ we set

$$N_b := (f^{\partial})^{\phi^{b-1}} \cdot (f^{\partial})^{\phi^{b-2}} \cdots (f^{\partial})^{\phi} \cdot f^{\partial} \in \mathbb{I}^b_{gg,\text{ord}}(-\phi^b, 1).$$

We also set $N_0 := 1_g$, the identity matrix.

Using the invertibility of f^{∂} we have the following consequences.

Corollary 3.4.5 *For all integers $a, b \geq 0$ and $r \geq \max\{a, b\}$, the following R-linear maps are isomorphisms:*

$$\mathbb{I}^r_{gg,ord}(\phi^a, \phi^b) \to \mathbb{I}^r_{gg,ord}(1, 1), \quad f \mapsto N^t_a \cdot f \cdot N_b,$$

$$\mathbb{I}^r_{gg,ord}(-\phi^a, \phi^b) \to \mathbb{I}^r_{gg,ord}(-1, 1), \ f \mapsto N^{-1}_a \cdot f \cdot N_b,$$

$$\mathbb{I}^r_{gg,ord}(\phi^a, -\phi^b) \to \mathbb{I}^r_{gg,ord}(1, -1), \ f \mapsto N^t_a \cdot f \cdot (N^t_b)^{-1}.$$

In particular, for each integer $a \geq 1$ we define recursively the form

$$f^{\langle a\rangle} = f^{\langle a\rangle}_{g,\text{crys}} := N^{\text{t}}_a \cdot (f^1_{g,\text{crys}})^{\phi^{a-1}} \cdot N_{a-1} \in \mathbb{I}^a_{gg}(1, 1) \tag{3.11}$$

and its transpose,

$$f^{\langle a\rangle\text{t}} := (f^{\langle a\rangle}_{g,\text{crys}})^{\text{t}} \in \mathbb{I}^a_{gg}(1, 1).$$

Also we define the form

$$f^{[a]} = f^{[a]}_{g,\text{crys}} := N^{\text{t}}_a \cdot f^a_{g,\text{crys}} \in \mathbb{I}^a_{gg}(1, 1) \tag{3.12}$$

and its transpose

$$f^{[a]\text{t}} := (f^{[a]}_{g,\text{crys}})^{\text{t}} \in \mathbb{I}^a_{gg}(1, 1).$$

Corollary 3.6.18(c) below will check that $f^{\langle a\rangle\text{t}} = f^{\langle a\rangle}$ and $f^{[a]} = f^{[a]\text{t}}$.

The following identity is a direct consequence of definitions:

$$f^{\langle a+1\rangle} = (f^{\partial})^{\text{t}} \cdot (f^{\langle a\rangle})^{\phi} \cdot f^{\partial}. \tag{3.13}$$

Similar to the case of δ-modular forms of size g, we have the following result for δ-modular forms of size 1.

Corollary 3.4.6 *Let $w \in W(r)$. Let $w' \in W$ be the unique element such that*

$$(\phi - 1)w' = w - deg(w).$$

Then the R-linear map

$$\mathbb{I}^r_{g,ord}(w) \to \mathbb{I}^r_{g,ord}(deg(w)), \quad f \mapsto \det(f^{\partial}_{g,crys})^{-w'} \cdot f,$$

is an isomorphism.

We end by recording the following facts proved in [Bu05] based on [Bar] (see [Bu05], Thm. 8.83 which relies on [Bu05], Prop. 8.22).

Lemma 3.4.7 *The following properties hold.*

(a) *We have* $\mathbb{I}^1_1(\phi - 1) = \mathbb{I}^1_1(1-\phi) = 0$. *In particular,*

$$f^{\partial}_{1,crys} \in \mathbb{I}^1_{1,ord}(\phi - 1)\backslash \mathbb{I}^1_1(\phi-1), \quad f_{\partial,1,crys} \in \mathbb{I}^1_{1,ord}(1-\phi)\backslash \mathbb{I}^1_1(1-\phi).$$

(b) *If* $a, b \in \mathbb{N}$ *are distinct, then the R-module* $\mathbb{I}_1^{\max\{a,b\}}(-\phi^a - \phi^b)$ *has rank* 1.

(c) *For each* $a \in \mathbb{N}$ *we have* $\mathbb{I}^a_1(-2\phi^a) = 0$.

3.5 Serre–Tate Expansions

Let $T = T^{(0)}, \dots, T^{(r)}$ and their entries be as in Sect. 2.3.4. Let

$$S^r_{\text{big-for}} := R[[T, \dots, T^{(r)}]] = R[[T^{(l)}_{ij} | 1 \le i \le j \le g, 0 \le l \le r]],$$

and define

$$S^r_{\text{for}} := \widehat{R[[T]][T', \dots, T^{(r)}]}$$

similarly. So $S^0_{\text{for}} = S^0_{\text{big-for}}$ is as in Corollary 3.1.2. The rings

$$S_{\text{big-for}} := \bigcup_{r \ge 0} S^r_{\text{big-for}} \quad \text{and} \quad S_{\text{for}} := \bigcup_{r\ge 0} S^r_{\text{for}}$$

are naturally in Ob($\widehat{\textbf{FilRing}_\delta}$) with δ defined by the rule: $\delta(T^{(l)}_{ij}) = T^{(l+1)}_{ij}$ for all integers $l \ge 0$ and $1 \le i \le j \le g$.

We consider the universal principally polarized abelian scheme $(A_{S^0_{\text{for}}}, \theta_{S^0_{\text{for}}})$ over Spec(S^0_{for}) that lifts (A_0, θ_0) and that is the algebraization of the formal principally polarized abelian scheme over $\mathcal{M}_{A_0,\theta_0}$ (see Corollary 3.1.2). Furthermore, we denote by $\zeta_i \in \text{Hom}(A^{\text{for}}_{S^0_{\text{for}}}, \mathbb{G}^{\text{for}}_m)$ the image via the isomorphism (cf. Diagram (3.6))

$$T_p(\check{A}_0) \simeq \text{Hom}(A^{\text{for}}_{S^0_{\text{for}}}, \mathbb{G}^{\text{for}}_m)$$

of the $i1$ entry of the column basis $T_p(\theta_0)(e)$ of $T_p(\check{A}_0)$. Defining $\omega_i := \zeta_i^*(dt/t)$, we get an R-basis $\omega_{S^0_{\text{for}}} = \{\omega_1, \dots, \omega_g\}$ of global 1-forms on $A^{\text{for}}_{S^0_{\text{for}}}$ which we identify with 1-forms on $A_{S^0_{\text{for}}}$.

Let $r \geq 0$ be an integer and let $w, w', w'' \in W$.

For $f \in M^r_g(w)$ let

$$\mathcal{E}(f) = \mathcal{E}_{A_0,\theta_0,e}(f) = F(T, T', \dots, T^{(r)}) := f(A_{S^0_{\text{for}}}, \theta_{S^0_{\text{for}}}, \omega_{S^0_{\text{for}}}, S_{\text{for}}) \in S^r_{\text{for}}.$$

We call $\mathcal{E}(f)$ the *Serre–Tate expansion* of f at (A_0, θ_0, e).

Similarly, for $f \in M^r_{gg}(w', w'')$ we let

$$\mathcal{E}(f) = \mathcal{E}_{A_0,\theta_0,e}(f) = F(T, T', \dots, T^{(r)}) \in \mathbf{Mat}_g(S^r_{\text{for}}) \subset \mathbf{Mat}_g(S^r_{\text{big-for}})$$

be the matrix of formal power series obtained by evaluating f at the quadruple $(A_{S^0_{\text{for}}}, \theta_{S^0_{\text{for}}}, \omega_{S^0_{\text{for}}}, S_{\text{for}})$ and we also call it the *Serre–Tate expansion* of f.

We obtain *Serre–Tate expansion R-linear maps*

$$\mathcal{E}_{A_0,\theta_0,e} : M^r_g(w) \to S^r_{\text{for}} \tag{3.14}$$

and

$$\mathcal{E}_{A_0,\theta_0,e} : M^r_{gg}(w', w'') \to \mathbf{Mat}_g(S^r_{\text{for}}). \tag{3.15}$$

As $A_{S^0_{\text{for}}}$ is ordinary, the R-linear maps of Diagrams (3.14) and (3.15) extend to R-linear maps

$$\mathcal{E}_{A_0,\theta_0,e} : M^r_{g,\text{ord}}(w) \to S^r_{\text{for}} \tag{3.16}$$

and

$$\mathcal{E}_{A_0,\theta_0,e} : M^r_{gg,\text{ord}}(w', w'') \to \mathbf{Mat}_g(S^r_{\text{for}}). \tag{3.17}$$

In what follows we would like to understand the restriction of these maps to Hecke covariant Siegel δ-modular forms.

For each matrix $M \in U_0$ (so $MM^\dagger = d \cdot 1_g$ with $d \in \mathbb{N}\backslash p\mathbb{N}$), and for every symmetric matrices $Q_1, Q_2 \in \mathbf{Mat}_g(R)$ that satisfy $Q_2 = M^{-1} \star Q_1 \star (M^\dagger)^{\text{t}}$, let (A_1, θ_1) and (A_2, θ_2) be the corresponding principally polarized abelian schemes over $\text{Spec}(R)$, let ω_1 and ω_2 be column bases of 1-forms on A_1 and A_2 (respectively) attached to the column basis e of $T_p(A_0)$, and let $u : A_1 \to A_2$ be the corresponding isogeny (see Sect. 3.3).

Let $f \in \mathbb{I}^r_{g,\text{ord}}(w)$ be an ordinary Hecke covariant Siegel δ-modular form of genus g, size 1, order $\leq r$ and weight $w \in W$ with $\frac{g \cdot \deg(w)}{2} \in \mathbb{Z}$; so Eq. (2.24) holds for it. As M has $\mathbb{Z}_p$-coefficients, from Lemma 3.3.2 we get that

$$\det([u])^{-w} = \det(M^\dagger)^{-\deg(w)}.$$

Also $\det(M)\det(M^{\dagger}) = d^g = \deg(u)$. Hence Eq. (2.24) becomes

$$f(A_2, \theta_2, \omega_2, R) = \det(M)^{\deg(w)} \cdot d^{-\frac{g \cdot \deg(w)}{2}} \cdot f(A_1, \theta_1, \omega_1, R). \qquad (3.18)$$

If $f \in \mathbb{I}^r_{gg,\mathrm{ord}}(\phi^a, \phi^b)$ is a Hecke covariant Siegel δ-modular form of genus g, size g, order $\leq r$ and weight (ϕ^a, ϕ^b), then by Eq. (2.28) we have:

$$f(A_2, \theta_2, \omega_2, R) = M^{\dagger} \cdot f(A_1, \theta_1, \omega_1, R) \cdot (M^{\mathrm{t}})^{-1}. \qquad (3.19)$$

If $f \in \mathbb{I}^r_{gg,\mathrm{ord}}(-\phi^a, \phi^b)$ is a Hecke covariant Siegel δ-modular form of genus g, size g, order $\leq r$ and weight $(-\phi^a, \phi^b)$, then by Eq. (2.28) we have:

$$f(A_2, \theta_2, \omega_2, R) = ((M^{\dagger})^{\mathrm{t}})^{-1} \cdot f(A_1, \theta_1, \omega_1, R) \cdot (M^{\dagger})^{\mathrm{t}}. \qquad (3.20)$$

If $f \in \mathbb{I}^r_{gg,\mathrm{ord}}(\phi^a, -\phi^b)$ is a Hecke covariant Siegel δ-modular form of genus g, size g, order $\leq r$ and weight $(\phi^a, -\phi^b)$, then by Eq. (2.28) we have:

$$f(A_2, \theta_2, \omega_2, R) = (M^{\dagger})^{\mathrm{t}} \cdot f(A_1, \theta_1, \omega_1, R) \cdot ((M^{\dagger})^{\mathrm{t}})^{-1}. \qquad (3.21)$$

If $f \in \mathbb{I}^r_{gg,\mathrm{ord}}(-\phi^a, -\phi^b)$ is a Hecke covariant Siegel δ-modular form of genus g, size g, order $\leq r$ and weight $(-\phi^a, -\phi^b)$, then by Eq. (2.28) we have:

$$f(A_2, \theta_2, \omega_2, R) = ((M^{\dagger})^{\mathrm{t}})^{-1} \cdot f(A_1, \theta_1, \omega_1, R) \cdot M. \qquad (3.22)$$

Let $\mathbb{Z}_p[[T]] := \mathbb{Z}_p[[T_{i,j} | 1 \leq i \leq j \leq g]]$ (recall that $T_{ij} = T_{ji}$). For each $M \in U_0$, $MM^{\dagger} = d \cdot 1_g$, $d \in \mathbb{N} \backslash p\mathbb{N}$, consider the $g \times g$ matrix with coefficients in $\mathbb{Z}_p[[T]]$ defined by

$$T^M := M^{-1} \star (\mathbb{E}_g + T) \star (M^{\dagger})^{\mathrm{t}} - \mathbb{E}_g.$$

Here the $\star$ operation is again defined in terms of the (left and right) $\mathbb{Z}_p$-module structure on the multiplicative abelian group $1 + \mathbb{Z}_p[[T]]$. Therefore if we write $M^{-1} = (m^{(-1)}_{ij})_{1 \leq i,j \leq g}$, then the ij entry of T^M is

$$(T^M)_{ij} = \prod_{\tilde{i}=1}^{g} \prod_{\tilde{j}=1}^{g} (1 + T_{\tilde{i}\tilde{j}})^{m^{(-1)}_{i\tilde{i}} m^{\dagger}_{j\tilde{j}}} - 1 \in \mathbb{Z}_p[[T]].$$

For every $G \in S^r_{\text{big-for}}$ we define

$$G^M := G(T^M, \delta(T^M), \ldots, \delta^r(T^M)) \in S^r_{\text{big-for}}.$$

Then for $F = (F_{ij})_{1 \leq i,j \leq g} \in \mathbf{Mat}_g(S^r_{\text{big-for}})$ we define

$$F^M := (F^M_{ij})_{1 \le i,j \le g} \in \mathbf{Mat}_g(S^r_{\text{big-for}}).$$

Using the injectivity of the R-algebra homomorphism in Diagram (2.3) it is easy to check that if a series $F = F(T, T', \dots, T^{(r)}) \in S^r_{\text{for}}$ has the property that

$$F(pX, \delta(pX), \dots, \delta^r(pX)) = 0$$

for all symmetric $g \times g$ matrices X with coefficients in R, then $F = 0$; indeed it is enough to note that the map

$$X \mapsto F(pX, \delta(pX), \dots, \delta^r(pX))$$

is a δ-function (see Definition 2.1.10) on the set of symmetric $g \times g$ matrices with entries in R. Using this remark we get that for $f \in \mathbb{I}^r_{g,\text{ord}}(w)$ we have an equality

$$\mathcal{E}(f)^M = \det(M)^{\deg(w)} \cdot d^{-\frac{g \cdot \deg(w)}{2}} \cdot \mathcal{E}(f); \tag{3.23}$$

indeed, in view of Lemma 3.3.1 and Eq. (3.18) the difference between the right-hand side and the left-hand side of Eq. (3.23) vanishes whenever $T, \dots, T^{(r)}$ are replaced by $pX, \dots, \delta^r(pX)$ where X is an arbitrary symmetric $g \times g$ matrix with entries in R.

Similarly, from Eq. (3.19) we get that for $f \in \mathbb{I}^r_{gg,\text{ord}}(\phi^a, \phi^b)$ we have

$$\mathcal{E}(f)^M = M^\dagger \cdot \mathcal{E}(f) \cdot (M^{\text{t}})^{-1}. \tag{3.24}$$

Similarly, from Eq. (3.20) we get that for $f \in \mathbb{I}^r_{gg,\text{ord}}(-\phi^a, \phi^b)$ we have

$$\mathcal{E}(f)^M = ((M^\dagger)^{\text{t}})^{-1} \cdot \mathcal{E}(f) \cdot (M^\dagger)^{\text{t}}. \tag{3.25}$$

Similarly, from Eq. (3.21) we get that for $f \in \mathbb{I}^r_{gg,\text{ord}}(\phi^a, -\phi^b)$ we have

$$\mathcal{E}(f)^M = (M^\dagger)^{\text{t}} \cdot \mathcal{E}(f) \cdot ((M^\dagger)^{\text{t}})^{-1}. \tag{3.26}$$

Similarly, from Eq. (3.22) we get that for $f \in \mathbb{I}^r_{gg,\text{ord}}(-\phi^a, -\phi^b)$ we have

$$\mathcal{E}(f)^M = ((M^\dagger)^{\text{t}})^{-1} \cdot \mathcal{E}(f) \cdot M. \tag{3.27}$$

Let $s \in \frac{1}{2}\mathbb{Z}$. If $gs \in \mathbb{Z}$ we consider the R-module

$$\mathcal{H}^r_g(U_0, s) = \mathcal{H}^r_g(U_0, s)_R := \{F \in S^r_{\text{big-for}} | F^M = \det(M)^{-2s} d^{gs} \cdot F \text{ for all } M \in U_0\}.$$

If $gs \notin \mathbb{Z}$ we set $\mathcal{H}^r_g(U_0, s) := 0$.

We define R-modules

$$\mathcal{H}^r_{gg}(U_0) = \mathcal{H}^r_{gg}(U_0)_R, \quad \mathcal{H}^r_{gg}(U_0)^\dagger = \mathcal{H}^r_{gg}(U_0)^\dagger_R, \quad \mathcal{H}^r_{gg}(U_0)^\ddagger = \mathcal{H}^r_{gg}(U_0)^\ddagger_R$$

as follows:

$$\mathcal{H}^r_{gg}(U_0) := \{F \in \mathbf{Mat}_g(S^r_{\text{big-for}}) | F^M = M^{-} \cdot F \cdot (M^{\mathrm{t}})^{-1} \text{ for all } M \in U_0\},$$

$$\mathcal{H}^r_{gg}(U_0)^\dagger := \{F \in \mathbf{Mat}_g(S^r_{\text{big-for}}) | F^M = ((M^{-})^{\mathrm{t}})^{-1} \cdot F \cdot (M^\dagger)^{\mathrm{t}} \text{ for all } M \in U_0\}.$$

$$\mathcal{H}^r_{gg}(U_0)^\ddagger := \{F \in \mathbf{Mat}_g(S^r_{\text{big-for}}) | F^M = ((M^\dagger)^{\mathrm{t}})^{-1} \cdot F \cdot M \text{ for all } M \in U_0\}.$$

Due to Eqs. (3.23)–(3.27) the rule $f \mapsto \mathcal{E}_{A_0,\theta_0,e}(f)$ defines R-linear maps

$$\mathcal{E}_{A_0,\theta_0,e} : \mathbb{I}^r_{g,\text{ord}}(w) \to \mathcal{H}^r_g(U_0, -\frac{\deg(w)}{2}), \tag{3.28}$$

$$\mathcal{E}_{A_0,\theta_0,e} : \mathbb{I}^r_{gg,\text{ord}}(\phi^a, \phi^b) \to \mathcal{H}^r_{gg}(U_0), \tag{3.29}$$

$$\mathcal{E}_{A_0,\theta_0,e} : \mathbb{I}^r_{gg,\text{ord}}(\phi^a, -\phi^b) \to \mathcal{H}^r_{gg}(U_0)^\dagger, \tag{3.30}$$

$$\mathcal{E}_{A_0,\theta_0,e} : \mathbb{I}^r_{gg,\text{ord}}(-\phi^a, \phi^b) \to \mathcal{H}^r_{gg}(U_0)^\dagger, \tag{3.31}$$

$$\mathcal{E}_{A_0,\theta_0,e} : \mathbb{I}^r_{gg,\text{ord}}(-\phi^a, -\phi^b) \to \mathcal{H}^r_{gg}(U_0)^\ddagger. \tag{3.32}$$

And by restriction we get R-linear maps

$$\mathcal{E}_{A_0,\theta_0,e} : \mathbb{I}^r_g(w) \to \mathcal{H}^r_g(U_0, -\frac{\deg(w)}{2}), \tag{3.33}$$

$$\mathcal{E}_{A_0,\theta_0,e} : \mathbb{I}^r_{gg}(\phi^a, \phi^b) \to \mathcal{H}^r_{gg}(U_0), \tag{3.34}$$

$$\mathcal{E}_{A_0,\theta_0,e} : \mathbb{I}^r_{gg}(-\phi^a, \phi^b) \to \mathcal{H}^r_{gg}(U_0)^\dagger, \tag{3.35}$$

$$\mathcal{E}_{A_0,\theta_0,e} : \mathbb{I}^r_{gg}(\phi^a, -\phi^b) \to \mathcal{H}^r_{gg}(U_0)^\dagger, \tag{3.36}$$

$$\mathcal{E}_{A_0,\theta_0,e} : \mathbb{I}^r_{gg}(-\phi^a, -\phi^b) \to \mathcal{H}^r_{gg}(U_0)^\ddagger. \tag{3.37}$$

The part (b) below is a 'Serre–Tate expansion principle'.

Proposition 3.5.1 **(a)** *The R-linear maps of Diagrams (2.41), (2.43), and (2.44) above are injective and have torsion free cokernels.*

(b) *The R-linear maps of Diagrams (3.14) to (3.17) and (3.28) to (3.37) above are injective and have torsion free cokernels.*

Proof For part (a) we only consider the case of the R-linear map

$$M^r_{g,\mathrm{ord}}(w) \to \mathcal{O}(J^r(X));$$

the other cases follow from this case.

Let $f \in M^r_{g,\mathrm{ord}}(w)$ be such that we have $f_X = 0 \in \mathcal{O}(J^r(X))$ (see Display (2.42)) for some affine open subscheme X of $\mathcal{X}_{\mathrm{ord}} = \mathcal{A}_{g,1,N,R,\mathrm{ord}}$ with $\overline{X}$ connected and with the property that there exists a column basis of 1-forms on the universal abelian scheme over X. As the geometric fibers of the stack $\mathcal{A}_{g,1}$ over $\mathrm{Spec}(\mathbb{Z})$ are irreducible (see [FC], Cor. 5.10), the geometric fibers of $\mathcal{X}_k$ have their connected components permuted transitively by the group $\mathbf{GSp}_{2g}(\mathbb{Z}/N\mathbb{Z})$. So if $\mathcal{C}$ is an arbitrary connected component of $\widehat{\mathcal{X}}$ (the p-adic completion of $\mathcal{X}$) it follows that $f_{\mathrm{Spf}(C)} = 0$ for some nonempty affine open formal subscheme $\mathrm{Spf}(C)$ of $\mathcal{C}$. Thus f evaluated at the formal principally polarized abelian scheme over $\mathrm{Spf}(C)$ endowed with a column basis of 1-forms vanishes, i.e., the element of $J^r(C)$ it defines is 0. From the connectivity (hence irreducibility) of $\mathcal{C}$ we get that f similarly vanishes at each formal principally polarized abelian scheme obtained from the universal one over $\mathcal{C}$ via a pullback and endowed with an arbitrary column basis of 1-forms.

Now let $B \in \mathrm{Ob}(\widehat{\mathbf{FilRing}_\delta})$. We want to prove that f vanishes at each abelian scheme of relative dimension g over $\mathrm{Spec}(B^0)$ which has a principal polarization and a column basis of 1-forms, i.e., the elements of B^r it defines are all 0. There exists a finite flat étale homomorphism $B^0 \to C^0$ such that the resulting principally polarized abelian scheme over $\mathrm{Spec}(C^0)$ has a symplectic similitude level-N structure. By [Bu05], Lem. 3.14, the union $C := \bigcup_{r \geq 0}(B^r \otimes_{B^0} C^0)$ has a natural structure of an object in $\widehat{\mathbf{FilRing}_\delta}$ and, as the B^r-homomorphisms

$$B^r \to B^r \otimes_{B^0} C^0 \tag{3.38}$$

are faithfully flat (hence injective), we can assume (by replacing B with C) that the principally polarized abelian scheme over $\mathrm{Spec}(B)$ has a symplectic similitude level-N structure. Thus the desired vanishing property of f (i.e., of the mentioned elements of B^r) follows from the previous paragraph as $\mathcal{C}$ was an arbitrary connected component of $\widehat{\mathcal{X}}$.

The torsion freeness property for the cokernel is proved similarly by using the fact that the reduction modulo p of the homomorphism of Diagram (3.38) is again faithfully flat and hence injective.

For part (b) we only consider the case of the R-linear map of Diagram (3.14): the cases of the other R-linear maps can be treated similarly. Let $f \in M^r_g(w)$ be such that $\mathcal{E}(f) = 0$. Let $P_0 \in \mathcal{X}(k)$ correspond to the principally polarized abelian scheme (A_0, θ_0) equipped with some symplectic similitude level-N structure. Let $\mathrm{Spf}(C) \subset \mathcal{C}$ be an affine open formal subscheme that contains P_0 and sufficiently small so that column bases of 1-forms on the corresponding formal abelian scheme over Spf(C) exist. As Spf(C) is connected and formally smooth over Spf(R), the R-algebra homomorphism $C \to S^0_{\text{for}}$ that corresponds to the point P_0 is injective with torsion free cokernel and thus also the R-algebra homomorphism

$$J^r(C) \to S^r_{\text{big-for}}$$

is injective with torsion free cokernel (see [Bu05], Prop. 4.43). Thus f evaluated at the formal principally polarized abelian scheme over Spf(C) endowed with a column basis of 1-forms, i.e., the element of $J^r(C)$ it defines is 0 as $\mathcal{E}(f)$ is so. From part (a) we get that $f = 0$. The assertion about the torsion free cokernel is proved similarly. □

Let $\mathbb{I} \in \{\mathbb{I}_g, \mathbb{I}_{g,\text{ord}}\}$ and similarly consider the rings $\mathbb{M}, \mathbb{M}'$ (see Display (2.45)). From Proposition 3.5.1(b), we get directly the following consequences.

Corollary 3.5.2 *The following properties hold.*

(a) *If $f_1 \in \mathbb{I}^r(w_1)$ and $f_2 \in \mathbb{I}^r(w_2)$ with $w_1, w_1 \in W$ are such that $f_1 f_2$ is 0 (respectively divisible by p) in $\mathbb{I}^r(w_1 + w_2)$, then either f_1 or f_2 is 0 (respectively is divisible by p in $\mathbb{I}^r(w_1)$ or $\mathbb{I}^r(w_2)$).*

(b) *The ring $\mathbb{I}$ is a p-adically separated integral domain and p is a prime element of it, hence the local ring $\mathbb{I}_{(p)}$ is a discrete valuation ring. Moreover, we have $\mathbb{I}^r \cap p\mathbb{I}^{r+1} = p\mathbb{I}^r$ for all integers $r \geq 0$.*

(c) *The endomorphism $\phi : \mathbb{I} \to \mathbb{I}$ is injective with injective reduction modulo p.*

(d) *The analog statements (a) to (c) hold for the rings $\mathbb{M}, \mathbb{M}', \mathbb{M}_X$ in place of $\mathbb{I}$. Thus the rings $\mathbb{I}, \mathbb{M}, \mathbb{M}', \mathbb{M}_X$ have natural structures of filtered integral δ-graded rings. Moreover, the morphisms in Diagrams (2.41) and (2.44) are injective with injective reduction modulo p.*

Property (a) of the next corollary says that no nonzero Siegel δ-modular form of size 1 can have an order smaller than the order of its weight.

Corollary 3.5.3 *For all $w \in W$ and integers $r \geq 0$ the following properties hold.*

(a) *For all integers s, r with $0 \leq s < ord(w) \leq r$ we have $M^r_g(w) \cap M^s_g = 0$.*

(b) *The R-algebra homomorphism $\mathbb{M}^r_g = \oplus_{w \in W(r)} \mathbb{M}^r_g(w) \to M^r_g$ given by summation is injective.*

(c) *For all $v \in W$, if $0 \neq F \in \mathbb{I}_g(w)$ and $G \in M^r_g$ are such that $FG \in \mathbb{I}_g(w + v)$, then $G \in \mathbb{I}^r_g(v)$.*

(d) *If $f \in \mathbb{I}^r_g(w) \backslash p\mathbb{I}^r_g(w)$, then*

$$(\mathbb{I}_{g,\langle f \rangle})^r = (\mathbb{M}_{g,\langle f \rangle})^r \cap \mathbb{I}_{g,\langle f \rangle}.$$

(e) *Properties (a) to (d) hold in the ordinary case as well.*

Proof We consider only the nonordinary case, i.e., the case of $\mathbb{I}_g$, etc., as the ordinary care, i.e., the case of $\mathbb{I}_{g,\mathrm{ord}}$, etc., is similar.

To prove part (a) we check that the assumption that there exists a nonzero $f \in M_g^r(w) \cap M_g^s$ leads to a contradiction. For each $\lambda \in (S^0_{\mathrm{for}})^\times$ we have

$$\mathcal{E}^\lambda(f) := f(A_{S^0_{\mathrm{for}}}, \theta_{S^0_{\mathrm{for}}}, \lambda\omega_{S^0_{\mathrm{for}}}, S_{\mathrm{for}}) \in S^s_{\mathrm{for}}, \tag{3.39}$$

$$\mathcal{E}(f) := f(A_{S^0_{\mathrm{for}}}, \theta_{S^0_{\mathrm{for}}}, \omega_{S^0_{\mathrm{for}}}, S_{\mathrm{for}}) \in S^s_{\mathrm{for}}. \tag{3.40}$$

As f has weight w, we have $\mathcal{E}^\lambda(f) = \lambda^{-gw}\mathcal{E}(f)$, see Eq. (2.22). As by the Serre–Tate expansion principle (Proposition 3.5.1 (b)) we have $\mathcal{E}(f) \neq 0$, we conclude that $\lambda^{gw} \in \mathrm{Frac}(S^s_{\mathrm{for}})$. But, as $s < \mathrm{ord}(w)$, this is clearly false, say, for $\lambda := 1_g + T_{11}$.

To prove part (b) we check that if $(f_w)_{w\in W(r)} \in \mathbb{M}^r_g = \oplus_{w\in W(r)} M^r_g(w)$ and $\sum_{w\in W(r)} f_w = 0$ in M^r_g, then $f_w = 0$ for all $w \in W(r)$.

Using the notation in Displays (3.39) and (3.40) we have $\sum_{w\in W(r)} \mathcal{E}^\lambda(f_w) = 0$ for all $\lambda \in R^\times$ hence $\sum_{w\in W(r)} \lambda^{-w}\mathcal{E}(f_w) = 0$ in S^r_{for}. For each monomial μ in the indeterminates $T_{ij}, T'_{ij}, \ldots, T^{(r)}_{ij}$ with $i, j \in \{1, \ldots, g\}$ let $a_{w,\mu} \in R$ be the coefficient of μ in $\mathcal{E}(f_w)$. We get that $\sum_{w\in W(r)} a_{w,\mu}\lambda^w = 0$ for all μ and all $\lambda \in R^\times$. The latter easily implies that $a_{w,\mu} = 0$ for all μ and $w \in W(r)$, hence $f_w = 0$ for all $w \in W(r)$.

For part (c) we use the fact that when checking the conditions defining weight and Hecke covariance it suffices to do so in the 'universal situation' in which case the filtered δ-rings under consideration are integral domains. So it is enough to check the conditions for every quadruple (A, θ, ω, S) with S an integral domain and $F(A, \theta, \omega, S) \neq 0$. For every such quadruple and each $\lambda \in (S^0)^\times$ we have

$$\begin{aligned} F(A, \theta, \lambda\omega, S)G(A, \theta, \lambda\omega, S) &= (FG)(A, \theta, \lambda\omega, S) \\ &= \det(\lambda)^{-w-v} F(A, \theta, \omega, S)G(A, \theta, \omega, S) \\ &= \det(\lambda)^{-v} F(A, \theta, \lambda\omega, S)G(A, \theta, \omega, S). \end{aligned}$$

As S is an integral domain we can cancel the nonzero factor $F(A, \theta, \lambda\omega, S)$ and get that G has weight v. A similar argument shows that G is Hecke covariant.

To prove part (d) we consider $F/f^v = G/f^u \in (\mathbb{M}_{g,\langle f\rangle})^r \cap \mathbb{I}_{g,\langle f\rangle}$ with $v, u \in W$, $F \in \mathbb{I}_g(wv)$, and $G, f^u \in \mathbb{M}^r_g(wu)$. Then $f^v G \in \mathbb{I}_g(wu + wv)$. By part (c) we have $G \in \mathbb{I}_g(wu)$, hence $G \in \mathbb{I}^r_g(wu)$ and therefore $G/f^u \in (\mathbb{I}_{g,\langle f\rangle})^r$.

The proofs of parts (a) to (d) in the ordinary case are similar. □

3.6 Comparison Maps: The Ordinary Case

In this section we first define some new modules and algebras and basic 'comparison maps' ♠, ♣, ♢, and ♡ between them. Then we will use these 'comparison maps', in conjunction with our Serre–Tate expansion maps $\mathcal{E}$ and with the geometric invariant theory of multiple quadratic forms (see Sects. 4.1–4.3), to prove our basic computations for different dimensions of K-algebras and dimensions of K-vector spaces of ordinary Hecke covariant Siegel δ-modular forms.

3.6.1 Basic Modules of δ-Modular Forms

We start by considering some variants of prior modules of formal power series (and new modules, consisting of polynomials) over rings other than R.

First, if in the definition of

$$\mathcal{H}^r_g(U_0, s),\ \mathcal{H}^r_{gg}(U_0),\ \mathcal{H}^r_{gg}(U_0)^\dagger,\ \mathcal{H}^r_{gg}(U_0)^\ddagger$$

we replace R by an arbitrary integral domain B that contains $\mathbb{Z}_p$, we get corresponding B-modules

$$\mathcal{H}^r_g(U_0, s)_B,\ \mathcal{H}^r_{gg}(U_0)_B,\ \mathcal{H}^r_{gg}(U_0)^\dagger_B,\ \mathcal{H}^r_{gg}(U_0)^\ddagger_B.$$

Next, for each polynomial

$$G \in B[T, \ldots, T^{(r)}] := B[T_{ij}, T'_{ij}, \ldots, T^{(r)}_{ij} | 1 \leq i, j \leq g]$$

(recall that we have $T_{ij} = T_{ji}$) and each $M \in U_0$ we set

$$G_M := G(M^{-1} \cdot T \cdot (M^\dagger)^{\mathrm{t}}, \ldots, M^{-1} \cdot T^{(r)} \cdot (M^\dagger)^{\mathrm{t}}).$$

Recall, here and below, $\cdot$ is matrix multiplication.

Let $s \in \frac{1}{2}\mathbb{Z}$. If $gs \notin \mathbb{Z}$, then we set $H^r_g(U_0, s)_B := 0$. If $gs \in \mathbb{Z}$, then we define B-modules

$$H^r_g(U_0, s)_B := \{G \in B[T, \ldots, T^{(r)}] | G_M = \det(M)^{-2s} d^{gs} \cdot G \ \text{ for all } \ M \in U_0\},$$

$$H^r_{gg}(U_0)_B := \{G \in \mathbf{Mat}_g(B[T, \ldots, T^{(r)}]) | G_M = M^\dagger \cdot G \cdot (M^{\mathrm{t}})^{-1} \ \text{ for all } \ M \in U_0\},$$

$$H^r_{gg}(U_0)^\dagger_B := \{G \in \mathbf{Mat}_g(B[T, \ldots, T^{(r)}]) | G_M = ((M^\dagger)^{\mathrm{t}})^{-1} \cdot G \cdot (M^\dagger)^{\mathrm{t}} \text{ for all } M \in U_0\},$$

$$H^r_{gg}(U_0)^\ddagger_B := \{G \in \mathbf{Mat}_g(B[T, \ldots, T^{(r)}]) | G_M = ((M^\dagger)^{\mathrm{t}})^{-1} \cdot G \cdot M \ \text{ for all } \ M \in U_0\}.$$

If $D \subset B$ is a multiplicative set, then we have

$$H^r_g(U_0, s)_{D^{-1}B} = D^{-1} H^r_g(U_0, s)_B,\ \ H^r_{gg}(U_0)_{D^{-1}B} = D^{-1} H^r_{gg}(U_0)_B,$$

$$H^r_{gg}(U_0)^\dagger_{D^{-1}B} = D^{-1}H^r_{gg}(U_0)^\dagger_B, \quad H^r_{gg}(U_0)^\ddagger_{D^{-1}B} = D^{-1}H^r_{gg}(U_0)^\ddagger_B.$$

Also, if $B \subset B'$ is a field extension, then we have

$$H^r_g(U_0, s)_{B'} = H^r_g(U_0, s)_B \otimes_B B', \quad H^r_{gg}(U_0)_{B'} = H^r_{gg}(U_0)_B \otimes_B B',$$

$$H^r_{gg}(U_0)^\dagger_{B'} = H^r_{gg}(U_0)^\dagger_B \otimes_B B', \quad H^r_{gg}(U_0)^\ddagger_{B'} = H^r_{gg}(U_0)^\ddagger_B \otimes_B B'.$$

Last two properties do not hold if H is replaced by $\mathcal{H}$ as we shall see below.

For an infinite integral domain B and for $s \in \frac{1}{2}\mathbb{Z}$ with $gs \in \mathbb{Z}$, let $\mathbb{H}^r_g(s)_B$ be the B-module of homogeneous polynomials G in the ring $B[T, \ldots, T^{(r)}]$ of degree gs that satisfy

$$G(X \cdot T \cdot X^{\mathrm{t}}, \ldots, X \cdot T^{(r)} \cdot X^{\mathrm{t}}) = \det(X)^{2s} G \tag{3.41}$$

for all $X \in \mathbf{Mat}_g(B)$. Here the entries of the matrices $T, \ldots, T^{(r)}$ are given degree 1. Note that $\mathbb{H}^r_g(s)_R$ coincides with the modules $\mathbb{H}^r_g(s)$ introduced in Eq. (2.34). If B is a field, then $\mathbb{H}^r_g(s)_B$ is the B-vector space of $\mathbf{SL}_g(B)$-invariant homogeneous polynomials of degree gs in $B[T, \ldots, T^{(r)}]$ under the left action defined by the rule:

$$(X, F) \mapsto F(X^{-1} \cdot T \cdot (X^{-1})^{\mathrm{t}}, \ldots, X^{-1} \cdot T^{(r)} \cdot (X^{-1})^{\mathrm{t}}).$$

Similarly, for an infinite integral domain B let $\mathbb{H}^r_{gg}(1)_B$ be the B-module of homogeneous matrices

$$G \in \mathbf{Mat}_g(B[T, \ldots, T^{(r)}])$$

of degree 1 such that we have

$$G(X \cdot T \cdot X^{\mathrm{t}}, X \cdot T' \cdot X^{\mathrm{t}}, \ldots, X \cdot T^{(r)} \cdot X^{\mathrm{t}}) = X \cdot G(T, T', \ldots, T^{(r)}) \cdot X^{\mathrm{t}} \tag{3.42}$$

for all $X \in \mathbf{Mat}_g(B)$. Again, for each multiplicative set $D \subset B$ we have

$$\mathbb{H}^r_g(s)_{D^{-1}B} = D^{-1}\mathbb{H}^r_g(s)_B \quad \text{and} \quad \mathbb{H}^r_{gg}(1)_{D^{-1}B} = D^{-1}\mathbb{H}^r_{gg}(1)_B.$$

Also, if $B \subset B'$ is an extension of infinite fields, then

$$\mathbb{H}^r_g(s)_{B'} = \mathbb{H}^r_g(s)_B \otimes_B B' \quad \text{and} \quad \mathbb{H}^r_{gg}(1)_{B'} = \mathbb{H}^r_{gg}(1)_B \otimes_B B'.$$

In what follows we will repeatedly use conditions (UAX1) and (UAX2) introduced in Sect. 3.2.

Lemma 3.6.1 *Assume (A_0, θ_0) satisfies condition (UAX1) and let B be an integral domain that contains $\mathbb{Z}_p$.*

(a) *Each polynomial in* $H_g^r(U_0, s)_B$ *is homogeneous of degree* gs. *In particular, we have* $H_g^r(U_0, s)_B = 0$ *for all* $s < 0$ *and* $H_g^r(U_0, 0)_B = B$.

(b) *Each matrix in* $H_{gg}^r(U_0)_B$ *is homogeneous of degree* 1.

(c) *We have* $H_{gg}^r(U_0)_B^\dagger \subset \mathbf{Mat}_g(B)$, *i.e., each matrix in* $H_{gg}^r(U_0)_B^\dagger$ *is homogeneous of degree* 0.

(d) *We have* $H_{gg}^r(U_0)_B^\ddagger = 0$.

Proof As (A_0, θ_0) satisfies condition (UAX1), there exists $\lambda \in \mathcal{U}_0$ such that $\lambda/\lambda^\dagger$ is not a root of unity. For $F \in H_g^r(U_0, s)$, we write $F = \sum_{n\geq 0} F_n$, with each F_n homogeneous of degree n, and we compute

$$F_{\lambda^2 \cdot 1_g} = \sum_{n\geq 0} (\lambda/\lambda^\dagger)^{-2n} \cdot F_n.$$

On the other hand, as $F \in H_g^r(U_0, s)$, we have

$$F_{\lambda^2 \cdot 1_g} = \lambda^{-4gs} d^{2gs} \cdot F = (\lambda/\lambda^\dagger)^{-2gs} \cdot F = \sum_{n\geq 0} (\lambda/\lambda^\dagger)^{-2gs} F_n.$$

From the two expressions of $F_{\lambda^2 \cdot 1_g}$ we get that $F_n = 0$ if $n \neq gs$. So part (a) holds.
Parts (b), (c), (d) are checked similarly. □

Proposition 3.6.2 *Assume* (A_0, θ_0) *satisfies conditions (UAX1) and (UAX2). Then*

$$H_g^r(U_0, s)_{\mathbb{Z}_p} = \mathbb{H}_g^r(s)_{\mathbb{Z}_p}, \quad H_{gg}^r(U_0)_{\mathbb{Z}_p} = \mathbb{H}_{gg}^r(1)_{\mathbb{Z}_p}, \quad H_{gg}^r(U_0)_{\mathbb{Z}_p}^\dagger = \mathbb{Z}_p \cdot 1_g$$

and hence for every integral domain B *that contains* $\mathbb{Z}_p$ *we have*

$$H_g^r(U_0, s)_B = \mathbb{H}_g^r(s)_B, \quad H_{gg}^r(U_0)_B = \mathbb{H}_{gg}^r(1)_B, \quad H_{gg}^r(U_0)_B^\dagger = B \cdot 1_g.$$

Proof Let $G \in H_g^r(U_0, s)_{\mathbb{Z}_p}$ and we prove that $G \in \mathbb{H}_g^r(s)_{\mathbb{Z}_p}$. (The converse is proved similarly.) We need to show that Eq. (3.41) holds for all $X \in \mathbf{Mat}_g(\mathbb{Z}_p)$. By Lemma 3.6.1(a), G is homogeneous of degree gs. As $\mathbb{Z}_{(p)}^\times \cdot U_0$ is Zariski dense in $\mathbf{GL}_{g,\mathbb{Z}_p}$, it is enough to show that Eq. (3.41) holds for all $X \in \mathbb{Z}_{(p)}^\times \cdot U_0$. By homogeneity of (3.41) it is enough to show that Eq. (3.41) holds for all $X \in U_0$. We know that we have

$$G(Y^{-1}T(Y^\dagger)^t, \ldots, Y^{-1}T^{(r)}(Y^\dagger)^t) = \det(Y)^{-2s} d^{gs} \cdot G(T, \ldots, T^{(r)}) \tag{3.43}$$

for all $Y \in U_0$. Write $YY^\dagger = d \cdot 1_g$. We get $Y^{-1} = d^{-1}Y^\dagger$. From this and the homogeneity part of Lemma 3.6.1 (a), as $\det(Y)\det(Y^\dagger) = d^g$, we get that the left-hand side of Eq. (3.43) equals

$$\det(Y^\dagger)^{2s} G(Y^\dagger T(Y^\dagger)^t, \ldots, Y^\dagger T^{(r)}(Y^\dagger)^t).$$

Setting $X = Y^\dagger$ we get that Eq. (3.41) holds for all $X \in U_0$.

Similarly, let $G \in H^r_{gg}(U_0)_{\mathbb{Z}_p}$ and let us prove that $G \in \mathbb{H}^r_{gg}(1)_{\mathbb{Z}_p}$; the converse is trivial (and does not need condition (UAX2)). Indeed, by Lemma 3.6.1(b), G is homogeneous of degree 1. We know that

$$G(M^{-1}T(M^\dagger)^{\mathrm{t}}, \ldots, M^{-1}T^{(r)}(M^\dagger)^{\mathrm{t}}) = M^\dagger \cdot G \cdot (M^{\mathrm{t}})^{-1}$$

holds for all $M \in U_0$. Using the fact that $MM^\dagger = d \cdot 1_g$ plus the homogeneity of G and writing $X = M^{-1}$ we get that Eq. (3.42) holds for all $X \in \mathbb{Z}^\times_{(p)} \cdot U_0$. Using condition (UAX2) we get that Eq. (3.42) holds for all $X \in \mathbf{GL}_g(\mathbb{Z}_p)$, so for all $X \in \mathbf{Mat}_g(\mathbb{Z}_p)$, hence $G \in \mathbb{H}^r_{gg}(1)_{\mathbb{Z}_p}$.

If $G \in H^r_{gg}(U_0)^\dagger_{\mathbb{Z}_p} \subset \mathbf{Mat}_g(\mathbb{Z}_p)$ (see Lemma 3.6.1(c)), we get that G commutes with all matrices in the set $\{(M^\dagger)^{\mathrm{t}} | M \in U_0\}$. By condition (UAX2) we get that G commutes with all the elements of $\mathbf{GL}_g(\mathbb{Z}_p)$, hence $G \in \mathbb{Z}_p \cdot 1_g$. □

In what follows we consider integers $s, r \geq 0$. Let

$$\Delta(g, r) := \{(m_0, \ldots, m_r) \in \mathbb{Z}^{r+1}_{\geq 0} | m_0 + \cdots + m_r = g\}.$$

Let B be an infinite integral domain. Note that $\mathbb{H}^r_g(s)_B$ contains all homogeneous polynomials of degree gs which are products of s polynomials of the form

$$\Theta_{m_0,\ldots,m_r} \in \mathbb{H}^r_g(1)_B \quad \text{for } (m_0, \ldots, m_r) \in \Delta(g, r) \tag{3.44}$$

that are defined by the identity

$$\det(y_0 T + y_1 T' + \cdots + y_r T^{(r)}) = \sum_{(m_0,\ldots,m_r)\in\Delta(g,r)} \Theta_{m_0,\ldots,m_r} y_0^{m_0} y_1^{m_1} \cdots y_r^{m_r}$$

in which $y_0, \ldots, y_r$ are indeterminates; so $\Theta_{m_0,\ldots,m_r}$ are partial polarizations of the invariant polynomial det. For $r = 1$ the classical notation is $\Theta_i = \Theta_{g-i,i}$ for $0 \leq i \leq g$; so $\Theta_0 = \det(T)$.

For an infinite field B of characteristic $\mathfrak{p} \geq 0$ and $s \in \frac{1}{2}\mathbb{Z}$, the dimension

$$D_{\mathfrak{p}}(g, r, s) := \dim_B(\mathbb{H}^r_g(s)_B), \tag{3.45}$$

depends (as we shall see) only on $g, r, s, \mathfrak{p}$, and equals 0 for $s < 0$. For B an infinite integral domain consider the integral extension of B-algebras

$$\mathbb{H}^r_{g,B} := \bigoplus_{s\in\mathbb{Z}} \mathbb{H}^r_g(s)_B \subset \mathbb{H}^r_{g,\mathrm{tot},B} := \bigoplus_{s\in\frac{1}{2}\mathbb{Z}} \mathbb{H}^r_g(s)_B.$$

For an infinite field B set

$$D(g, r) := \dim(\mathbb{H}^r_{g,B}) = \dim(\mathbb{H}^r_{g,\mathrm{tot},B}),$$

which, as we shall see shortly, does indeed depend only on g and r.

We have the following result (see Theorem 4.2.8(a)) which, at least in characteristic 0, is classical.

Lemma 3.6.3 *Let B be an infinite field of characteristic $\mathfrak{p} \geq 0$, $\mathfrak{p} \neq 2$. The B-algebra $\mathbb{H}^1_{g,tot,B}$ is generated by the algebraically independent elements $\Theta_0, \ldots, \Theta_g$. Hence, for $s \in \mathbb{N} \cup \{0\}$, we have*

$$D_{\mathfrak{p}}(g, 1, s) = \frac{(g+s)!}{g!s!} \quad \textit{and} \quad D_{\mathfrak{p}}(g, 0, s) = 1.$$

In addition $D_{\mathfrak{p}}(g, 1, s) = 0$ if g is even and $s \notin \mathbb{Z}$.

Also we have the following properties.

Lemma 3.6.4 *The following properties hold.*

(a) *Let B be either an infinite field or R. Then for each integer $r \geq 1$, the B-algebra $\mathbb{H}^r_{g,tot,B}$ is finitely generated and a unique factorization domain.*

(b) *For B an infinite field we have*

$$D(g, r) = (r+1) \cdot \frac{g(g+1)}{2} - g^2 + 1.$$

(c) *For a field B of characteristic 0 there exists a polynomial $D_{g,r}(x) \in \mathbb{Q}[x]$ of degree*

$$D(g, r) - 1 = \frac{g(rg + r - g + 1)}{2}$$

such that for all $s \in \mathbb{Z}_{\geq 0}$ big enough we have

$$D_0(g, r, s) = D_{g,r}(s).$$

Moreover, if g is even, then there exists a polynomial $D_{g,odd,r}(x) \in \mathbb{Q}[x]$ of degree at most $D(g, r) - 1$ such that for all $s \in \mathbb{N}$ big enough we have

$$D_0(g, r, \frac{1}{2} + s) = D_{g,odd,r}(s).$$

For detailed information on the numbers $D_0(g, r, s)$ (not needed for the present chapter) we refer to results in Chap. 4 (e.g., Theorem 4.2.8, Proposition 4.2.10, and Corollaries 4.2.14 and 4.2.15).

Proof If B is an infinite field or R, then, by Lemma 4.2.3(a) and (b) of Sect. 4.2.2, $\mathbb{H}^r_{g,\mathrm{tot},B}$ is the B-algebra of invariants $B[T_{ij} | 1 \leq i \leq j \leq g]^{\mathbf{SL}_g}$. To prove part (b) or part (a) in case B is a field, we can assume B is algebraically closed in which case we conclude by Corollary 4.2.4 of Sect. 4.2.3. It remains to prove part (a) in case $B = R$. Recalling that the upper bar signifies the reduction modulo p (or in the present context, tensoring with k over R), note that the homomorphism $\overline{\mathbb{H}^r_{g,\mathrm{tot},R}} \to \mathbb{H}^r_{g,\mathrm{tot},k}$ is injective because if p times a polynomial is invariant then the polynomial is invariant. Hence p is a prime element in $\mathbb{H}^r_{g,\mathrm{tot},R}$.

As $\mathbb{H}^r_{g,\mathrm{tot},R} \otimes_R K = \mathbb{H}^r_{g,\mathrm{tot},K}$ is a unique factorization domain it follows that $\mathbb{H}^r_{g,\mathrm{tot},R}$ is a unique factorization domain (see [Mat], Thm. 20.2), and from a general result on finite generation of rings of invariants over universally Japanese rings such as R (see [Se], Thm. 2; cf. also Sect. 4.2.3) we get that $\mathbb{H}^r_{g,\mathrm{tot},R}$ is a finitely generated R-algebra. However, for convenience we include a direct argument.

Let $\mathbb{H}^{r,\Theta}_{g,\mathrm{tot},R}$ be the R-subalgebra of $\mathbb{H}^r_{g,\mathrm{tot},R}$ generated by the elements in Display (3.44): it is of finite type over R. As R is universally Japanese, the normalization $\mathbb{B}$ of $\mathbb{H}^{r,\Theta}_{g,\mathrm{tot},R}$ in $\mathrm{Frac}(\mathbb{H}^r_{g,\mathrm{tot},R})$ is a finitely generated R-algebra. The rings $\mathbb{H}^{r,\Theta}_{g,\mathrm{tot},R}$ and $\mathbb{H}^r_{g,\mathrm{tot},R}$ are $\frac{1}{2}\mathbb{Z}_{\geq 0}$-graded. Then $\mathbb{B}$ is a $\frac{1}{2}\mathbb{Z}_{\geq 0}$-graded R-subalgebra of $\mathbb{H}^r_{g,\mathrm{tot},R}$ as the latter is normal. From Theorem 4.2.13(a) of Sect. 4.2.8 we get that $\mathbb{B} \otimes_R K = \mathbb{H}^r_{g,\mathrm{tot},K} = \mathbb{H}^r_{g,\mathrm{tot},R} \otimes_R K$ and the image $\tilde{\mathbb{B}}$ of $\overline{\mathbb{B}}$ in $\overline{\mathbb{H}^r_{g,\mathrm{tot},R}} \subset \mathbb{H}^r_{g,\mathrm{tot},k}$ is such that the k-homomorphism $\tilde{\mathbb{B}} \to \mathbb{H}^r_{g,\mathrm{tot},k}$ is integral. Thus $\overline{\mathbb{H}^r_{g,\mathrm{tot},R}}$ is a finitely generated k-algebra (being a submodule of the finite module $\mathbb{H}^r_{g,\mathrm{tot},k}$ over the finitely generated k-algebra $\tilde{\mathbb{B}}$). We pick a set of homogeneous generators for this k-algebra, lift them to a set of homogeneous elements of $\mathbb{H}^r_{g,\mathrm{tot},R}$, and enlarge this set to generate a $\frac{1}{2}\mathbb{Z}_{\geq 0}$-graded R-subalgebra $\mathbb{A}$ of $\mathbb{H}^r_{g,\mathrm{tot},R}$ with the properties that we have $\mathbb{A} \otimes_R K = \mathbb{H}^r_{g,\mathrm{tot},K} = \mathbb{H}^r_{g,\mathrm{tot},R} \otimes_R K$ and the k-homomorphism $\overline{\mathbb{A}} \to \overline{\mathbb{H}^r_{g,\mathrm{tot},R}}$ is surjective. We claim that $\mathbb{A} = \mathbb{H}^r_{g,\mathrm{tot},R}$ and this will complete the argument that $\mathbb{H}^r_{g,\mathrm{tot},R}$ is a finitely generated R-algebra. To check the claim, let $s \in \frac{1}{2}\mathbb{Z}_{\geq 0}$ and let $\mathbb{A}(s)$ and $\mathbb{H}^r_{g,\mathrm{tot},R}(s)$ be the corresponding graded R-module pieces. Then $\mathbb{A}(s) \subset \mathbb{H}^r_{g,\mathrm{tot},R}(s)$ becomes an isomorphism after tensoring with K and it becomes surjective after tensoring with k. As $\mathbb{A}(s)$ is a finitely generated R-module, we get that $\mathbb{A}(s) = \mathbb{H}^r_{g,\mathrm{tot},R}(s)$. As this is true for all $s \in \frac{1}{2}\mathbb{Z}_{\geq 0}$, we conclude that our claim holds.

Part (c) follows from Corollary 4.2.14 of Sect. 4.2.10. □

For explicit formulas for $D_{2,2}(t)$, $D_{2,3}(t)$ and $D_{2,4}(t)$ see Corollary 4.3.8 (a), (b) and (c) (respectively) of Sect. 4.3.4.

We continue by providing a construction of elements in some of the modules introduced above. Let

$$\psi(t, t') := \frac{1}{p} \log \frac{1 + t^p + pt'}{(1+t)^p} \in R[[t, t']],$$

where t, t' as above are two indeterminates. Let

$$\Psi \in \mathbf{Mat}_g(R[[t, t']])$$

be such that its ij entry is

$$\Psi_{ij} := \psi(T_{ij}, T'_{ij}) = \frac{\log(\frac{1+T_{ij}^{\phi}}{(1+T_{ij})^p})}{p}$$

and set

$$\mathcal{F} := \det(\Psi) \in R[[t, t']].$$

Then $\mathcal{F}$ satisfies

$$\mathcal{F}^M = \det(M)^{-2} d^g \cdot \mathcal{F}$$

and thus $\mathcal{F} \in \mathcal{H}_g^1(U_0, 1) = \mathcal{H}_g^1(U_0, 1)_R$. On the other hand, recall that $K = \mathrm{Frac}(R)$ and consider the logarithm of the multiplicative formal group

$$\mathfrak{f}(t) := \log(1+t) = \sum_{n=1}^{\infty} (-1)^{n+1} \frac{t^n}{n} \in K[[t]].$$

Let $\mathfrak{f}(T) := (\mathfrak{f}(T_{ij}))_{1 \leq i,j \leq g} \in \mathbf{Mat}_g(K[[t]])$. Then $\tilde{\mathcal{F}} := \det(\mathfrak{f}(T))$ satisfies

$$\tilde{\mathcal{F}}^M = \det(M)^{-2} d^g \cdot \tilde{\mathcal{F}}$$

for all $M \in U_0$ and thus $\tilde{\mathcal{F}} \in \mathcal{H}_g^0(U_0, 1)_K \subset \mathcal{H}_g^r(U_0, 1)_K$ for all integers $r \geq 0$. But for all integers $r \geq 0$ we have $\tilde{\mathcal{F}} \notin \mathcal{H}_g^r(U_0, 1)_R$.

More generally, consider the formal power series

$$\Psi_{m_0, m_1, \ldots, m_r} \in K[[T, \ldots, T^{(r)}]] := K[[T_{ij}^{(l)} | 1 \leq i \leq j \leq g, 0 \leq l \leq r]]$$

defined by the equality

$$\det(z_0 \mathfrak{f}(T) + z_1 \Psi + z_2 \Psi^{\phi} + \cdots + z_r \Psi^{\phi^{r-1}})$$

$$= \sum_{(m_0, \ldots, m_r) \in \Delta(g,r)} z_0^{m_0} z_1^{m_1} \ldots z_r^{m_r} \Psi_{m_0, m_1, \ldots, m_r}$$

in which $z_0, \ldots, z_r$ are $r+1$ indeterminates. In other words

$$\Psi_{m_0, m_1, \ldots, m_r} := \Theta_{m_0, m_1, \ldots, m_r}(\mathfrak{f}(T), \Psi, \ldots, \Psi^{\phi^{r-1}}),$$

where $\Theta_{m_0, m_1, \ldots, m_r} \in \mathbb{H}_g^r(1)$ was defined right after Display (3.44). Then clearly

$$\Psi_{m_0, m_1, \ldots, m_r} \in \mathcal{H}_g^r(U_0, 1)_K.$$

Note that, in the notation above, we also have

$$\Psi, \Psi^{\phi}, \ldots, \Psi^{\phi^{r-1}} \in \mathcal{H}^r_{gg}(U_0)_R, \tag{3.46}$$

$$\mathfrak{f}(T) \in \mathcal{H}^r_{gg}(U_0)_K \backslash \mathcal{H}^r_{gg}(U_0)_R. \tag{3.47}$$

3.6.2 The Maps ♠, ♣, and ♡

Proposition 3.6.5 *There exist well-defined K-linear maps*

$$\spadesuit : \mathbb{H}^r_g(s)_K \to \mathcal{H}^r_g(U_0, s)_K$$

given by the formula

$$F \mapsto F^{\spadesuit} := F(\mathfrak{f}(T), \Psi, \ldots, \Psi^{\phi^{r-1}}).$$

Proof This follows using Displays (3.46) and (3.47) via a computation similar to the one in the proof of Proposition 3.6.2. □

In order to define our next 'comparison map' ♣ we need some preparation.

Lemma 3.6.6 *Let $n \in \mathbb{N}$. Let $a_1, \ldots, a_n \in \mathbb{Z}_p$ and $x = \{x_1, \ldots, x_n\}$ be a set of indeterminates. Then for every integer $r \geq 0$ the following identity holds*

$$\delta^r(\sum_{m=1}^{n} a_m x_m) - \sum_{m=1}^{n} a_m x_m^{(r)} \in (x, x', \ldots, x^{(r)})^2$$

in the ring $\mathbb{Z}_p[x, x', \ldots, x^{(r)}] := \mathbb{Z}_p[x_m, x'_m, \ldots, x_m^{(r)} | 1 \leq m \leq n]$.

The proof of the lemma is an easy induction on $n \in \mathbb{N}$.

For the universal symmetric $g \times g$ matrix T and an integer $r \geq 0$ we get the following result.

Lemma 3.6.7 *For all $M \in U_0$ we have an equality of $g \times g$ matrices*

$$\delta^r(T^M) = M^{-1} \cdot T^{(r)} \cdot (M^{\dagger})^t + C_2^M,$$

where the 'correction factor' C_2^M is such that all its entries belong to the squared ideal $(T_{ij}^{(l)} | 1 \leq i \leq j \leq g, 0 \leq l \leq r)^2$ of the $\mathbb{Z}_p$-subalgebra $\mathbb{Z}_p[[T, \ldots, T^{(r)}]]$ (i.e., $\mathbb{Z}_p[[T_{ij}^{(l)} | 1 \leq i \leq j \leq g, 0 \leq l \leq r]]$) of $S^r_{big\text{-}for}$.

For each commutative ring B, integer $a \geq 0$, and formal power series F in $B[[T, \ldots, T^{(r)}]] := B[[T_{ij}^{(l)} | 1 \leq i \leq j \leq g, 0 \leq l \leq r]]$, let the degree a component F_a of F be the sum of all terms of F of total degree a. If $F \neq 0$, the initial component $F^{\clubsuit}$ of F will be the nonzero component of F which is of smallest total degree, and we set $0^\clubsuit := 0$. More generally, for F a matrix whose entries are formal power series in $B[[T, \ldots, T^{(r)}]]$, let F_a be the matrix whose entries are the a-components of the entries of the matrix F, and we define $F^\clubsuit$ similarly.

Directly from Lemma 3.6.6 we get the following properties.

Lemma 3.6.8 *For an integral domain B that contains $\mathbb{Z}_p$ and all $s \in \frac{1}{2}\mathbb{Z}$ with $gs \in \mathbb{Z}$ the following properties hold.*

(a) *If $F \in \mathcal{H}^r_g(U_0, s)_B$, then $F^\clubsuit \in H^r_g(U_0, s)_B$.*

(b) *If $F \in \mathcal{H}^r_{gg}(U_0)_B$, then $F^\clubsuit \in H^r_{gg}(U_0)_B$.*

(c) *If $F \in \mathcal{H}^r_{gg}(U_0)^\dagger_B$, then $F^\clubsuit \in H^r_{gg}(U_0)^\dagger_B$.*

(d) *If $F \in \mathcal{H}^r_{gg}(U_0)^\ddagger_B$, then $F^\clubsuit \in H^r_{gg}(U_0)^\ddagger_B$.*

From Lemmas 3.6.1 and 3.6.8 we get the following properties.

Corollary 3.6.9 *Assume (A_0, θ_0) satisfies condition (UAX1) and let B be an integral domain that contains $\mathbb{Z}_p$. Then for all $s \in \frac{1}{2}\mathbb{Z}$ with $gs \in \mathbb{Z}$ the following properties hold.*

(a) *For each nonzero $F \in \mathcal{H}^r_g(U_0, s)_B$, the polynomial $F^\clubsuit \in H^r_g(U_0, s)_B$ is the component of degree gs of F (i.e., $\deg(F^\clubsuit) = gs$). Thus the B-linear map*

$$\clubsuit : \mathcal{H}^r_g(U_0, s)_B \to H^r_g(U_0, s)_B, \quad F \mapsto F^\clubsuit$$

is injective and hence $\mathcal{H}^r_g(U_0, s)_B$ has finite rank bounded by the rank of $H^r_g(U_0, s)_B$; in particular, we have $\mathcal{H}^r_g(U_0, s)_B = 0$ for $s < 0$ and $\mathcal{H}^r_g(U_0, 0)$ has rank ≤ 1. Moreover, for every $F_1 \in \mathcal{H}^r_g(U_0, s_1)_B$ and $F_2 \in \mathcal{H}^r_g(U_0, s_2)_B$ we have the properties that $F_1 F_2 \in \mathcal{H}^r_g(U_0, s_1 + s_2)_B$ and $(F_1 F_2)^\clubsuit = F_1^\clubsuit F_2^\clubsuit$.

(b) *For each nonzero $F \in \mathcal{H}^r_{gg}(U_0)_B$, the matrix $F^\clubsuit \in H^r_{gg}(U_0)_B$ is the component of degree 1 of F (i.e., we have $\deg(F^\clubsuit) = 1$). Thus the B-linear map*

$$\clubsuit : \mathcal{H}^r_{gg}(U_0)_B \to H^r_{gg}(U_0)_B, \quad F \mapsto F^\clubsuit$$

is injective and hence $\mathcal{H}^r_{gg}(U_0)_B$ has finite rank bounded by the rank of $H^r_{gg}(U_0)_B$.

(c) *For each $F \in \mathcal{H}^r_{gg}(U_0)^\dagger_B$, the matrix $F^\clubsuit \in H^r_{gg}(U_0)^\dagger_B$ is the component of degree 0 of F. Thus the B-linear map*

$$\clubsuit : \mathcal{H}^r_{gg}(U_0)^\dagger_B \to H^r_{gg}(U_0)^\dagger_B, \quad F \mapsto F^\clubsuit$$

is injective.

(d) *We have $\mathcal{H}^r_{gg}(U_0)^\ddagger_B = 0$.*

As we shall see later, the B-linear map $\mathcal{H}^r_g(U_0, s)_B \to H^r_g(U_0, s)_B$ is not surjective in general (even if $g = 1$).

From Propositions 3.5.1(b) and 3.6.2 and Corollary 3.6.9(a) we get directly the following consequence.

Corollary 3.6.10 *Let $w \in W$. If $deg(w) > 0$, then $\mathbb{I}^r_{g,ord}(w) = 0$. If $deg(w) = 0$, then $\mathbb{I}^r_{g,ord}(w)$ has rank at most 1. For $w \in W(r)$, if $s := -deg(w)/2 \in \frac{1}{2}\mathbb{N}$ and $gs \in \mathbb{Z}$, then we have a natural injective R-linear map*

$$\heartsuit : \mathbb{I}^r_{g,ord}(w) \xrightarrow{\mathcal{E}} \mathcal{H}^r_g(U_0, s) \xrightarrow{\clubsuit} \mathbb{H}^r_g(s)$$

and thus the rank of the free R-module $\mathbb{I}^r_{g,ord}(w)$ is less or equal to $D_0(g, r, s)$.

We recall that, by definition, $\mathbb{I}^r_{g,\mathrm{ord}}(w) = 0$ if $\frac{g \cdot \deg(w)}{2} \notin \mathbb{Z}$, see Remarks 2.3.9 (d) and 2.3.16.

Lemma 3.6.11 *For $s \in \frac{1}{2}\mathbb{N}$ with $gs \in \mathbb{Z}$ and $r \geq 1$ an integer, the composite K-linear map*

$$\mathbb{H}^r_g(s)_K \xrightarrow{\spadesuit} \mathcal{H}^r_g(U_0, s)_K \xrightarrow{\clubsuit} \mathbb{H}^r_g(s)_K$$

is given by the formula

$$F \mapsto (F^{\spadesuit})^{\clubsuit} = F(T, T' - T, p(T'' - T'), \ldots, p^{r-1}(T^{(r)} - T^{(r-1)})).$$

In particular, this composite is injective, hence a K-linear isomorphism. Thus the K-linear maps $\clubsuit$ and $\spadesuit$ are isomorphisms.

Proof This follows from the fact that the linear part of $\mathfrak{f}(T)$ is T and the linear part of Ψ^{ϕ^i} equals $p^i(T^{(i+1)} - T^{(i)})$. □

The following lemma on 'elimination of logarithms' will play a key role later on.

Lemma 3.6.12 *Let $r \geq 1$ be an integer. Let $F \in R[T, \ldots, T^{(r)}]$ be a homogeneous polynomial such that we have*

$$F(\mathfrak{f}(T), \Psi, \Psi^{\phi}, \ldots, \Psi^{\phi^{r-1}}) \in S^r_{big\text{-}for} \otimes_R K. \tag{3.48}$$

Then $F \in R[T', \ldots, T^{(r)}] := R[T^{(l)}_{ij} | 1 \leq i \leq j \leq g, 1 \leq l \leq r]$.

Proof Let $\mathcal{S}_r$ be the set of all homogeneous polynomials

$$F \in R[T, \ldots, T^{(r)}] \backslash R[T', \ldots, T^{(r)}]$$

such that Display (3.48) holds. We want to show $\mathcal{S}_r = \emptyset$. Assume $\mathcal{S}_r \neq \emptyset$ for some $r \geq 1$ and seek a contradiction. Let r be the minimum of all positive integers for which $\mathcal{S}_r \neq \emptyset$ and let $F \in \mathcal{S}_r$ have degree

$$\deg(F) = c := \min\{\deg(G) | G \in \mathcal{S}_r\}.$$

Write $F = F_0 + F_1$ where $F_0 \in R[T', \ldots, T^{(r)}]$ and F_1 is in the ideal (T) of $R[T, \ldots, T^{(r)}]$ generated by the entries of T. As $F \in \mathcal{S}_r$ we get that $F_1 \in \mathcal{S}_r$. Replacing F by F_1 we can assume that, in addition, $F \in (T)$. For $1 \leq i, j \leq g$ we compute

$$\frac{\partial}{\partial T_{ij}^{(r)}}(\Psi_{ij}^{\phi^{r-1}}) = \frac{1}{p}\cdot\frac{\partial}{\partial T_{ij}^{(r)}}\left(\log\left(\frac{1+T_{ij}^{\phi^r}}{(1+T_{ij}^{\phi^{r-1}})^p}\right)\right)$$

$$= \frac{1}{p}\cdot\left(\frac{1+T_{ij}^{\phi^r}}{(1+T_{ij}^{\phi^{r-1}})^p}\right)^{-1}\cdot\frac{\partial}{\partial T_{ij}^{(r)}}\left(\frac{1+T_{ij}^{\phi^r}}{(1+T_{ij}^{\phi^{r-1}})^p}\right)$$

$$= \frac{1}{p}\cdot(1+T_{ij}^{\phi^r})^{-1}\cdot\frac{\partial}{\partial T_{ij}^{(r)}}(T_{ij}^{\phi^r})$$

$$= p^{r-1}\cdot(1+T_{ij}^{\phi^r})^{-1}.$$

Hence

$$\frac{\partial}{\partial T_{ij}^{(r)}}(F(\mathfrak{f}(T),\Psi,\ldots,\Psi^{\phi^{r-1}})) = \frac{\partial F}{\partial T_{ij}^{(r)}}(\mathfrak{f}(T),\Psi,\ldots,\Psi^{\phi^{r-1}})\cdot p^{r-1}\cdot(1+T_{ij}^{\phi^r})^{-1}.$$

As the right hand side of the above equality belongs to $S^r_{\text{big-for}}\otimes_R K$ it follows that

$$\frac{\partial F}{\partial T_{ij}^{(r)}}(\mathfrak{f}(T),\Psi,\ldots,\Psi^{\phi^{r-1}}) \in S^r_{\text{big-for}}\otimes_R K.$$

As $\frac{\partial F}{\partial T_{ij}^{(r)}}$ is homogeneous of degree $c-1$, by the minimality of F, we must have

$$\frac{\partial F}{\partial T_{ij}^{(r)}} \in R[T',\ldots,T^{(r)}].$$

On the other hand, as $r \geq 1$ and $F \in (T)$ we also have

$$\frac{\partial F}{\partial T_{ij}^{(r)}} \in (T) \subset R[T,\ldots,T^{(r)}].$$

We conclude that for all integers $1 \leq i, j \leq g$ we have

$$\frac{\partial F}{\partial T_{ij}^{(r)}} = 0.$$

If $r \geq 2$ we get $F \in \mathcal{S}_{r-1}$, contradicting the minimality of the positive integer r. If $r = 1$ we get $F \in R[T]$. As $F(\mathfrak{f}(T)) \in R[[T]]\otimes_R K$ it follows that

$$\frac{\partial}{\partial T_{ij}}(F(\mathfrak{f}(T))) = \frac{\partial F}{\partial T_{ij}}(\mathfrak{f}(T))\cdot(1+T_{ij})^{-1} \in R[[T]]\otimes_R K.$$

Hence

$$\frac{\partial F}{\partial T_{ij}}(\mathfrak{f}(T)) \in R[[T]] \otimes_R K.$$

By the minimality of c we get

$$\frac{\partial F}{\partial T_{ij}} \in R[T'] \cap R[[T]] = R,$$

hence $c = 1$. So $F(\mathfrak{f}(T)) = \sum \lambda_{ij}\mathfrak{f}(T_{ij}) = \sum \lambda_{ij} \log(1 + T_{ij}) \in R[[T]] \otimes_R K$, with $\lambda_{ij} \in R$, hence $F = 0$, a contradiction. □

3.6.3 Applications to δ-Modular Forms of Size g

Next we briefly turn our attention to δ-modular forms of size g. For B an infinite integral domain we recall that $\mathbb{H}^0_{gg}(1)_B$ is the B-module of all matrices $F \in \mathbf{Mat}_g(B[T])$ such that for every $X \in \mathbf{Mat}_g(B)$, inside $\mathbf{Mat}_g(B[T])$ we have an identity

$$F(XTX^{\mathrm{t}}) = X \cdot F(T) \cdot X^{\mathrm{t}}. \tag{3.49}$$

Lemma 3.6.13 *Let B be a field of characteristic 0. Then the B-vector space $\mathbb{H}^0_{gg}(1)_B$ has dimension 1 with basis the matrix T.*

Proof We can assume that B is algebraically closed. Setting $T = 1_g$ and taking $X \in \mathbf{O}_g(B) = \{X \in \mathbf{Mat}_g(B) | XX^{\mathrm{t}} = 1_g\}$ in Eq. (3.49), we get

$$F(1_g) = X \cdot F(1_g) \cdot X^{-1}.$$

So $F(1_g)$ is a matrix that commutes with all matrices in $\mathbf{O}_g(B)$. We claim that $F(1_g)$ must be a scalar matrix. Indeed the matrix $F(1_g)$ induces an endomorphism of the $\mathbf{O}_g(B)$-module B^g. This module is known to be simple; this is trivial to check for $g \in \{1, 2\}$ while for $g \geq 3$ this follows from the fact that the $\boldsymbol{so}_{g,B}$-module B^g is simple (for instance, see [FH], Thms. 19.2 and 19.4). Then our claim follows from Schur's Lemma.

In view of our claim we can write $F(1_g) = \lambda \cdot 1_g$ with $\lambda \in B$. Setting $T = 1_g$ in Eq. (3.49), we get that for all $X \in \mathbf{GL}_g(B)$ we have

$$F(XX^{\mathrm{t}}) = \lambda XX^{\mathrm{t}}.$$

By Sylvester's theorem, each symmetric matrix in $\mathbf{GL}_g(B)$ can be written in the form XX^{t} for some $X \in \mathbf{GL}_g(B)$. It follows that $F(Y) = \lambda Y$ for every symmetric matrix $Y \in \mathbf{GL}_g(B)$ hence $F = \lambda \cdot T$. □

Proposition 3.6.14 *Assume B is a field of characteristic 0. Then $\mathbb{H}^r_{gg}(1)_B$ has a basis consisting of the matrices $T, T', \dots, T^{(r)}$.*

Proof Let $G \in \mathbb{H}^r_{gg}(1)_B$; so, for all $M \in \mathbf{GL}_g(B)$, Eq. (3.42) holds. As G is homogeneous of degree 1 we have

$$G(T, T', \dots, T^{(r)}) = G(T, 0, \dots, 0) + G(0, T', 0, \dots, 0) + \cdots + G(0, 0, \dots, T^{(r)}). \quad (3.50)$$

Setting all but one of the matrices $T, T', \dots, T^{(r)}$ in Eq. (3.42) equal to 0, from Lemma 3.6.13 we get that the terms in the right-hand side of Equation (3.50) have the form $b_0T, b_1T', \dots, b_rT^{(r)}$ respectively, with $b_0, \dots, b_r \in B$, and we are done by Eq. (3.50). □

Corollary 3.6.15 *Assume conditions (UAX1) and (UAX2) hold. Then for every integer $r \geq 0$ the R-module $\mathcal{H}^r_{gg}(U_0)_R$ has rank r and hence for each element F in it there exists a nonzero element $\lambda \in R$ such that λF is an R-linear combination of $\Psi, \Psi^{\phi}, \dots, \Psi^{\phi^{r-1}}$.*

Proof The elements $\mathfrak{f}(T), \Psi, \dots, \Psi^{\psi^{r-1}}$ of the K-vector space $\mathcal{H}^r_{gg}(U_0)_K$ are linearly independent. By Corollary 3.6.9(b) we have

$$\dim_K(\mathcal{H}^r_{gg}(U_0)_K) \leq \dim_K(H^r_{gg}(U_0)_K).$$

By Proposition 3.6.2 we have

$$\dim_K(H^r_{gg}(U_0)_K) = \dim_K(\mathbb{H}^r_{gg}(1)_K).$$

By Proposition 3.6.14 we have

$$\dim_K(\mathbb{H}^r_{gg}(1)_K) = r + 1.$$

So $\mathfrak{f}(T), \Psi, \dots, \Psi^{\psi^{r-1}}$ is a basis of $\mathcal{H}^r_{gg}(U_0)_K$. Let $F \in \mathcal{H}^r_{gg}(U_0)_R$. Then there exists a nonzero $\lambda \in R$ and there exist $\lambda_0, \dots, \lambda_r \in R$ such that

$$\lambda F = \lambda_0 \mathfrak{f}(T) + \lambda_1 \Psi + \cdots + \lambda_r \Psi^{\phi^{r-1}}.$$

So $\lambda_0 \mathfrak{f}(T) \in S^0_{\text{for}}$. This clearly implies $\lambda_0 = 0$ which ends our proof. □

Assumption on notation. *From now on throughout the monograph, we will assume, for simplicity, that (A_0, θ_0, e) is as in Display* (3.3) *and e is chosen such that $\mathcal{E}_{E_0,\theta_0,e}(f^1_{1,\text{crys}}) = \psi$. This is possible by* [Bu05], *Sect. 8.1, pp. 239–240. In particular conditions* (UAX1) *and* (UAX2) *hold.*

For the next result we note that, if t is an indeterminate, then the base change of the universal formal abelian scheme of relative dimension g over the formal spectrum $\text{Spf}(R_g/(T_{ij} - T_{ji} | 1 \leq i < j \leq g))$ which lifts (A_0, θ_0) via the morphism

$$\text{Spf}(R[[t]]) \to \text{Spf}(R_g/(T_{ij} - T_{ji} | 1 \leq i < j \leq g))$$

induced by $T_{ij} \mapsto t \cdot \delta_{ij}$ is the g-fold product of the universal formal elliptic curve over $\mathrm{Spf}(R[[t]])$ which lifts E_0 with its product polarization. Then substituting

$$(T, T', \dots, T^{(r)}) \text{ by } (t \cdot 1_g, t' \cdot 1_g, \dots, t^{(r)} \cdot 1_g) \tag{3.51}$$

in the Serre–Tate expansion matrix $\mathcal{E}_{A_0,\theta_0,e}(f^r_{g,\mathrm{crys}})$, where $t', t'', \dots$ are indeterminates that play the role of $T', T'', \dots$ for $g = 1$, we get

$$\mathcal{E}_{E_0,\theta_0,e}(f^r_{1,\mathrm{crys}}) \cdot 1_g \in \mathbf{Mat}_g(R[[t, t', \dots, t^{(r)}]]).$$

Theorem 3.6.16 *For all integers $r \geq 1$ the following properties hold.*

(a) *The form $f^r_{g,crys} \in \mathbb{I}^r_{gg}(\phi^r, 1)$ has the Serre–Tate expansion*

$$\mathcal{E}_{A_0,\theta_0,e}(f^r_{g,crys}) = \Psi^{\phi^{r-1}} + p\Psi^{\phi^{r-2}} + \dots + p^{r-1}\Psi.$$

(b) *The free R-modules $\mathbb{I}^1_{gg}(\phi, 1)$ and $\mathbb{I}^1_{gg,ord}(\phi, 1)$ are equal and have rank 1, with basis $f^1_{g,crys}$.*

Proof Part (a) was proved for $g = 1$ in [Bu05], Prop. 8.61. For arbitrary $g \geq 1$, from Corollary 3.6.15 we get that there exists $\lambda_0, \dots, \lambda_r \in R$, such that we have an equality

$$\mathcal{E}_{A_0,\theta_0,e}(f^r_{g,\mathrm{crys}}) = \lambda_0 \Psi + \dots + \lambda_r \Psi^{\phi^{r-1}}.$$

Making the substitution of Display (3.51) in this equality we get that

$$\mathcal{E}_{A_0,\theta_0,e}(f^r_{1,\mathrm{crys}}) = \lambda_0 \psi + \dots + \lambda_r \psi^{\phi^{r-1}}.$$

By the genus 1 case of the theorem we get that

$$(\lambda_0, \dots, \lambda_r) = (p^{r-1}, \dots, p, 1),$$

from which part (a) follows.

To prove part (b) we note that, by Corollary 3.6.15, $\mathcal{H}^1_{gg}(U_0)$ has rank 1 and we conclude by Proposition 3.5.1(b) and part (a). □

Similarly we have the following basic properties.

Theorem 3.6.17 *The following properties hold.*

(a) *The Serre–Tate expansion of* $f^{\partial}_{g,crys} \in \mathbb{I}^1_{gg,ord}(-\phi, 1)$ *equals* 1_g.

(b) *The free* R*-module* $\mathbb{I}^1_{gg,ord}(-\phi, 1)$ *has basis* $f^{\partial}_{g,crys}$.

(c) *Moreover for all* $a, b \in \{0, 1, \ldots, r\}$ *we have the following rank results:*

(c.i) *the* R*-module* $\mathbb{I}^r_{gg,ord}(\phi^a, \phi^b)$ *has rank* r*;*

(c.ii) *the* R*-modules* $\mathbb{I}^r_{gg,ord}(-\phi^a, \phi^b)$ *and* $\mathbb{I}^r_{gg,ord}(\phi^a, -\phi^b)$ *have rank* 1*;*

(c.iii) *the* R*-module* $\mathbb{I}^r_{gg,ord}(-\phi^a, -\phi^b)$ *is* 0.

Proof From Propositions 3.5.1(b) and 3.6.2 and Corollary 3.6.15 we get that the free R-module $\mathbb{I}^1_{gg,\text{ord}}(-\phi, 1)$ has rank at most 1. As it contains $f^{\partial}_{g,\text{crys}}$, by the torsion freeness part in Proposition 3.5.1(b) we get that it is equal to $\mathbb{I}^1_{gg}(-\phi, 1)$ and this latter R-module has rank 1. Also, as $\mathcal{H}^1_{gg}(U_0)^{\dagger}$ contains 1_g, by Propositions 3.5.1(b) and 3.6.2 plus Corollary 3.6.9(c), we have $\mathcal{H}^1_{gg}(U_0)^{\dagger} = R1_g$. So the Serre–Tate expansion of $f^{\partial}_{g,\text{crys}}$ equals $\eta \cdot 1_g$ for some $\eta \in R$. We claim that $\eta = 1$. It suffices to check this for $g = 1$, and this was checked in [Bu05], Prop. 8.59. So parts (a) and (b) are proved.

Part (c.i) can be proved by a similar argument: one notes that, by Corollary 3.6.15, $\mathcal{H}^r_{gg}(U_0)$ has rank r and one concludes by Proposition 3.5.1(b), Corollary 3.4.5, and Theorem 3.6.16(a).

Part (c.ii) follows similarly by combining part (a), Propositions 3.5.1(b) and 3.6.2, Corollary 3.6.9(c), and Lemma 3.4.4.

Part (c.iii) follows similarly by combining Proposition 3.5.1(b) and Corollary 3.6.9(d). □

Corollary 3.6.18 *For all integers* $a \geq 1$*, with the forms* $f^{\langle a \rangle}$ *and* $f^{[a]}$ *as in Displays (3.11) and (3.12), the following properties hold.*

(a) *The Serre–Tate expansion of* $f^{\langle a \rangle}$ *equals* $\Psi^{\phi^{a-1}}$.

(b) *The Serre–Tate expansion of* $f^{[a]}$ *equals* $\Psi^{\phi^{a-1}} + p\Psi^{\phi^{a-2}} + \cdots + p^{a-1}\Psi$.

(c) *The identities* $f^{\langle a \rangle t} = f^{\langle a \rangle}$ *and* $f^{[a]t} = f^{[a]}$ *hold.*

(d) *If* $a \geq 2$*, then we have an equality*

$$f^{[a]} = f^{\langle a \rangle} + pf^{\langle a-1 \rangle} + \cdots + p^{a-1} f^{\langle 1 \rangle}.$$

Proof Parts (a) and (b) follow from Theorems 3.6.16(a) and 3.6.17(a).

To check part (c), note that the forms $f^{\langle a \rangle \text{t}}$ and $f^{\langle a \rangle}$ both belong to $\mathbb{I}^a_{gg}(1, 1)$ and have the same Serre–Tate expansion due to part (a) and the symmetry of the matrix $\Psi = (\Psi_{ij})_{1 \leq i,j \leq g}$. We conclude that part (c) holds by the Serre–Tate expansion principle (Proposition 3.5.1(b)). The same argument applies to $f^{[a]}$.

Part (d) follows from parts (a) and (b) and the Serre–Tate expansion principle. □

3.6.4 The Maps $\diamondsuit$ and Ω

For all integers $r \geq 0$ we recall the R-algebra of homogeneous invariants $\mathbb{H}^r_{g,\text{tot}} = \bigoplus_{s\in\frac{1}{2}\mathbb{Z}} \mathbb{H}^r_g(s)$, and we consider the R-algebra

$$\mathbb{I}^{r,\mathbb{Z}}_{g,\text{ord}} := \bigoplus_{s\in\frac{1}{2}\mathbb{Z}} \mathbb{I}^r_{g,\text{ord}}(-2s) = \bigoplus_{m\in\mathbb{N}\cup\{0\}} \mathbb{I}^r_{g,\text{ord}}(-m).$$

Denoting $R[T, T', T'', \ldots] := \cup_{r\geq 0} R[T, \ldots, T^{(r)}]$, we consider the R-algebra monomorphism

$$\sigma : R[T, T', T'', \ldots] \to R[T, T', T'', \ldots], \quad F \mapsto F^\sigma,$$

where

$$T^\sigma := T',\ (T')^\sigma := T'',\ \text{etc.}, \quad a^\sigma := a, \quad a \in R,$$

and the $\mathbb{Z}_p$-algebra automorphism

$$\tau : R[T, T', T'', \ldots] \to R[T, T', T'', \ldots], \quad F \mapsto F^\tau,$$

where

$$T^\tau := T,\ (T')^\tau := T',\ \text{etc.}, \quad a^\tau := \phi(a), \ a \in R;$$

so for all integers $l \geq 0$ and $1 \leq i \leq j \leq g$ we have identities $\sigma(T^{(l)}_{ij}) = T^{(l+1)}_{ij}$ and $\tau(T^{(l)}_{ij}) = T^{(l)}_{ij}$. Thus $\sigma\tau = \tau\sigma$. Moreover, for $r \geq 1$ we have induced $\mathbb{Z}_p$-linear maps

$$\mathbb{H}^{r-1}_g(s) \to \mathbb{H}^r_g(s), \quad F \mapsto F^\sigma, \tag{3.52}$$

$$\mathbb{H}^{r-1}_g(s) \to \mathbb{H}^{r-1}_g(s), \quad F \mapsto F^\tau. \tag{3.53}$$

and a $\mathbb{Z}_p$-bilinear map

$$\mathbb{H}^{r-1}_g(s) \times \mathbb{H}^{r-1}_g(s) \to \mathbb{H}^r_g((p+1)s),$$

defined by

$$(F, G) \mapsto \{F, G\}_{\sigma,\tau} := G^{\sigma\tau} F^p - F^{\sigma\tau} G^p.$$

Proposition 3.6.19 *For all integers $r \geq 1$ and all $s \in \frac{1}{2}\mathbb{Z}$ the following properties hold.*

(a) *We have a well-defined R-linear map*

$$\diamondsuit : \mathbb{H}^{r-1}_g(s) \to \mathbb{I}^r_{g,ord}(-2s), \quad F \mapsto F^\diamondsuit := F(f^{\langle 1\rangle}, \ldots, f^{\langle r\rangle}).$$

(b) *For* $F \in \mathbb{H}_g^{r-1}(s)$ *we have the following formula:*

$$(F^\diamondsuit)^\phi = \det(f^\partial)^{-2s} \cdot (F^{\sigma\tau})^\diamondsuit. \tag{3.54}$$

(c) *For* $F, G \in \mathbb{H}_g^{r-1}(s)$ *we have the following formula:*

$$(\{F, G\}_{\sigma,\tau})^\diamondsuit = p \cdot \det(f^\partial)^{2s} \cdot \{F^\diamondsuit, G^\diamondsuit\}_\delta \tag{3.55}$$

(see the part after Definition 2.1.15 of Sect. 2.1 for the notation $\{F^\diamondsuit, G^\diamondsuit\}_\delta$*).*

Proof The expression $F(f^{\langle 1\rangle}, \ldots, f^{\langle r\rangle})$ is well-defined as $f^{\langle 1\rangle}, \ldots, f^{\langle r\rangle}$ are symmetric, see Corollary 3.6.18(c). This expression is trivially seen to define an element in $\mathbb{I}_{g,\mathrm{ord}}^r(-2s)$. Thus part (a) holds.

Equation (3.54) reads

$$F(f^{\langle 1\rangle}, \ldots, f^{\langle r\rangle})^\phi = \det(f^\partial)^{-2s} \cdot F^\tau(f^{\langle 2\rangle}, \ldots, f^{\langle r+1\rangle}); \tag{3.56}$$

as Eq. (3.56) follows from Eq. (3.13), part (b) holds.

Part (c) holds as Eq. (3.55) follows directly from Eq. (3.54). □

Theorem 3.6.20 *For all* $r \in \mathbb{N}$ *and all* $s \in \frac{1}{2}\mathbb{Z}$ *the* R*-linear map*

$$\diamondsuit : \mathbb{H}_g^{r-1}(s) \to \mathbb{I}_{g,ord}^r(-2s)$$

is injective and has a torsion cokernel. In particular, the following properties hold.

(a) *For* $w \in W$, *if* $s := -deg(w)/2$, *then we have*

$$rank_R(\mathbb{I}_{g,ord}^r(w)) = D_0(g, r-1, s).$$

(b) *The* R*-algebra homomorphism*

$$\diamondsuit : \mathbb{H}_{g,tot}^{r-1} \to \mathbb{I}_{g,ord}^{r,\mathbb{Z}}$$

is injective and it induces a K*-algebra isomorphism*

$$\diamondsuit : \mathbb{H}_{g,tot}^{r-1} \otimes_R K \to \mathbb{I}_{g,ord}^{r,\mathbb{Z}} \otimes_R K.$$

Proof We have the following diagram of R-linear maps:

$$\begin{array}{ccccccc}
\mathbb{H}_g^{r-1}(s)_K & \stackrel{\sigma}{\to} & \mathbb{H}_g^r(s)_K & \stackrel{\spadesuit}{\to} & \mathcal{H}_g^r(U_0, s)_K & \stackrel{\clubsuit}{\to} & \mathbb{H}_g^r(s)_K \\
\alpha \uparrow & & & & \beta \uparrow & & \gamma \uparrow \\
\mathbb{H}_g^{r-1}(s) & \stackrel{\diamondsuit}{\to} & \mathbb{I}_{g,\mathrm{ord}}^r(-2s) & \stackrel{\mathcal{E}}{\to} & \mathcal{H}_g^r(U_0, s) & \stackrel{\clubsuit}{\to} & \mathbb{H}_g^r(s)
\end{array}$$

where α, γ and β are inclusions. As the R-linear maps on the upper row are injective it follows that $\diamondsuit$ on the lower row is injective.

Now for each $f \in \mathbb{I}^r_{g,\mathrm{ord}}(-2s)$ we have that

$$\beta(\mathcal{E}(f)) \in S^r_{\text{big-for}}.$$

As ♠ is an R-linear isomorphism (see Lemma 3.6.11), there exists $F \in \mathbb{H}^r_g(s)_K$ such that $F^{♠} = \beta(\mathcal{E}(f))$. By Lemma 3.6.12 we must have $F \in R[T', \dots, T^{(r)}]$. But then, clearly $F = F_0^\sigma$ for some $F_0 \in \mathbb{H}^{r-1}_g(s)_K$. We conclude that the image of the injective R-linear map $\beta \circ \mathcal{E}$ is contained in the image of the R-linear map ♠ $\circ\, \sigma$. Hence we have an inequality

$$\dim(\mathbb{I}^r_{g,\mathrm{ord}}(-2s) \otimes_R K) \leq \dim(\mathbb{H}^{r-1}_g(s)_K)$$

which, as $\diamondsuit$ is injective, is an equality. So the cokernel of $\diamondsuit$ is torsion. This proves part (b). Part (a) follows from part (b), Display (3.45), and Corollary 3.4.6. □

Corollary 3.6.21 *For each $n \in \mathbb{N}$, the R-modules $\mathbb{I}^1_g(-n\phi - n)$ and $\mathbb{I}^1_{g,ord}(-n\phi - n)$ coincide and are free of rank* 1 *with basis* $(\det(f^1_{g,crys}))^n$.

Proof Theorem 3.6.20(a) and Lemma 3.6.3 imply that the statement is true after tensorization with K. So to prove the corollary it is enough to check that $(\det(f^1_{g,\mathrm{crys}}))^n$ is not divisible by p in $\mathbb{I}^1_{g,\mathrm{ord}}(-n\phi - n)$. The latter condition follows from the torsion freeness part of the Serre–Tate expansion principle (see Proposition 3.5.1(b)) and the fact that the Serre–Tate expansion of $(\det(f^1_{g,\mathrm{crys}}))^n$ equals 1. □

For ordinary Hecke covariant Siegel δ-modular forms of size g we have the following consequence.

Corollary 3.6.22 *For all integers $a, b \geq 0$ and $r \geq \max\{a, b\}$, the R-module $\mathbb{I}^r_{gg,ord}(\phi^a, \phi^b)$ has rank r.*

Proof Recall that by Corollary 3.6.15 the R-module $\mathcal{H}^r_{gg}(U_0)_R$ has rank r. Hence by Proposition 3.5.1(b) the R-module $\mathbb{I}^r_{gg,\mathrm{ord}}(1, 1)$ has rank $\leq r$. As $f^{\langle 1\rangle}, \dots, f^{\langle r\rangle}$ belong to the latter module and are R-linearly independent (as their Serre–Tate expansions are, see Corollary 3.6.18(a)) it follows that $\mathbb{I}^r_{gg,\mathrm{ord}}(1, 1)$ has rank r. We conclude by Corollary 3.4.5. □

By Theorem 3.6.20(a) the R-linear maps $\diamondsuit : \mathbb{H}^{r-1}_g(s) \to \mathbb{I}^r_{g,\mathrm{ord}}(-2s)$ become surjective after inverting p; but they are not surjective in general (see Proposition 3.6.24(b) below for a more precise statement).

Indeed let us consider the torsion R-modules

$$\mathbb{V}^r_g(s) := \frac{\mathbb{I}^r_{g,\mathrm{ord}}(-2s)}{\diamondsuit(\mathbb{H}^{r-1}_g(s))}, \quad \mathbb{V}^r_g := \frac{\mathbb{I}^{r,\mathbb{Z}}_{g,\mathrm{ord}}}{\diamondsuit(\mathbb{H}^{r-1}_{g,\mathrm{tot}})} = \bigoplus_{s \in \frac{1}{2}\mathbb{Z}_{\geq 0}} \mathbb{V}^r_g(s).$$

Denote by $\mathbb{V}^r_g(s)[p]$ the set of all elements of $\mathbb{V}^r_g(s)$ annihilated by p. So $\mathbb{V}^r_g(s)[p]$ has a natural structure of a k-vector space. We denote by $\mathbb{V}^r_g(s)[p]^\tau$ the k-vector space obtained from $\mathbb{V}_g^r(s)[p]$ by restricting the scalars via the Frobenius automorphism of k.

Consider the $\mathbb{Z}_p$-bilinear map

$$\Omega : \mathbb{H}_g^{r-1}(s) \times \mathbb{H}_g^{r-1}(s) \to \mathbb{V}_g^{r+1}((p+1)s)[p]$$

defined (see Proposition 3.6.19(c)) by:

$$\Omega(F, G) := (\det(f^{\partial}))^{2s}\{F^{\diamondsuit}, G^{\diamondsuit}\}_{\delta} + \diamondsuit(\mathbb{H}_g^r((p+1)s))$$

$$= \frac{1}{p}(\{F, G\}_{\sigma,\tau})^{\diamondsuit} + \diamondsuit(\mathbb{H}_g^r((p+1)s)).$$

Clearly, the map Ω induces an antisymmetric k-bilinear map

$$\overline{\Omega} : \overline{\mathbb{H}_g^{r-1}(s)} \times \overline{\mathbb{H}_g^{r-1}(s)} \to \mathbb{V}_g^{r+1}((p+1)s)[p]^{\tau}.$$

For instance, additivity in the first argument follows from the formula (see Proposition 3.6.19(b)):

$$\{F_1^{\diamondsuit} + F_2^{\diamondsuit}, G^{\diamondsuit}\}_{\delta} = \{F_1^{\diamondsuit}, G^{\diamondsuit}\}_{\delta} + \{F_2^{\diamondsuit}, G^{\diamondsuit}\}_{\delta} - \det(f^{\partial})^{-2s}(G^{\sigma\tau} \cdot C_p(F_1, F_2))^{\diamondsuit}.$$

To state our next result and for future use we introduce the following definition.

Definition 3.6.23 Let U, V be two vector spaces over an arbitrary field. A bilinear antisymmetric map $\Omega : U \times U \to V$ is called *totally nondegenerate* if for all $x, y \in U$, we have $\Omega(x, y) = 0$ if and only if x and y are linearly dependent.

Proposition 3.6.24 *For an integer $r \geq 2$ and $s \in \frac{1}{2}\mathbb{Z}_{\geq 0}$ the following properties hold.*

(a) *The above bilinear map $\overline{\Omega}$ is totally nondegenerate.*

(b) *We have an inequality*

$$\dim_k(\mathbb{V}_g^{r+1}((p+1)s)[p]) \geq 2D_0(g, r-1, s) - 3.$$

In particular, if $D_0(g, r-1, s) \geq 2$, then the map

$$\diamondsuit : \mathbb{H}_g^r((p+1)s) \to \mathbb{I}_{g,ord}^{r+1}(-2(p+1)s)$$

is not surjective.

(c) *The annihilator of the R-module $\mathbb{V}_g^r$ is 0.*

Proof To prove part (a) assume $F, G \in \mathbb{H}_g^{r-1}(s)$ and $\overline{\Omega}(\overline{F}, \overline{G}) = 0$; we need to show that $\overline{F}$ and $\overline{G}$ are k-linearly dependent. We have $\Omega(F, G) = 0$ so there exists $H \in \mathbb{H}_g^r((p+1)s)$ such that $(\{F, G\}_{\sigma,\tau})^{\diamondsuit} = pH^{\diamondsuit}$. By the injectivity of $\diamondsuit$ we have

$\{F, G\}_{\sigma,\tau} = pH$. This implies that either $\overline{G} = 0$ (in which case we are done) or $\overline{G} \neq 0$ and $(\overline{F}/\overline{G})^{\sigma\tau} = (\overline{F}/\overline{G})^p$ in the field $k(T, \dots, T^{(r)})$. The latter clearly implies $\overline{F}/\overline{G} \in k$ and we are done again.

To check part (b) note that by part (a) plus a general fact of linear algebra (see Lemma 3.6.25 below) the k-linear span of the image of $\overline{\Omega}$ has dimension at least $2\dim_k(\overline{\mathbb{H}_g^{r-1}(s)}) - 3$. We then conclude by the fact that

$$\dim_k(\overline{\mathbb{H}_g^{r-1}(s)}) = \dim_K(\mathbb{H}_g^{r-1}(s)_K) = D_0(g, r-1, s).$$

To check part (c) note that, by part (b), $\mathbb{V}_g^r[p] \neq 0$ hence there exists a prime element F of the unique factorization domain $\mathbb{H}_{g,\text{tot}}^{r-1}$ with $\frac{F}{p} \in \mathbb{I}_{g,\text{ord}}^{r,\mathbb{Z}} \backslash \diamondsuit(\mathbb{H}_{g,\text{tot}}^{r-1})$. For each $n \in \mathbb{N}$ the image of $\frac{F^n}{p^n}$ in $\mathbb{V}_g$ has annihilator $p^n R$ in R. Thus the annihilator of $\mathbb{V}_g$ in R is contained in $\bigcap_{n \geq 1} p^n R = 0$, hence it is 0. □

In the above proof we applied over k the following general fact.

Lemma 3.6.25 *Let U and V be finite dimensional vector spaces over an algebraically closed field B and let $b : U \times U \to V$ be a B-bilinear antisymmetric map. Assume b is totally nondegenerate. Then the following inequality holds:*

$$\dim_B(Span(Im(b))) \geq 2\dim_B(U) - 3.$$

Proof Let $u_1, \dots, u_n$ be a basis of U and set $v_{ij} := b(u_i, u_j) \in V$ for $1 \leq i < j \leq n$. Let L be the vector space over B of all $(n^2 - n)/2$-tuples

$$c := (c_{ij})_{(i,j) \in \{1,\dots,n\}^2 \backslash \{(l,l) | l \in \{1,\dots,n\}\}, i<j} \in B^{(n^2-n)/2}$$

and let

$$L' := \{c \in L | \sum_{1 \leq i < j \leq n} c_{ij} v_{ij} = 0\}.$$

For $c \in L \backslash \{0\}$ let $[c] \in \mathbb{P}(L) := (L \backslash \{0\})/B^{\times}$ be the corresponding point in the projective space; similarly, for $\lambda := (\lambda_i)_{1 \leq i \leq n} \in B^n \backslash \{0\}$ let $[\lambda] \in \mathbb{P}(B^n)$ be the corresponding point. Let

$$\varphi : (\mathbb{P}(B^n) \times \mathbb{P}(B^n)) \backslash \Delta \to \mathbb{P}(L)$$

be the morphism

$$\varphi([\lambda], [\mu]) := [(\lambda_i \mu_j - \mu_i \lambda_j)_{1 \leq i < j \leq n}],$$

where Δ is the diagonal of $\mathbb{P}(B^n) \times \mathbb{P}(B^n)$. The image $\text{Im}(\varphi)$ is the Grassmannian of 2-planes in an n-dimensional vector space, so $\dim(\text{Im}(\varphi)) = 2n - 4$. If there exists $[(c_{ij})] \in \text{Im}(\varphi) \cap \mathbb{P}(L')$, then we have $c_{ij} = \lambda_i \mu_j - \lambda_j \mu_i$ for some $(\lambda_i)_{1 \leq i \leq n}, (\mu_i)_{1 \leq i \leq n} \in B^n \backslash \{0\}$ and for $x := \sum_{i=1}^n \lambda_i u_i$ and $y := \sum_{i=1}^n \mu_i u_i$ we get that x and y are B-linearly independent and $b(x, y) = 0$, a contradiction. Thus we have $\text{Im}(\varphi) \cap \mathbb{P}(L') = \emptyset$ and hence

$$\dim_B(L') \leq \dim_B(L) - 1 - \dim(\mathrm{Im}(\varphi)) \leq \frac{n(n-1)}{2} - 2n + 3,$$

which implies that the lemma holds. □

Remark 3.6.26 Lemma 3.6.25 is optimal in the sense that for every vector space U over B of dimension at least 2 there exists a vector space V and a totally non-degenerate antisymmetric bilinear map $b : U \times U \to V$ such that the inequality of Lemma 3.6.25 is an equality. To construct such a V and b fix a basis $u_1, u_2, \ldots, u_n$ of U with $n \geq 2$, let L and φ be as in the above proof, let $L' \subset L$ be a subspace of dimension $\frac{n(n-1)}{2} - 2n + 3$ such that $\mathbb{P}(L') \cap \mathrm{Im}(\varphi) = \emptyset$ (such an L' exists because $\dim(\mathrm{Im}(\varphi)) + \frac{n(n-1)}{2} - 2n + 3 \leq \dim_B(L) - 1$) and take $V := (\wedge^2 U)/U'$ where U' is the subspace of $\wedge^2 U$ consisting of all expressions of the form $\sum_{i<j} c_{ij} u_i \wedge u_j$ with $(c_{ij})_{1 \leq i < j \leq n} \in L'$. Finally take b to be the natural quotient composition $U \times U \to \wedge^2 U \to V$.

In what follows we consider the rings:

$$\mathbb{H}_{g,\mathrm{tot}} := \bigcup_{r=0}^{\infty} \mathbb{H}^r_{g,\mathrm{tot}}, \quad \mathbb{I}^{\mathbb{Z}}_{g,\mathrm{ord}} := \bigcup_{r=0}^{\infty} \mathbb{I}^{r,\mathbb{Z}}_{g,\mathrm{ord}}.$$

In view of Lemma 3.6.4(b), we get the following consequence.

Corollary 3.6.27 *For all integers $r \geq 2$ we have*

$$\mathrm{tr.deg}_R(\mathbb{I}^{r,\mathbb{Z}}_{g,ord}) = r\frac{g(g+1)}{2} - g^2 + 1.$$

From the last corollary we get directly the following consequence.

Corollary 3.6.28 *We have*

$$\mathrm{asy.tr.deg}(\mathbb{I}^{\mathbb{Z}}_{g,ord}) = \frac{g(g+1)}{2}.$$

On the other hand recall the rings $\mathbb{I}^r_{g,\mathrm{ord}}$ and their union $\mathbb{I}_{g,\mathrm{ord}}$. Also recall the filtered δ-ring $\mathbb{I}_{g,\mathrm{ord},((p))}$. For $r \in \mathbb{N}$ let $z_{<r} := \{z_0, \ldots, z_{r-1}\}$ be a set of indeterminates and we consider the R-algebra monomorphism between rings of Laurent polynomials in the set of indeterminates $z_{<r}$,

$$\mathbb{H}^{r-1,\mathrm{torus}}_{g,\mathrm{tot}} := \mathbb{H}^{r-1}_{g,\mathrm{tot}}[z_{<r}, \frac{1}{z_0 \cdots z_{r-1}}] \to \mathbb{I}^{r,\mathbb{Z}}_{g,\mathrm{ord}}[z_{<r}, \frac{1}{z_0 \cdots z_{r-1}}],$$

that is induced by $\diamond$ and sends z_i to z_i for all $i \in \{0, \ldots, r-1\}$. The ring of Laurent polynomials

$$\mathbb{H}^{\mathrm{torus}}_{g,\mathrm{tot}} := \cup_{r \geq 1} \mathbb{H}^{r-1,\mathrm{torus}}_{g,\mathrm{tot}}$$

in the set of indeterminates $z := \cup_{r\in\mathbb{N}} z_{<r}$ is a filtered W-graded ring with the grading defined by letting the elements of $\mathbb{H}^r_g(s)$ have weight $-2s$ and letting z_i have weight $\phi^{i+1} - \phi^i$ for each integer $i \geq 0$, see Sect. 2.3.4.

Theorem 3.6.29 *The following properties hold.*

(a) *For all integers $r \geq 1$ we have an $\mathbb{I}^{r,\mathbb{Z}}_{g,ord}$-algebra isomorphism*

$$\mathbb{I}^{r,\mathbb{Z}}_{g,ord}[z_{<r}, \frac{1}{z_0 \cdots z_{r-1}}] \simeq \mathbb{I}^r_{g,ord},$$

and hence monomorphisms of R-algebras,

$$\diamondsuit : \mathbb{H}^{r-1,torus}_{g,tot} \to \mathbb{I}^r_{g,ord}, \quad \diamondsuit : \mathbb{H}^{torus}_{g,tot} \to \mathbb{I}_{g,ord},$$

whose cokernels are torsion.

(b) *For all integers $r \geq 2$ we have*

$$\text{tr.deg}_R(\mathbb{I}^r_{g,ord}) = r\frac{g(g+1)}{2} - g^2 + 1 + r.$$

(c) *We have*

$$\text{asy.tr.deg}(\mathbb{I}_{g,ord}) = \frac{g(g+1)}{2} + 1.$$

(d) *We have an equality $\mathbb{I}_{g,ord,((p))} = \mathbb{I}^{\mathbb{Z}}_{g,ord,((p))}$. In particular, we have*

$$\text{asy.tr.deg}(\mathbb{I}_{g,ord,((p))}) = \frac{g(g+1)}{2}.$$

(e) *We have*

$$\phi\text{−tr.deg}(\mathbb{I}_{g,ord}) = \frac{g(g+1)}{2} + 1.$$

(f) *We have*

$$\phi\text{−tr.deg}(\mathbb{I}_{g,ord,((p))}) = \frac{g(g+1)}{2}.$$

(g) *The field Frac$(\mathbb{I}_{g,ord})$ is ϕ-generated over K by a finite set of elements in $\mathbb{I}^4_{g,ord}$.*

(h) *For each integer $r \geq 1$, the form $\det(f^r_{g,crys})$ belongs to the image of the R-algebra monomorphism $\diamondsuit : \mathbb{H}^{r-1,torus}_g \to \mathbb{I}^r_{g,ord}$.*

Proof To check part (a) we define a $\mathbb{I}^{r,\mathbb{Z}}_{g,\text{ord}}$-algebra homomorphism

$$\mathbb{I}^{r,\mathbb{Z}}_{g,\text{ord}}[z_{<r}, \frac{1}{z_0 \cdots z_{r-1}}] \to \mathbb{I}^r_{g,\text{ord}}$$

by the rule

$$z_i \mapsto \det(f^{\partial}_{g,\mathrm{crys}})^{\phi^i} \in \mathbb{I}^r_{g,\mathrm{ord}}(\phi^i(\phi-1)), \quad i \in \{0, \ldots, r-1\}.$$

Hence, for $s \in \frac{1}{2}\mathbb{Z}$, $f \in \mathbb{I}^r_{g,\mathrm{ord}}(-2s)$, and $w' = \sum_{i=0}^{r-1} a_i\phi^i \in W$ we have

$$f \cdot z_0^{a_0} \cdots z_{r-1}^{a_{r-1}} \mapsto f \cdot (f^{\partial}_{g,\mathrm{crys}})^{w'} \in \mathbb{I}^r_{g,\mathrm{ord}}(w),$$

where

$$w := -2s + (\phi - 1)w' \in W, \quad \deg(w) = -2s.$$

By Corollary 3.4.6, we have an induced R-linear isomorphism

$$\mathbb{I}^r_{g,\mathrm{ord}}(-2s) \cdot z_0^{a_0} \cdots z_{r-1}^{a_{r-1}} \simeq \mathbb{I}^r_{g,\mathrm{ord}}(w).$$

Based on this, as w determines the pair (s, w'), we conclude that the part (a) and the first sentence of part (d) hold.

Part (b) follows from part (a) and Corollary 3.6.21.

Part (c) follows from part (b).

The second sentence of part (d) follows from part (c).

For the proof of parts (e) and (f) recall our convention (see the beginning of Sect. 2.2) on algebraically independent families. Furthermore, by the cardinality of a family $(x_i)_{i\in\mathcal{I}}$ we will mean the cardinality of the set $\{x_i | i \in \mathcal{I}\}$. Also, by the union of two families $(x_i)_{i\in\mathcal{I}}$ and $(y_j)_{j\in\mathcal{J}}$ we will understand the family $(z_k)_{k\in\mathcal{I}\coprod\mathcal{J}}$, indexed by the coproduct $\mathcal{I}\coprod\mathcal{J}$ in the category of sets where, if $\kappa_1 : \mathcal{I} \to \mathcal{I}\coprod\mathcal{J}$ and $\kappa_2 : \mathcal{J} \to \mathcal{I}\coprod\mathcal{J}$ are the natural injections, then $z_{\kappa_1(i)} := x_i$ for all $i \in \mathcal{I}$ and $z_{\kappa_2(j)} := y_j$ for all $j \in \mathcal{J}$.

In order to check parts (e) and (f), based on part (d), Inequality (2.7) and Remark 2.2.9(a) it suffices to check that $\mathbb{I}_{g,\mathrm{ord}}$ contains $\frac{g(g+1)}{2} + 1$ elements that are ϕ-algebraically independent. To check this we use the fact that, by Corollary 4.2.19(a) of Sect. 4.2, the ring $\mathbb{H}_g$ contains homogeneous polynomials $F_1, \ldots, F_{\frac{g(g+1)}{2}}$ of degree gs, for some $s \in \mathbb{N}$, with coefficients in $\mathbb{Z}$, such that the family

$$(\sigma^j F_i)_{(i,j)\in\{1,\ldots,\frac{g(g+1)}{2}\}\times(\mathbb{N}\cup\{0\})}$$

is algebraically independent over $\mathbb{Q}$, therefore over K. In particular, σ and τ act as the identity on the coefficients of these polynomials. Hence, for each integer $N \geq 0$, the family

$$((F_i^{\sigma^j})^{\diamond})_{(i,j)\in\{1,\ldots,\frac{g(g+1)}{2}\}\times\{0,\ldots,N\}}$$

of elements in $\mathbb{I}_{g,\mathrm{ord}}$ is algebraically independent over K. By Proposition 3.6.19(b), for each integer $j \geq 1$, we have

$$(F_i^{\sigma^j})^\diamond = (\det(f^\partial))^{2s(1+\phi+\cdots+\phi^{j-1})} \cdot ((F_i)^\diamond)^{\phi^j}. \tag{3.57}$$

Thus the family of elements of weight $-2s$

$$\left((\det(f^\partial))^{2s(1+\phi+\cdots+\phi^{j-1})} \cdot ((F_i)^\diamond)^{\phi^j}\right)_{(i,j)\in\{1,\ldots,\frac{g(g+1)}{2}\}\times\{0,\ldots,N\}} \tag{3.58}$$

is algebraically independent over K. By weight considerations one easily sees that the union of Family (3.58) with Family

$$(\det(f^\partial)^{\phi^j})_{j\in\{0,\ldots,N-1\}} \tag{3.59}$$

is algebraically independent over K. We claim that the union of Family (3.59) with Family

$$(((F_i)^\diamond)^{\phi^j})_{(i,j)\in\{1,\ldots,\frac{g(g+1)}{2}\}\times\{0,\ldots,N\}} \tag{3.60}$$

is algebraically independent over K. This follows from the following two facts.

(i) The K-algebra generated by the members of the union of Families (3.58) and (3.59) is contained in the K-algebra generated by the members of the union of Families (3.60) and (3.59).

(j) The cardinality of the union of Families (3.58) and (3.59) is the same as the cardinality of the union of Families (3.60) and (3.59).

Hence Family

$$(b_i)_{i\in\{0,\ldots,\frac{g(g+1)}{2}\}} \tag{3.61}$$

where

$$b_0 := \det(f^\partial),\ b_1 := (F_1)^\diamond, \ldots, b_{\frac{g(g+1)}{2}} := (F_{\frac{g(g+1)}{2}})^\diamond,$$

is ϕ-algebraically independent and parts (e) and (f) follow.

To check part (g), we note that by Corollary 4.2.19 (b) and (d) of Sect. 4.2, the field $\mathrm{Frac}(\mathbb{H}^3_{g,\mathrm{tot},\mathbb{K}})$ contains elements $G_1, \ldots, G_{N_3}$ such that Family

$$(G_i^{\sigma^j})_{(i,j)\in\{1,\ldots,N_3\}\times(\mathbb{N}\cup\{0\})} \tag{3.62}$$

generates the field $\mathrm{Frac}(\mathbb{H}_{g,\mathrm{tot},\mathbb{K}})$ over $\mathbb{K}$. By enlarging $N_3 \in \mathbb{N}$ we can assume that $G_1, \ldots, G_{N_3}$ are homogeneous elements of $\mathbb{H}^3_{g,\mathrm{tot},\mathbb{Q}}$. Then Family (3.62) generates the field $\mathrm{Frac}(\mathbb{H}_{g,\mathrm{tot},\mathbb{Q}})$ over $\mathbb{Q}$ and hence the field $\mathrm{Frac}(\mathbb{H}_{g,\mathrm{tot}})$ over K. As σ acts trivially on the coefficients of G_i we get that Eq. (3.57) holds with F_i replaced by G_i. We conclude that the ordinary Hecke covariant Siegel δ-modular forms

$$\det(f^\partial), (G_1)^\diamond, \ldots, (G_{N_3})^\diamond \in \mathbb{I}^4_{g,\mathrm{ord}}$$

and their images via ϕ^j, $j \geq 0$, generate the field $\mathrm{Frac}(\mathbb{I}_{g,\mathrm{ord}})$ over K.

Part (h) follows from the fact that $\det(f^r_{g,\text{crys}})$ equals a W-power of $\det(f^\partial_{g,\text{crys}})$ times $\det(f^{[r]})$ and thus (see Corollary 3.6.18(d) and Display (3.44)) times a $\mathbb{Z}$-linear combination of elements of the form $\Theta_{m_0,\dots,m_r}(f^{\langle 1\rangle},\dots,f^{\langle r\rangle})$. □

Corollary 3.6.30 *We have an inequality*

$$\text{tr.deg}_k(\mathbb{I}^r_{g,ord}\otimes_R k)\leq r\frac{g(g+1)}{2}-g^2+1+r.$$

Proof As p is a prime element in the ring $\mathbb{I}^r_{g,\text{ord}}$ (see Corollary 3.5.2(a)), it follows by Cohen's 'dimension inequality' in [Ma], Thm. 15.5 applied to the extension $R\subset\mathbb{I}^r_{g,\text{ord}}$ and to the prime ideal $p\mathbb{I}^r_{g,\text{ord}}$ that $\text{tr.deg}_k(\mathbb{I}^r_{g,\text{ord}}\otimes_R k)\leq\text{tr.deg}_R(\mathbb{I}^r_{g,\text{ord}})$. We conclude by Theorem 3.6.29(b). □

Corollary 3.6.31 *The following properties hold.*

(a) *For each integer $r\geq 1$, the K-algebra $\mathbb{I}^r_{g,ord}\otimes_R K$ is finitely generated, Cohen–Macauley, unique factorization domain, hence Gorenstein, and moreover $\mathbb{I}^r_{g,ord}$ is a unique factorization domain.*

(b) *If g is odd, then for all $w\in W$ with $deg(w)$ odd we have $\mathbb{I}^r_{g,ord}(w)=0$.*

(c) *If g is even, then there exists $r\in\{0,1,2,3,4\}$ and $w\in W(r)$ such that $deg(w)$ is odd and $\mathbb{I}^r_{g,ord}(w)\neq 0$.*

(d) *We have $\mathbb{I}^r_{g,ord}(w)=0$ for all $w\in W$ with $deg(w)=-1$.*

(e) *Let $w\in W(r)$ with $deg(w)=0$, so $w\in(\phi-1)W$. Then the R-module $\mathbb{I}^r_{g,ord}(w)$ is free of rank 1 with basis $\det(f^\partial_{g,crys})^{\frac{w}{\phi-1}}$.*

(f) *All nonzero elements in $\mathbb{I}_{g,ord}(-2)\otimes_R K$ generate prime ideals in the ring $\mathbb{I}_{g,ord}\otimes_R K$.*

(g) *Every two K-linearly independent elements in $\mathbb{I}_{g,ord}(-2)\otimes_R K$ generate two distinct ideals in $\mathbb{I}_{g,ord}\otimes_R K$.*

Proof As we have a K-algebra isomorphism $\mathbb{H}^{r-1,\text{torus}}_{g,\text{tot}}\otimes_R K\simeq\mathbb{I}^r_{g,\text{ord}}\otimes_R K$ (see Theorem 3.6.29(a)) to prove the K-algebra statement of part (a) it suffices to show that $\mathbb{H}^{r-1}_{g,\text{tot}}\otimes_R K$ has all the desired properties and this follows from Lemma 4.2.3.

To prove that $\mathbb{I}^r_{g,\text{ord}}$ is a unique factorization domain, we first recall that p is a prime element of $\mathbb{I}^r_{g,\text{ord}}$ (see Corollary 3.5.2(b)) and we will use this to check that we have an identity $\mathbb{I}^r_{g,\text{ord}}=\mathbb{I}^r_{g,\text{ord}}[\frac{1}{p}]\cap(\mathbb{I}^r_{g,\text{ord}})_{(p)}$ between subrings of $\text{Frac}(\mathbb{I}^r_{g,\text{ord}})$. The inclusion '$\subset$' is clear and for the reversed inclusion we consider an element $x\in\mathbb{I}^r_{g,\text{ord}}[\frac{1}{p}]\cap(\mathbb{I}^r_{g,\text{ord}})_{(p)}$. Let $i\in\mathbb{N}\cup\{0\}$ be the smallest integer such that for $y:=p^ix$ we have $y\in\mathbb{I}^r_{g,\text{ord}}$. It suffices to show that the assumption that $i>0$ leads to a contradiction. As $x\in(\mathbb{I}^r_{g,\text{ord}})_{(p)}$, there exists $a\in\mathbb{I}^r_{g,\text{ord}}\setminus p\mathbb{I}^r_{g,\text{ord}}$ such that $z:=sx\in\mathbb{I}^r_{g,\text{ord}}$. From the identity $p^iz=sy$ between elements of $\mathbb{I}^r_{g,\text{ord}}$, as $i>0$ we get that p divides sy. So, as p is a prime element of $\mathbb{I}^r_{g,\text{ord}}$ which does not divide s, we get that p divides y and therefore $p^{i-1}x=\frac{y}{p}\in\mathbb{I}^r_{g,\text{ord}}$ which contradicts the minimality property of $i>0$.

As $\mathbb{I}^r_{g,\mathrm{ord}}[\frac{1}{p}] \simeq \mathbb{I}^r_{g,\mathrm{ord}} \otimes_R K$ is a unique factorization domain, it is a Krull ring. Moreover, $(\mathbb{I}^r_{g,\mathrm{ord}})_{(p)}$ is a discrete valuation ring (see Corollary 3.5.2(b)). From the last two sentences and the prior paragraph we get that $\mathbb{I}^r_{g,\mathrm{ord}}$ is a Krull ring (see [Sa], Cor. (a) after Prop. 4.1). Based on this and [Sa], Cor. after Thm. 6.3 we get that $\mathbb{I}^r_{g,\mathrm{ord}}$ is a unique factorization domain. Thus part (a) holds.

Part (b) follows from Lemma 4.2.3(a) of Sect. 4.2.2, part (c) follows from Corollary 4.2.19(c) of Sect. 4.2.11 and part (d) follows from Proposition 4.2.6 of Sect. 4.2.4, these references being applied over an algebraic closure of K.

Part (e) follows directly from the proof of Theorem 3.6.29(a).

To check part (f), let $f \in \mathbb{I}^r_{g,\mathrm{ord}}(-2) \otimes_R K$ be a nonzero element. By the unique factorization property in part (a), it is enough to check that f is irreducible in $\mathbb{I}^r_{g,\mathrm{ord}} \otimes_R K$. But if f is reducible in $\mathbb{I}^r_{g,\mathrm{ord}} \otimes_R K$, it is easy to see that we have a product decomposition $f = f'f''$ with $f' \in \mathbb{I}^r_{g,\mathrm{ord}}(w') \otimes_R K$ and $f'' \in \mathbb{I}^r_{g,\mathrm{ord}}(w'') \otimes_R K$ for some $w', w'' \in W$ and f', f'' nonunits of $\mathbb{I}^r_{g,\mathrm{ord}} \otimes_R K$. Therefore we have $\deg(w') + \deg(w'') = -2$. As $\deg(w')$ and $\deg(w'')$ are ≤ 0 (see Corollary 3.6.10) it follows by part (d) that either $\deg(w')$ or $\deg(w'')$ is 0. But for $\deg(w) = 0$ the K-vector space $\mathbb{I}_{g,\mathrm{ord}}(w) \otimes_R K$ is one dimensional with basis an invertible element of $\mathbb{I}_{g,\mathrm{ord}} \otimes_R K$, see part (e). So either f' or f'' is invertible in $\mathbb{I}^r_{g,\mathrm{ord}} \otimes_R K$, a contradiction.

Finally part (g) follows from the fact that $\mathbb{I}_{g,\mathrm{ord}}(0) \otimes_R K = K$. □

Remark 3.6.32 **(a)** Now we can explicitly compute the Serre–Tate expansions of all elements of the R-module $\mathbb{I}^r_{g,\mathrm{ord}}(w)$, $w \in W$. Indeed, by the proof of Theorem 3.6.29(a), each element in $\mathbb{I}^r_{g,\mathrm{ord}}(w)$ is a K-multiple of a product

$$f := F^{\diamond} \cdot \det(f^{\partial})^{w'} \tag{3.63}$$

for some $F \in \mathbb{H}^{r-1}_g(s)$, where $s := -\deg(w)/2$ and $w' := \frac{w+2s}{\phi-1} \in W$. On the other hand it follows by Theorems 3.6.16(a) and 3.6.17(a) that the Serre–Tate expansion of f is given by the formula:

$$\mathcal{E}_{A_0,\theta_0,e}(f) = F(\Psi, \Psi^{\phi}, \dots, \Psi^{\phi^{r-1}}).$$

(b) For $w \in W$, each element of $\mathbb{I}_{g,\mathrm{ord}}(w)$ vanishes at every ordinary CM_R-point of $\mathcal{X}_{\mathrm{ord}}$: it is an f as in Eq. (3.63) and $F^{\diamond}$ vanishes at every ordinary CM_R-point by Proposition 3.6.19(a) and Remark 3.4.3. In particular, for all linear systems $\mathbf{f}$ of weight w in $\mathbb{I}_{g,\mathrm{ord}}$ the set of $\mathbf{f}$-unstable points $\mathcal{X}_{\mathrm{ord}}(R)^{\mathrm{u}}_{\mathbf{f}}$ in $\mathcal{X}_{\mathrm{ord}}(R)$ contains the set of ordinary CM_R-points. More generally, we can describe the set of $\mathbf{f}$-unstable points $\mathcal{X}_{\mathrm{ord}}(R)^{\mathrm{u}}_{\mathbf{f}}$ in terms of the unstable locus of $\mathbf{SL}_g$ acting on spaces of multiple quadratic forms as follows. The K-algebra $\mathbb{H}^{r-1}_{g,\mathrm{tot}} \otimes_R K$ is known (by classical geometric invariant theory) to be finitely generated. Let $s \in \frac{1}{2}\mathbb{N}$ be such that if $F_1, \dots, F_{D_0(g,r-1,s)}$ is a K-basis of the homogeneous component $\mathbb{H}^{r-1}_g(s)_K$, then the radical ideal generated by $F_1, \dots, F_{D_0(g,r-1,s)}$ in $\mathbb{H}^{r-1}_g \otimes_R K$ contains a set of generators of the K-algebra $\mathbb{H}^{r-1}_g \otimes_R K$.

We can assume $F_1, \dots, F_{D_0(g,r-1,s)} \in \mathbb{H}_g^r(s)$, i.e., the coefficients of the F_is are in R. With the notation of Sect. 2.3.4, let $\mathbf{V}_{g,R}^r := \mathrm{Spec}(R[T, T', \dots, T^{(r-1)}])$ and we consider the closed subscheme $\Sigma_{\mathbf{SL}_g}^r$ of $\mathbf{V}_{g,R}^r$ defined by $F_1, \dots, F_{D_0(g,r-1,s)}$; for an algebraic closure $\mathbb{K}$ of K, the $\mathbb{K}$-valued points of $\Sigma_{\mathbf{SL}_g}^r$, are precisely the unstable points for the action of $\mathbf{SL}_g(\mathbb{K})$ on $\mathbf{V}_{g,R}^r(\mathbb{K})$ in the sense of [MFK], App. to Ch. I, Sect. B, Def. By the proof of Theorem 3.6.29(a), a basis of the K-linear space $\mathbb{I}_{g,\mathrm{ord}}^r(-2s)$ is given by the forms

$$F_1^{\diamond}, \dots, F_{D_0(g,r-1,s)}^{\diamond}.$$

Let $X \subset \mathcal{X}_{\mathrm{ord}}$ be an affine open subscheme with $\overline{X}$ connected and such that the universal polarized abelian scheme (A_X, θ_X) has a column basis ω_X of 1-forms on A_X and consider the map

$$f^{\langle 1,\dots,r\rangle} : X(R) \to \mathbf{V}_{g,R}^r(R)$$

which maps $P \in X(R)$ to

$$(f^{\langle 1\rangle}(P), \dots, f^{\langle r\rangle}(P)) := (f^{\langle 1\rangle}(A_P, \theta_P, \omega_P, R), \dots, f^{\langle r\rangle}(A_P, \theta_P, \omega_P, R))$$

where A_P, θ_P, ω_P over $\mathrm{Spec}(R)$ are as usual induced by A_X, θ_X, ω_X via base change by $P : \mathrm{Spec}(R) \to X$. Then, one can check, by directly using the definitions, that the set $\mathcal{X}_{\mathrm{ord}}(R)_{\mathbf{f}}^{\mathrm{u}}$ of $\mathbf{f}$-unstable points for $\mathbf{f}$ defined by an arbitrary R-basis of $\mathbb{I}_{g,\mathrm{ord}}^r(-2s)$, such as the one above, is given by the formula

$$\mathcal{X}_{\mathrm{ord}}(R)_{\mathbf{f}}^{\mathrm{u}} = (f^{\langle 1,\dots,r\rangle})^{-1}(\Sigma_{\mathbf{SL}_g}^r(R)).$$

(c) In the proof of Theorem 3.6.29(e) and (f) we considered elements of $\mathbb{I}_{g,\mathrm{ord}}$, $b_0 = \det(f^{\partial})$ of weight $\phi - 1$ and $b_i = (F_i)^{\diamond}$ for $i \in \{1, \dots, \frac{g(g+1)}{2}\}$ of weight $-2s$ for some $s \in \mathbb{N}$, and we checked that the Family (3.61) is ϕ-algebraically independent over R. Based on this, the argument in the proof of Remark 2.2.9(a) implies that the family

$$(\tilde{b}_i)_{i \in \{0,\dots,\frac{g(g+1)}{2}\}}$$

of elements of the same weight $-sg(g+1)$,

$$\tilde{b}_0 := b_1 b_2 \cdots b_{\frac{g(g+1)}{2}}, \quad \tilde{b}_1 := b_0^{2s} b_1^{\phi} b_2 \cdots b_{\frac{g(g+1)}{2}}, \dots, \tilde{b}_{\frac{g(g+1)}{2}} := b_0^{2s} b_1 b_2 \cdots b_{\frac{g(g+1)}{2}}^{\phi},$$

is ϕ-algebraically independent.

The following is a direct consequence of Lemma 3.6.3 and Theorem 3.6.29(a).

Corollary 3.6.33 *We have a $K[\Theta_0]$-algebra isomorphism*

$$K[\Theta_0, z_0, z_0^{-1}] \simeq \mathbb{I}_{g,ord}^1 \otimes_R K.$$

that maps z_0 *to* $\det(f^{\partial}_{g,crys})$ *and that extends to a* $K[\Theta_0, \ldots, \Theta_g]$*-algebra isomorphism*

$$K[\Theta_0, \ldots, \Theta_g, z_0, z_1, \frac{1}{z_0 z_1}] \simeq \mathbb{I}^2_{g,ord} \otimes_R K,$$

which maps z_1 *to* $\det(f^{\partial}_{g,crys})^{\phi}$.

The structure of $\mathbb{I}^r_{g,\mathrm{ord}} \otimes_R K$ is more complicated as $r \geq 3$ increases. A 'good picture' of what happens in this case can be obtained by combining Theorem 3.6.29 (a) with the results of Sects. 4.2 and 4.3; we leave this to the reader.

In view of Theorem 3.6.29(a) (and the work in Sect. 4.2) the structure of the R-algebra $\mathbb{I}^r_{g,\mathrm{ord}}$ is more or less well understood. By contrast, the structure of the R-algebras $\mathbb{I}^r_g$ remains somewhat mysterious. We will address these rings in the next two sections. Before that, we record the following generalization of Lemma 3.4.7(a).

Corollary 3.6.34 *We have* $\mathbb{I}^1_g(\phi - 1) = \mathbb{I}^1_g(1 - \phi) = 0$ *and thus* $\mathbb{I}_g \neq \mathbb{I}_{g,ord}$.

Proof We only show that the assumption $\mathbb{I}^1_g(\phi - 1) \neq 0$ leads to contradiction; the case of $\mathbb{I}^1_g(1 - \phi)$ is similar (in fact easier) and is left to the reader. As the R-module $\mathbb{I}^1_{g,\mathrm{ord}}(\phi - 1)/\mathbb{I}^1_g(\phi - 1)$ is torsion free (see Proposition 3.5.1(b)), from Corollary 3.6.31(e) and our assumption we get that the following relation holds: $\det(f^{\partial}_{g,\mathrm{crys}}) \in \mathbb{I}^1_g(\phi - 1) = \mathbb{I}^1_{g,\mathrm{ord}}(\phi - 1)$.

Let $Y \subset Y_1(N)$ be an affine open subset of the affine modular curve of level $\Gamma_1(N)$ over Spec(R) with $N \geq 3$ not divisible by p, such that the reduction modulo p of Y is not contained in the ordinary locus of the reduction modulo p of $Y_1(N)$.

Let E_Y be the universal elliptic curve over Y, let θ_{E_Y} be the Abel–Jacobi map (polarization) of E_Y and let ω_{E_Y} be a column basis of 1-forms on E_Y.

Let $(E_{1,R}, \theta_1), \ldots, (E_{g-1,R}, \theta_{g-1})$ be ordinary elliptic curves over Spec(R) endowed with their Abel–Jacobi maps (polarizations) such that by evaluating we have $f^{\partial}_{1,\mathrm{crys}}(E_{i,R}, \theta_i, \omega_i, R) \in R^{\times}$ for $i \in \{1, \ldots, g-1\}$, where each ω_i is a column basis of 1-forms on $E_{i,R}$.

Let $(E_{i,Y}, \theta_{i,Y}, \omega_{i,Y}) := (E_{i,R}, \theta_i, \omega_i) \times_{\mathrm{Spec}(R)} Y$. Let the following quadruple $(A_Y, \theta_{A_Y}, \omega_{A_Y}) \in \mathbf{M}_g(\mathcal{O}(Y))$ be defined by

$$(A_Y, \theta_{A_Y}) := (E_Y, \theta_{E_Y}) \times_Y (E_{1,Y}, \theta_{1,Y}) \times_Y \cdots \times_Y (E_{g-1,Y}, \theta_{g-1,Y})$$

and by the column basis of 1-forms on A_Y whose 11 entry is the pullback of ω_Y and whose $i1$ entry for $i \in \{2, \ldots, g\}$ is the pullback of $\omega_{i-1,Y}$. We have that

$$\det(f^{\partial}_{g,\mathrm{crys}})(A_Y, \theta_{A_Y}, \omega_{A_Y}, \mathcal{O}^{\infty}(Y)) \in \mathcal{O}(J^1(Y)).$$

On the other hand the above element equals

$$f^{\partial}_{1,\text{crys}}(E_Y, \theta_{E_Y}, \omega_{E_Y}, \mathcal{O}^{\infty}(Y)) \cdot \prod_{i=1}^{g-1} f^{\partial}_{1,\text{crys}}(E_{i,R}, \theta_i, \omega_i, R).$$

So we have

$$f_Y := f^{\partial}_{1,\text{crys}}(E_Y, \theta_{E_Y}, \omega_{E_Y}, \mathcal{O}^{\infty}(Y)) \in \mathcal{O}(J^1(Y)).$$

For a prime ℓ that does not divide pN, let $Y_1(N, \ell)$ be the curve over Spec(R) that parametrizes isogenies of degree ℓ between elliptic curves with level $\Gamma_1(N)$ structure that respect the level $\Gamma_1(N)$ structures. We consider the two canonical projections $\pi_1, \pi_2 : Y_1(N, \ell) \to Y_1(N)$ and the intersection $Y_\ell := \pi_1^{-1}(Y) \cap \pi_2^{-1}(Y)$.

Let $Y_{\text{ord}} \subset Y$ and $Y_{\ell,\text{ord}} \subset Y_\ell$ be the open subschemes which over K are isomorphisms and whose fibers over k are ordinary loci. For $i \in \{1, 2\}$ let $\pi_i^* E_Y/Y_\ell$ be the pullback of E_Y/Y via $\pi_i^* : Y_\ell \to Y$ and consider the 1-forms $\pi_i^*\omega_{E_Y}$ on $\pi_i^* E_Y$. Let

$$\lambda := \frac{u^*(\pi_1^*\omega_{E_Y})}{\pi_2^*\omega_{E_Y}} \in \mathcal{O}(Y_\ell)^\times$$

with $u : \pi_2^* E_Y \to \pi_1^* E_Y$ the 'universal isogeny' over Y_ℓ; hence $\lambda^{\phi-1} \in \mathcal{O}(J^1(Y_\ell))^\times$. By the Hecke covariance of $f^{\partial}_{1,\text{crys}}$ we have the following equality in $\mathcal{O}(J^1(Y_{\ell,\text{ord}}))$:

$$\pi_2^* f_Y = \lambda^{\phi-1} \cdot \pi_1^* f_Y. \tag{3.64}$$

By the injectivity of the R-algebra homomorphism $\mathcal{O}(J^1(Y_\ell)) \to \mathcal{O}(J^1(Y_{\ell,\text{ord}}))$, Eq. (3.64) holds in $\mathcal{O}(J^1(Y_\ell))$. The existence of such a function f_Y in $\mathcal{O}(J^1(Y))$ contradicts [Bu05], Thm. 8.83 3). □

Remark 3.6.35 It may be reasonable to expect that $\mathbb{I}^r_g(w) = 0$ for all $w \neq 0$ with $\deg(w) = 0$. By Corollary 3.6.31(e) this is equivalent to the statement that

$$\det(f^{\partial}_{g,\text{crys}})^{\frac{w}{\phi-1}} \notin \mathbb{I}^r_g(w) \tag{3.65}$$

for all such w. Corollary 3.6.34 gives a partial result in this direction. One can show that Statement (3.65) holds for $g = 1$ provided $w = \sum_{i=0}^r a_i\phi^i \in W$ with $a_0, \ldots, a_r$ integers such that $\sum_{i=0}^r a_i = 0$ and $\sum_{i=0}^r a_i p^i < 0$. This follows from the fact that the reduction modulo p of $f^{\partial}_{1,\text{crys}}$ is the Hasse invariant, see [Bu05], Prop. 8.57. On the other hand note that we do not know, for instance, if Statement (3.65) holds even for $g = 1$ and $w = n(\phi - 1)$, with $n \geq 2$ an integer. This is so as, for $n \geq 2$, from the fact that $f^{\partial}_{1,\text{crys}} \notin \mathbb{I}^1_1(\phi - 1)$ it does not follow a priori that $(f^{\partial}_{1,\text{crys}})^n \notin \mathbb{I}^1_1(n(\phi - 1))$. Concretely, the following phenomenon occurs. Let S be a smooth R-algebra, let $h \in S\backslash pS$, and let $f \in \widehat{S_h}$ be such that $f^n \in \widehat{S}$ for some $n \geq 2$; then in general it does not follow that $f \in \widehat{S}$. We include a simple example. Let $S := R[x]$ with x an indeterminate, let $n \geq 2$ be an integer that does not divide p, and let $h := x$. The equation $y^n = x^{n-1}(x - p)$ in y has a solution

$$z := x - p/n + \sum_{m=2}^{\infty} p^m z_m \in \widehat{R[x, x^{-1}]},$$

with each $z_m \in R[x, x^{-1}]$. We have $z^n = x^{n-1}(x - p) \in \widehat{R[x]}$ but $z \notin \widehat{R[x]}$.

3.7 Comparison Maps: The Nonordinary Case

In this section we define some further new modules and algebras and we introduce some further 'comparison maps', $\triangle$, $\Box$, between them. We will use these maps, in conjunction with our Serre–Tate expansion maps $\mathcal{E}$ and with the geometric invariant theory of multiple endomorphisms (see Sects. 4.1.1, 4.1.5 and 4.1.6), to derive our basic estimates from below for the dimensions of K-vector spaces of Hecke covariant Siegel δ-modular forms.

3.7.1 *The Maps* $\Box$

Let $n \in \mathbb{N}$ and let $X_1 = (X_{1,jl})_{1 \leq j,l \leq g}, \ldots, X_n = (X_{n,jl})_{1 \leq j,l \leq g}$ be universal $g \times g$ matrices whose entries are indeterminates. Let B be an integral domain which is a $\mathbb{Z}_p$-algebra. Consider the left action of $\mathbf{GL}_n(B)$ on the polynomial B-algebra $B[X_1, \ldots, X_n] := B[X_{i,jl} | 1 \leq i \leq n, 1 \leq j, l \leq g]$ defined by the rule:

$$(M, F) \mapsto F(M^{-1} X_1 M, \ldots, M^{-1} X_n M).$$

We denote by $\mathbb{H}^n_{g,\mathrm{conj},B}$ the B-subalgebra of $B[X_1, \ldots, X_n]$ formed by all $\mathbf{GL}_g(B)$-invariant polynomials. Clearly

$$\mathbb{H}^n_{g,\mathrm{conj},D^{-1}B} = \mathbb{H}^n_{g,\mathrm{conj}} \otimes_B D^{-1} B$$

for each multiplicative set $D \subset B$. Also

$$\mathbb{H}^n_{g,\mathrm{conj},B'} = \mathbb{H}^n_{g,\mathrm{conj}} \otimes_B B'$$

for each extension $B \subset B'$ of infinite fields. Moreover, we have an identity

$$\mathbb{H}^n_{g,\mathrm{conj},R} = \mathbb{H}^n_{g,\mathrm{conj}}.$$

For B an algebraically closed field and $n \geq 2$ we have the formula (see Theorem 4.1.2 of Sect. 4.11):

$$\mathrm{tr.deg}_R(\mathbb{H}^n_{g,\mathrm{conj},B}) = D_{\mathrm{conj}}(g, n) := (n-1)g^2 + 1.$$

Also recall that the ‘first main theorem’ in this setting (see Theorem 4.1.20 of Sect. 4.1.6) implies that if B an infinite field, then the B-algebra $\mathbb{H}^n_{g,\mathrm{conj},B}$ is generated by finitely many polynomials of the form

$$\tau_{j;i_1,i_2,\dots,i_N} := c_j(X_{i_1}X_{i_2}\cdots X_{i_N})$$

where $j \in \{1,\dots,g\}$, $N \in \mathbb{N}$, and $i_1,\dots,i_N \in \{1,\dots,n\}$, and where for a matrix $M \in \mathbf{Mat}_g(C)$ with C a ring, we let $c_0(M),\dots,c_g(M) \in C[t]$ be such that

$$\det(t\cdot 1_g - M) = \sum_{j=0}^{g}(-1)^j c_j(M)t^{g-j} \in C[t]$$

is its characteristic polynomial; henceforth $c_0(M) = 1$, $c_1(M)$ is the trace $\mathrm{Tr}(M)$ of M, and $c_g(M) = \det(M)$. The relations between these generators provided by the ‘second main theorem’ (see [P], Introd.) will not be used in what follows.

Note that the natural k-algebra homomorphisms

$$\mathbb{H}^n_{g,\mathrm{conj},R} \otimes_R k \to \mathbb{H}^n_{g,\mathrm{conj},k}$$

are isomorphisms. Indeed they are surjective by the first main theorem and they are injective because for $F \in R[X_1,\dots,X_n]$, if pF is $\mathbf{GL}_g(R)$-invariant, then F itself is $\mathbf{GL}_g(R)$-invariant. By Nakayama’s Lemma applied to R-modules of homogeneous components of $\mathbb{H}^n_{g,\mathrm{conj},R}$ it follows that the latter R-algebra is generated by finitely many elements of the form $\tau_{j;i_1,i_2,\dots,i_N}$; we will refer to this statement as the ‘first main theorem over R’.

More generally, let $x_1,\dots,x_{n+2},\xi_1,\dots,\xi_{n+2}$ be indeterminates and view

$$R[X_1,\dots,X_n,x_1,\xi_1,\dots,x_{n+2},\xi_{n+2}] := R[X_1,\dots,X_n][x_1,\xi_1,\dots,x_{n+2},\xi_{n+2}] \tag{3.66}$$

as a W-graded R-algebra by letting

$$\deg(x_i) := -\phi^{i-1} - \phi^i \ \text{ and } \ \deg(\xi_i) := -1 - \phi^i \ \text{ for all } \ i \in \{1,\dots,n+2\}$$

and

$$\deg(X_{i,jl}) := -1 - \phi^i - \phi^{i+1} - \phi^{i+2} \ \text{ for all } \ i \in \{1,\dots,n\} \ \text{ and all } \ j,l \in \{1,\dots,g\}.$$

Then by definition

$$\mathbb{H}^n_{g,\mathrm{conj}}[x_1,\dots,x_{n+2},\xi_1,\dots,\xi_{n+2}]$$

is a W-graded R-subalgebra of Display (3.66). We will prove the following result.

Theorem 3.7.1 *For all integers $n \geq 2$ there exists a natural homomorphism*

$$\square = \square_n : \mathbb{H}^n_{g,conj}[x_1,\dots,x_{n+2},\xi_1,\dots,\xi_{n+2}] \to \mathbb{I}^{n+2}_g,$$

of W-graded R-algebras whose image satisfies

$$\text{tr.deg}_R(Im(\Box)) \geq (n-1)\frac{g(g+1)}{2} + n + g + 1.$$

The homomorphisms $\Box_n$ in the above theorem will be compatible as $n \geq 2$ varies and will be entirely explicit, as we will see shortly.

Corollary 3.7.2 *The following properties hold.*

(a) *For all integers $r \geq 4$ we have inequalities*

$$(r-3)\frac{g(g+1)}{2} + r + g - 1 \leq \text{tr.deg}_R(\mathbb{I}_g^r) \leq r \cdot \frac{g(g+1)}{2} - g^2 + 1 + r.$$

(b) *We have an inequality*

$$\text{tr.deg}_{\mathbb{I}_g}(\mathbb{I}_{g,ord}) \leq \frac{g(g+1)}{2} + 2.$$

(c) *We have equalities*

$$\text{asy.tr.deg}(\mathbb{I}_g) = \phi\text{−tr.deg}(\mathbb{I}_g) = \frac{g(g+1)}{2} + 1.$$

(d) *We have an equality*

$$\phi\text{−tr.deg}(\mathbb{I}_{g,((p))}) = \frac{g(g+1)}{2}.$$

Proof Part (a) follows from Theorems 3.7.1 and 3.6.29(b).

To check part (b) note that the difference between the upper and the lower bounds for $\text{tr.deg}_R(\mathbb{I}_g^r)$ in part (a) equals $\frac{g(g+1)}{2} + 2$. From this and Theorem 3.6.29(b) we get that

$$\text{tr.deg}_{\mathbb{I}_g^r}(\mathbb{I}_{g,\text{ord}}^r) = \text{tr.deg}_R(\mathbb{I}_{g,\text{ord}}^r) - \text{tr.deg}_R(\mathbb{I}_g^r) \leq \frac{g(g+1)}{2} + 2$$

for all $r \geq 4$. Thus part (b) follows from the part of Remark 2.2.6 after Display (2.8).

The equality $\text{asy.tr.deg}(\mathbb{I}_g) = \frac{g(g+1)}{2} + 1$ in part (c) follows from part (a). The equality ϕ-$\text{tr.deg}(\mathbb{I}_g) = \frac{g(g+1)}{2} + 1$ in part (c) follows from the implication in Display (2.8) plus Theorem 3.6.29(c); alternatively one can use part (b) plus Theorem 3.6.29(c) and the additivity of the ϕ-transcendence degree (see Eq. (2.6)).

Part (d) follows from part (c) and Remark 2.2.9 (a). □

Remark 3.7.3 **(a)** By an argument similar to the one in the proof of Corollary 3.6.30 we have that

$$\text{tr.deg}_k(\mathbb{I}^r_g \otimes_R k) \leq \text{tr.deg}_R(\mathbb{I}^r_g).$$

(b) By Remark 3.6.32 it follows that for all linear systems f weight $w \in W$ in $\mathbb{I}_{g,\text{ord}}$ the set of f-unstable points $\mathcal{X}(R)^{\text{u}}_f$ in $\mathcal{X}(R)$ contains the set of ordinary CM$_R$-points.

3.7.2 Cyclic Products and the Maps Δ

In order to proceed with our proof we need to recall (and slightly extend) the *cyclic product construction* from [BB], Sect. 4.2. Recall our *basic* δ-modular forms $f^r = f^r_{g,\text{crys}} \in \mathbb{I}^r_{gg}(\phi^r, 1)$, $r \in \mathbb{N}$. For integers $a > b \geq 0$ we set

$$f^{ab} := \phi^b f^{a-b} \in \mathbb{I}^a_{gg}(\phi^a, \phi^b) \quad \text{and} \quad f^{ba} := (f^{ab})^{\text{t}} \in \mathbb{I}^a_{gg}(\phi^b, \phi^a). \tag{3.67}$$

Let Cyc^r be the set of all $2s$-tuples $(a_1, a_2, \ldots, a_{2s-1}, a_{2s})$ of arbitrary even length $2s \in 2\mathbb{N}$, such that the following properties hold:

(i) for all $i \in \{1, \ldots, 2s\}$, we have $a_i \in \{0, 1, \ldots, r\}$;

(ii) for all $i \in \{1, \ldots, 2s-1\}$, we have $a_i \neq a_{i+1}$;

(iii) we have $a_{2s} \neq a_1$.

For each $(a_1, a_2, \ldots, a_{2s-1}, a_{2s}) \in \text{Cyc}^r$ we define the *matrix cyclic product*

$$F^{a_1,a_2,\ldots,a_{2s-1},a_{2s}} \in M^r_{gg} = \mathbf{Mat}_g(M^r_g)$$

by the formula

$$F^{a_1,a_2,\ldots,a_{2s-1},a_{2s}} := f^{a_1a_2}(f^{a_3a_2})^* f^{a_3a_4}(f^{a_5a_4})^* \cdots f^{a_{2s-1}a_{2s}}(f^{a_1a_{2s}})^* \tag{3.68}$$

(recall that for a square matrix M, M^* is its adjugate).

Note that for each $S \in \text{Ob}(\widehat{\mathbf{FilRing}_\delta})$ and every triple $(x_1, x_2, u) \in \mathbf{M}'_g(S^0)$, where $x_1 = (A_1, \theta_1, \omega_1), x_2 = (A_2, \theta_2, \omega_2), uu^{\text{t}} = d \in \mathbb{N}\backslash p\mathbb{N}$, the following equality holds (see Eq. (2.28)):

$$F^{a_1,\ldots,a_{2s}}(x_2, S) = d^{-gs}(\det([u]))^{-w}[u]^{\phi^{a_1}} F^{a_1,\ldots,a_{2s}}(x_1, S)([u]^{\phi^{a_1}})^{-1}, \tag{3.69}$$

where $w := -\sum_{i=1}^{2s} \phi^{a_i} \in W$ and $[u] \in \mathbf{GL}_g(R)$ is defined by the usual identity $u^*\omega_2 = [u]\omega_1$. Then, for $j \in \{1, \ldots, g\}$ we define the (scalar) *cyclic products*

$$f_{j;a_1,a_2,\ldots,a_{2s-1},a_{2s}} := c_j(F^{a_1,a_2,\ldots,a_{2s-1},a_{2s}}) \in M^r_g.$$

For $j = 1$ we abbreviate

$$f_{a_1,a_2,\ldots,a_{2s-1},a_{2s}} := f_{1;a_1,a_2,\ldots,a_{2s-1},a_{2s}} := \mathrm{Tr}(F^{a_1,a_2,\ldots,a_{2s-1},a_{2s}}) \in M^r_g. \tag{3.70}$$

Note that, for $a \in \mathbb{N}$ we have

$$f_{0,a} = \mathrm{Tr}(F^{0,a}) = \mathrm{Tr}(f^{0a}(f^{0a})^*) = \mathrm{Tr}(\det(f^{0a}) \cdot 1_g) = g \cdot \det(f^a).$$

Clearly the K-algebra generated by the members of the family

$$(f_{j;a_1,a_2,\ldots,a_{2s-1},a_{2s}})_{(j,(a_1,a_2,\ldots,a_{2s-1},a_{2s}))\in\{1,\ldots g\}\times \mathrm{Cyc}^r}$$

is the same as the K-algebra generated by the members of the family

$$(f_{a_1,a_2,\ldots,a_{2s-1},a_{2s}})_{(a_1,a_2,\ldots,a_{2s-1},a_{2s})\in \mathrm{Cyc}^r}$$

(as this last family is stable under powers); however, the R-algebras generated by the members of these two families are distinct if $p \leq g$. By Eq. (3.69) we have

$$f_{j;a_1,a_2,\ldots,a_{2s-1},a_{2s}} \in \mathbb{I}^r_g(-j(\sum_{i=1}^{2s} \phi^{a_i})). \tag{3.71}$$

On the other hand, for each integer $r \geq 1$ we consider the R-subalgebra

$$R[Q^{(1)}, \ldots, Q^{(r)}] \subset S^r_{\mathrm{for}}$$

generated by the entries of the matrices

$$Q^{(s)} := \Psi^{\phi^{s-1}} \in \mathbf{Mat}_g(S^r_{\mathrm{for}}), \quad s \in \mathbb{N} \tag{3.72}$$

(see Displays (2.35) and (2.36)). The entries of these symmetric matrices that are above and on the diagonal are (distinct and) algebraically independent over R (because, by definition, the same is true for the matrices $\Psi, \delta\Psi, \ldots, \delta^{r-1}\Psi$ and, on the other hand, the K-algebra generated by the entries of $Q^{(1)}, \ldots, Q^{(r)}$ coincides with the K-algebra generated by the entries of $\Psi, \delta\Psi, \ldots, \delta^{r-1}\Psi$); we will view these entries of $Q^{(1)}, \ldots, Q^{(r)}$ as indeterminates over R and also as indeterminates in the ring $R[Q^{(1)}, \ldots, Q^{(r)}]/(p)$.

We consider the reduction modulo p homomorphism

$$\Delta : R[Q^{(1)}, \ldots, Q^{(r)}] \to k[Q^{(1)}, \ldots, Q^{(r)}] := R[Q^{(1)}, \ldots, Q^{(r)}]/(p), \ \ G \mapsto G^{\Delta}.$$

We note that the natural k-algebra homomorphism

$$k[Q^{(1)}, \ldots, Q^{(r)}] \to S^r_{\mathrm{for}} \otimes_R k = k[[T]][T', \ldots, T^{(r)}]$$

is not injective.

For each $(a_1, a_2, \ldots, a_{2s-1}, a_{2s}) \in \mathrm{Cyc}^r$ we define a matrix with entries in the ring $R[Q^{(1)}, \ldots, Q^{(r)}]$ by the formula

$$Y_{a_1,a_2,\ldots,a_{2s-1},a_{2s}} := Q^{(m_1)}(Q^{(m_2)})^* Q^{(m_3)}(Q^{(m_4)})^* \ldots Q^{(m_{2s-1})}(Q^{(m_{2s})})^*,$$

where

$$m_1 := \max\{a_1, a_2\}, \quad m_2 := \max\{a_2, a_3\}, \ldots, m_{2s} := \max\{a_{2s}, a_1\}.$$

For each such matrix and $j \in \{1, \ldots, g\}$ we define a polynomial

$$y_{j;a_1,a_2,\ldots,a_{2s-1},a_{2s}} := c_j(Y_{a_1,a_2,\ldots,a_{2s-1}\ a_{2s}}) \in R[Q^{(1)}, \ldots, Q^{(r)}]$$

and we set

$$y_{a_1,a_2,\ldots,a_{2s-1},a_{2s}} := y_{1;a_1,a_2,\ldots,a_{2s-1},a_{2s}}.$$

Lemma 3.7.4 *For each $(a_1, a_2, \ldots, a_{2s-1}, a_{2s}) \in \mathrm{Cyc}^r$ we have*

$$(\mathcal{E}_{A_0,\theta_0,e}(f_{j;a_1,a_2,\ldots,a_{2s-1},a_{2s}}))^\Delta = y^\Delta_{j;a_1,a_2,\ldots,a_{2s-1},a_{2s}} \neq 0.$$

Proof This follows directly from Theorem 3.6.16(a). □

With $\Lambda_0, \ldots, \Lambda_r$ as indeterminates, let the polynomial algebras

$$R[Q^{(1)}, \ldots, Q^{(r)}, \Lambda_0, \ldots, \Lambda_r] := R[Q^{(1)}, \ldots, Q^{(r)}][\Lambda_0, \ldots, \Lambda_r],$$

$$k[Q^{(1)}, \ldots, Q^{(r)}, \Lambda_0, \ldots, \Lambda_r] := k[Q^{(1)}, \ldots, Q^{(r)}][\Lambda_0, \ldots, \Lambda_r].$$

We extend the homomorphism Δ above to a reduction modulo p homomorphism between these polynomial algebras

$$\Delta : R[Q^{(1)}, \ldots, Q^{(r)}, \Lambda_0, \ldots, \Lambda_r] \to k[Q^{(1)}, \ldots, Q^{(r)}, \Lambda_0, \ldots, \Lambda_r],$$

by setting $\Lambda_j^\Delta := \Lambda_j$.

Remark 3.7.5 For $F_1, \ldots, F_m \in R[Q^{(1)}, \ldots, Q^{(r)}, \Lambda_0, \ldots, \Lambda_r]$, if $F_1^\Delta, \ldots, F_m^\Delta$ are algebraically independent in $k[Q^{(1)}, \ldots, Q^{(r)}, \Lambda_0, \ldots, \Lambda_r]$, then $F_1, \ldots, F_m$ are algebraically independent in $R[Q^{(1)}, \ldots, Q^{(r)}, \Lambda_0, \ldots, \Lambda_r]$.

We define

$$z_{j;a_1,a_2,\ldots,a_{2s-1},a_{2s}} := y_{j;a_1,a_2,\ldots,a_{2s-1},a_{2s}} \Lambda^j_{a_1} \Lambda^j_{a_2} \cdots \Lambda^j_{a_{2s-1}} \Lambda^j_{a_{2s}},$$

$$z_{a_1,a_2,\ldots,a_{2s-1},a_{2s}} := z_{1;a_1,a_2,\ldots,a_{2s-1},a_{2s}}.$$

So for $\mathcal{E} = \mathcal{E}_{A_0,\theta_0,e}$ we have

$$(\mathcal{E}(f_{j;a_1,a_2,\dots,a_{2s-1},a_{2s}})\Lambda^j_{a_1}\cdots\Lambda^j_{a_{2s}})^\triangle = z^\triangle_{j;a_1,a_2,\dots,a_{2s-1},a_{2s}}. \tag{3.73}$$

Lemma 3.7.6 *Consider the unique R-algebra homomorphism*

$$\square : R[X_1,\dots,X_n,x_1,\dots,x_{n+2},\xi_1,\dots,\xi_{n+2}] \to M_g^{n+2},\quad F \mapsto F^\square,$$

defined by the rules

$$X_i^\square := F^{0,i,i+1,i+2},\ i \in \{1,\dots,n\},$$

$$x_i^\square := f_{i-1,i}, \qquad i \in \{1,\dots,n+2\},$$

$$\xi_i^\square := f_{0,i}, \qquad i \in \{1,\dots,n+2\}.$$

Then $\square$ induces a W-graded R-algebra homomorphism

$$\square : \mathbb{H}^n_{g,conj}[x_1,\dots,x_{n+2},\xi_1,\dots,\xi_{n+2}] \to \mathbb{I}_g^{n+2}.$$

Proof By the 'first main theorem over R' it is enough to check that for all $i_1,\dots,i_N \in \{1,\dots,g\}$ the element $\tau^\square_{j;i_1,\dots,i_N}$ belongs to $\mathbb{I}_g^{n+2}(w)$, where $w \in \mathbb{Z}[\phi]$ is the sum of the degrees of $X_{i_1},\dots,X_{i_N}$; but this follows from the fact that

$$\tau^\square_{j;i_1,\dots,i_N} = f_{j;0,i_1,i_1+1,i_1+2,0,i_2,i_2+1,i_2+2,\dots,0,i_N,i_N+1,i_N+2}.$$

$\square$

Proof of Theorem 3.7.1. Consider the polynomial k-algebra

$$k[X_1,\dots,X_n,x_1,\dots,x_{n+2}]$$

in the indeterminates $x_1,\dots,x_{n+2}$ and the entries of the matrices $X_1,\dots,X_n$. Consider the k-algebra homomorphism

$$k[X_1,\dots,X_n,x_1,\dots,x_{n+2}] \to k[Q^{(1)},\dots,Q^{(n+2)}],\quad F \mapsto \widetilde{F}, \tag{3.74}$$

defined by

$$\widetilde{X}_i := Q^{(i)}(Q^{(i+1)})^* \in \mathbf{Mat}_g(S^{i+1}_{\text{for}}),\ \ i \in \{1,\dots,n\},$$

$$\widetilde{x}_i := \det(Q^{(i)}),\ \ i \in \{1,\dots,n+2\}.$$

Let $\widetilde{\mathbb{H}^n_{g,\text{conj},k}}$ be the image of $\mathbb{H}^n_{g,\text{conj},k}$ under the k-algebra homomorphism of Display (3.74). For $n \geq 2$, by Theorem 4.1.4 applied to the following correspondance $(Q^{(1)},\dots,Q^{(n+1)}) \longleftrightarrow (Q_0,\dots,Q_n)$ we have

$$\dim(\widetilde{\mathbb{H}^n_{g,\text{conj},k}}) = (n-1)\frac{g(g+1)}{2} + g. \tag{3.75}$$

By Eq. (3.75) there exists a set

$$I \subset \bigcup_{N=1}^{\infty} (\{1, \ldots, g\} \times \{1, \ldots, n\}^N)$$

whose cardinality $|I|$ is given by

$$|I| = (n-1)\frac{g(g+1)}{2} + g \tag{3.76}$$

such that Family

$$(c_j(Q^{(i_1)}(Q^{(i_1+1)})^* \cdots Q^{(i_N)}(Q^{(i_N+1)})^*))_{(j,i_1,\ldots,i_N)\in I}, \tag{3.77}$$

is algebraically independent over k in the ring $k[Q^{(1)}, \ldots, Q^{(n+1)}]$. But the members of Family (3.77) can be written as

$$\frac{c_j(Q^{(i_1)}(Q^{(i_1+1)})^* Q^{(i_1+2)}(Q^{(i_1+2)})^* \cdots Q^{(i_N)}(Q^{(i_N+1)})^* Q^{(i_N+2)}(Q^{(i_N+2)})^*)}{\det(Q^{(i_1+2)})^j \cdots \det(Q^{(i_N+1)})^j \det(Q^{(i_N+2)})^j}$$

which in their turn can be written as

$$\left(\frac{y_{j;0,i_1,i_1+1,i_1+2,0,i_1,i_1+1,i_1+2,\ldots,0,i_N,i_N+1,i_N+2}}{(y_{i_1+1,i_1+2} \cdots y_{i_N+1,i_N+2})^j}\right)^{\triangle}, \quad (j, i_1, \ldots, i_N) \in I.$$

So if we consider the two Families

$$(y^{\triangle}_{i-1,i})_{i\in\{1,\ldots,n+2\}}, \tag{3.78}$$

$$(y^{\triangle}_{j;0,i_1,i_1+1,i_1+2,\ldots,0,i_N,i_N+1,i_N+2})_{(j,i_1,\ldots,i_N)\in I}, \tag{3.79}$$

then the members of Family (3.78)∪(3.79) generate a field extension of k of transcendence degree $\geq |I|$. If we consider Family

$$(\Lambda_i)_{i\in\{0,\ldots,n+2\}}, \tag{3.80}$$

we get that the field extension of k generated by the members of Family (3.78)∪(3.79)∪(3.80) has transcendence degree $\geq |I| + n + 3$. Consider the four Families

$$(z^{\triangle}_{j;0,i_1,i_1+1,i_1+2,\ldots,0,i_N,i_N+1,i_N+2})_{(j,i_1,\ldots,i_N)\in I}, \tag{3.81}$$

$$(z^{\triangle}_{i-1,i})_{i\in\{1,\ldots,n+2\}}, \tag{3.82}$$

$$(z^{\triangle}_{0,i})_{i\in\{2,\ldots,n+2\}}, \tag{3.83}$$

$$(\Lambda_i)_{i\in\{0,n+2\}}. \tag{3.84}$$

We claim that the field extension of k generated by the members of Family (3.81)∪(3.82)∪(3.83)∪(3.84) has transcendence degree $\geq |I|+n+3$. To check this, we note that the field extension of k generated by the members of Family (3.78)∪(3.79)∪(3.80) is contained in the field extension of k generated by the members of Family (3.81)∪(3.82)∪(3.83)∪(3.84); if $i\in\{1,\dots,n+2\}$ then we have $\max\{0,i\}=\max\{i-1,i\}=i$, so $Y_{0,i}=Q^{(i)}(Q^{(i)})^*=Y_{i-1,i}$, hence $y_{0,i}=y_{i-1,i}$, and thus

$$\Lambda_{i-1}=\Lambda_0\frac{z_{i-1,i}}{z_{0,i}}.$$

So the field extension of k generated by the members of Family (3.81)∪(3.82)∪(3.83) has transcendence degree $\geq |I|+n+1$. Consider the three Families

$$(\mathcal{E}(f_{j;0,i_1,i_1+1,i_1+2,\dots,0,i_N,i_N+1,i_N+2})\Lambda_0^j\cdots\Lambda_{i_N+2}^j)_{(j,i_1,\dots,i_N)\in I}, \tag{3.85}$$

$$(\mathcal{E}(f_{i-1,i})\Lambda_{i-1}\Lambda_i)_{i\in\{1,\dots,n+2\}}, \tag{3.86}$$

$$(\mathcal{E}(f_{0,i})\Lambda_0\Lambda_i)_{i\in\{2,\dots,n+2\}}. \tag{3.87}$$

By Eq. (3.73) and Remark 3.7.5 it follows that the field extension of K generated by the members of Family (3.85)∪(3.86)∪(3.87) has transcendence degree $\geq |I|+n+1$. By the Serre–Tate expansion principle (Proposition 3.5.1(b)) the R-algebra homomorphism

$$\bigoplus_{w\leq 0}\mathbb{I}_g^{n+2}(w)\to R[[T,\dots,T^{(n+2)}]][\Lambda_0,\dots,\Lambda_{n+2}]$$

defined by

$$f\mapsto \mathcal{E}(f)\cdot\Lambda_0^{c_0}\cdots\Lambda_{n+2}^{c_{n+2}},\quad f\in\mathbb{I}_g^{n+2}(w),\quad w=-\sum_{j=0}^{n+2}c_j\phi^j\leq 0,$$

is injective. Consider the three Families

$$(f_{j;0,i_1,i_1+1,i_1+2,\dots,0,i_N,i_N+1,i_N+2})_{(j,i_1,\dots,i_N)\in I}, \tag{3.88}$$

$$(f_{i-1,i})_{i\in\{1,\dots,n+2\}}, \tag{3.89}$$

$$(f_{0,i})_{i\in\{2,\dots,n+2\}}. \tag{3.90}$$

The field extension of K generated by the members of Family (3.88)∪(3.89)∪(3.90) has transcendence degree $\geq |I|+n+1$ and we conclude by Eq. (3.76). □

3.7.3 Applications to Some Loci in $\mathcal{A}_{g,1,N,R}$

We end by showing that certain remarkable Zariski dense loci in $\mathcal{A}_{g,1,N,R}$ are contained in zero loci of Hecke covariant Siegel δ-modular forms.

We say that two matrices $M_1, M_2 \in \mathbf{Mat}_g(R)$ *pseudo-commute* if there exists a unit $c \in R^\times$ such that $M_1M_2 = cM_2M_1$. For $(A, \theta, \omega) \in \mathbf{M}_g(R)$, we recall that $\mathrm{End}(A, \theta)^{(p)}$ is the subring of $\mathrm{End}(A)$ generated by all polarized isogenies $u : A \to A$. Consider the map

$$[\ ]_\omega : \mathrm{End}(A, \theta)^{(p)} \to \mathbf{Mat}_g(R)$$

defined by the rule $u \mapsto [u]_\omega$, where the matrix $[u]_\omega$ is uniquely determined by the identity $u^*\omega = [u]_\omega\omega$. We claim that $[\ ]_\omega$ is a ring monomorphism. Indeed, embedding R into the complex field $\mathbb{C}$ we are reduced to checking the claim for R replaced by $\mathbb{C}$; however, in this case, our claim follows from the fact that for each abelian variety A over $\mathbb{C}$, the endomorphism ring $\mathrm{End}(A)$ is the ring of $\mathbb{C}$-linear endomorphisms of $H^0(A(\mathbb{C}), \Omega)^*$ that send the lattice $H_1(A(\mathbb{C}), \mathbb{Z})$ into itself.

Lemma 3.7.7 *Let $F^{a_1,\dots,a_{2s}}$ be an arbitrary matrix cyclic product and recall that $f_{a_1,\dots,a_{2s}} \in \mathbb{I}^r_g(-(\sum_{i=1}^{2s}\phi^{a_i}))$ is its trace (see Display (3.71)). Then for every $(A, \theta, \omega) \in \boldsymbol{M}_g(R)$ the following properties hold.*

(a) *The matrix $F^{a_1,\dots,a_{2s}}(A, \theta, \omega) \in \boldsymbol{Mat}_g(R)$ pseudo-commutes with each matrix of the form $M^{\phi^{a_1}}$ with $M \in Im([\ ]_\omega)$.*

(b) *If $f_{a_1,\dots,a_{2s}}(A, \theta, \omega) \neq 0$, then part (a) holds with 'pseudo-commutes' replaced by 'commutes'.*

Proof Part (a) follows from Eq. (3.69).

Part (b) also follows by taking traces in Eq. (3.69) and using part (a). □

Theorem 3.7.8 *For each integer $s \geq 2$ and every nonnegative distinct integers $a_1, \dots, a_{2s}$ there exists a nonzero δ-modular form*

$$\mathcal{D}^{a_1,\dots,a_{2s}} \in \mathbb{I}_g(-(g^2 - g + 1)\sum_{i=1}^{2s}\phi^{a_i})$$

such that for all $(A, \theta, \omega) \in \boldsymbol{M}_g(R)$, if we have $\mathcal{D}^{a_1,\dots,a_{2s}}(A, \theta, \omega) \neq 0$, then the ring $End(A, \theta)^{(p)}$ is commutative.

Proof For a monic polynomial $C(t) = t^g + c_1t^{g-1} + \dots + c_g \in B[t]$ with coefficients in a ring B let $\mathrm{Disc}^0_g(C(t)) := \mathrm{Disc}_g(1, c_1, \dots, c_g) \in B$, where Disc_g is the discriminant of the generic binary (i.e., in 2 indeterminates) form of degree g. Also for a matrix $M \in \mathbf{Mat}_g(B)$ we let $\chi_M(t) := \det(t \cdot 1_g - M) \in B[t]$ be its characteristic polynomial. We can assume that $a_1 < \dots < a_{2s}$.

Define

$$D^{a_1,\ldots,a_{2s}} := \mathrm{Disc}_g^0(\chi_{F^{a_1,\ldots,a_{2s}}}(t)) \in M_g$$

where $F^{a_1,\ldots,a_{2s}}$ is the corresponding matrix cyclic product. Note that the coefficients $c_1, \ldots, c_g$ of $\chi_{F^{a_1,\ldots,a_{2s}}}(t)$ belong to M_g and, by Newton's formulas, for each $j \in \{1, \ldots, g\}$, c_j multiplied by a suitable power of p is a weighted homogeneous polynomial of degree j, with integer coefficients, in the forms $\mathrm{Tr}((F^{a_1,\ldots,a_{2s}})^a)$, $a \in \{1, \ldots, j\}$, with respect to the weights $1, \ldots, j$. Thus for each j we have $c_j \in \mathbb{I}_g(-j\sum_{i=1}^{2s}\phi^{a_i})$. Also Disc_g^0 is a weighted homogeneous polynomial of degree $g(g-1)$, with respect to the weights $1, \ldots, g$. Therefore, it follows that $D^{a_1,\ldots,a_{2s}} \in \mathbb{I}_g(-g(g-1)\sum_{i=1}^{2s}\phi^{a_i})$. Set

$$\mathcal{D}^{a_1,\ldots,a_{2s}} := f_{a_1,\ldots,a_{2s}} D^{a_1,\ldots,a_{2s}}.$$

Note that, by Lemma 3.7.7(b), if $\mathcal{D}^{a_1,\ldots,a_{2s}}$ does not vanish at some triple $(A, \theta, \omega) \in \mathbf{M}_g(R)$, then every element of $\phi^{a_1}(\mathrm{Im}([\ ]_\omega))$ commutes with the matrix $F^{a_1,\ldots,a_{2s}}(A, \theta, \omega)$ that has distinct eigenvalues. Hence $\phi^{a_1}(\mathrm{Im}([\ ]_\omega))$ is contained in an abelian R-subalgebra of $\mathbf{Mat}_g(R)$. In order to conclude the proof we need to check that $D^{a_1,\ldots,a_{2s}} \neq 0$. It suffices to show that the image $\mathcal{E}(D^{a_1,\ldots,a_{2s}})^\Delta$ of $\mathcal{E}(D^{a_1,\ldots,a_{2s}})$ via the reduction modulo p homomorphism between rings of polynomials

$$\Delta : R[Q^{(1)}, \ldots, Q^{(2s)}] \to k[Q^{(1)}, \ldots, Q^{(2s)}]$$

is nonzero. But, by Theorem 3.6.16(a), we have

$$\mathcal{E}(f^r_{g,\mathrm{crys}}) \equiv \Psi^{\phi^{r-1}} \mod p$$

hence, using Displays (3.67), (3.68), (3.70), and (3.72) and the equalities

$$\max\{a_1, a_2\} = a_2, \ \ldots, \max\{a_{2s-1}, a_{2s}\} = \max\{a_{2s}, a_1\} = a_{2s}$$

we get that the mentioned image equals the polynomial

$$\mathrm{Disc}_g^0(\chi_{Q^{(a_2)}(Q^{(a_3)})^* \cdots Q^{(a_{2s-2})}(Q^{(a_{2s-1})})^* Q^{(a_{2s})}(Q^{(a_{2s})})^*}(t)).$$

As $s \geq 2$, evaluating the matrices of indeterminates $Q^{(a_3)}, \ldots, Q^{(a_{2s})}$ at 1_g we get the nonzero polynomial $\mathrm{Disc}_g^0(\chi_{Q^{(a_2)}}(t))$, so $D^{a_1,\ldots,a_{2s}} \neq 0$. □

Theorem 3.7.9 *Let $s \geq 2$ be an integer and let*

$$a_1 < \cdots < a_{2s} \quad \text{and} \quad b_1 < \cdots < b_{2s}$$

be nonnegative integers such that $a_1 = b_1$ and $\{a_2, a_3\} \cap \{b_2, b_3\} = \emptyset$. Define

$$m_g := \sum_{j=1}^{g-1} j\binom{g}{j}, \quad w_a := \sum_{i=1}^{2s} \phi^{a_i}, \quad w_b := \sum_{i=1}^{2s} \phi^{b_i}.$$

Then there exists a nonzero δ-modular form

$$\mathcal{G} \in \mathbb{I}_g(-(m_g + 1)(w_a + w_b))$$

such that for all $(A, \theta, \omega) \in \boldsymbol{M}_g(R)$ *with* $\mathcal{G}(A, \theta, \omega) \neq 0$, *the polarized isogeny class of* (A, θ) *is indecomposable.*

Proof For a ring C, matrix $M \in \mathbf{Mat}_g(C)$ and integer $j \in \{1, \dots, g-1\}$ let $M^{\wedge j} \in \mathbf{Mat}_{\binom{g}{j}}(C)$ be the matrix representation of the j-th wedge power of the C-linear endomorphism of C^g defined by M with respect to the standard basis

$$\{e_{i_1} \wedge \cdots \wedge e_{i_j} | 1 \leq i_1 < \cdots < i_j \leq g\}$$

of $\bigwedge^j C^g$ ordered lexicographically, where $\{e_1, \dots, e_g\}$ is the standard basis of C^g. For $M_1, M_2 \in \mathbf{Mat}_g(C)$ we have $(M_1 M_2)^{\wedge j} = M_1^{\wedge j} M_2^{\wedge j}$ and we set

$$\Phi_j(M_1, M_2) := \det(M_1^{\wedge j} M_2^{\wedge j} - M_2^{\wedge j} M_1^{\wedge j}) \in C.$$

Consider the matrix cyclic products

$$F_a := F^{a_1,\dots,a_{2s}}, \quad F_b := F^{b_1,\dots,b_{2s}}.$$

As $a_1 = b_1$ the products $F_a F_b$ and $F_b F_a$ are also matrix cyclic products:

$$F_a F_b = F^{a_1,\dots,a_{2s},b_1,\dots,b_{2s}}, \quad F_b F_a = F^{b_1,\dots,b_{2s},a_1,\dots,a_{2s}}.$$

Define $G_j := \Phi_j(F_a, F_b) \in \mathbb{I}_g$, $j \in \{1, \dots, g-1\}$. Using Eq. (2.28) one easily checks that

$$G_j \in \mathbb{I}_g\left(-j\binom{g}{j}(w_a + w_b)\right).$$

Thus $\prod_{j=1}^{g-1} G_j \in \mathbb{I}_g(-m_g(w_a + w_b))$. From this and Display (3.71) we get:

$$\mathcal{G} := f_{a_1,\dots,a_{2s}} f_{b_1,\dots,b_{2s}} \prod_{j=1}^{g-1} G_j \in \mathbb{I}_g(-(m_g + 1)(w_a + w_b)).$$

We claim that $\mathcal{G}$ has the desired properties. We first prove that $\mathcal{G}$ is nonzero. It is enough to check that each G_j is nonzero.

As in the proof of Theorem 3.7.8 it is enough to show that the image $\mathcal{E}(G_j)^{\triangle}$ of $\mathcal{E}(G_j)$ in the ring $k[Q^{(1)}, \dots, Q^{(\max\{a_{2s}, b_{2s}\})}]$ is nonzero. Evaluating the matrices of indeterminates $Q^{(a_4)}, \dots, Q^{(a_{2s})}$ and $Q^{(b_4)}, \dots, Q^{(b_{2s})}$ at 1_g one gets the expression

$$\widetilde{G}_j := \Phi_j(Q^{(a_2)}(Q^{(a_3)})^*, Q^{(b_2)}(Q^{(b_3)})^*).$$

As $\{a_2, a_3\} \cap \{b_2, b_3\} = \emptyset$ the entries of the symmetric matrices $Q^{(a_2)}$, $(Q^{(a_3)})^*$, $Q^{(b_2)}$, $(Q^{(b_3)})^*$ that are above and on the diagonal are algebraically independent. So it is enough to find symmetric matrices $A_2, A_3, B_2, B_3 \in \mathbf{Mat}_g(k)$ such that $\Phi_j(A_2 A_3^*, B_2 B_3^*) \neq 0$. Each diagonalizable matrix in $\mathbf{GL}_g(k)$ is a product of two symmetric matrices (see Remark 4.1.19). So it is enough to find two diagonalizable matrices $D_1, D_2 \in \mathbf{GL}_g(k)$ such that $\Phi_j(D_1, D_2) \neq 0$. This follows from Lemma 4.1.14 and from the fact that the set of diagonalizable matrices is Zariski dense in the set of all matrices.

To end our proof we need to show that if the polarized isogeny class (A, θ) is decomposable then $\mathcal{G}(A, \theta, \omega) = 0$ for all ω. By Hecke covariance we may assume $(A, \theta) = (A_1, \theta_1) \times_{\mathrm{Spec}(R)} (A_2, \theta_2)$ and we may take $\omega = (\mathrm{pr}_1^*(\omega_1), \mathrm{pr}_2^*(\omega_2))^{\mathrm{t}}$ where $\mathrm{pr}_1, \mathrm{pr}_2$ are the projections of A onto A_1, A_2 (respectively) and ω_1, ω_2 are column bases of 1-forms on A_1, A_2 (respectively). For $m \in \mathbb{Z}$ let $[m]_i : A_i \to A_i, i \in \{1, 2\}$ be the multiplication by m on A_i. Then we may consider the polarized isogenies,

$$[1]_1 \times [1]_2, [1]_1 \times [-1]_2 : A \to A.$$

Their sum $u := [2]_1 \times [0]_2$ belongs to $\mathrm{End}(A, \theta)^{(p)}$. Let g_1 and g_2 be the relative dimensions of A_1 and A_2 so the relative dimension of A is $g = g_1 + g_2$. Note that the matrix $[u]_\omega$ of u with respect to ω has the block form $\begin{pmatrix} 2 \cdot 1_{g_1} & 0_{g_1 \times g_2} \\ 0_{g_2 \times g_1} & 0_{g_2 \times g_2} \end{pmatrix}$ where, for $i, j \in \{1, 2\}$, we denoted by $0_{g_i \times g_j}$ the $g_i \times g_j$ matrix with zero entries. By Lemma 3.7.7(b) this matrix commutes with $F_a(A, \theta, \omega)$ and $F_b(A, \theta, \omega)$ hence the latter matrices have entries in R and a block form $\begin{pmatrix} \bigstar & 0 \\ 0 & \bigstar \end{pmatrix}$ so the endomorphisms over K defined by these matrices have a common invariant subspace of dimension $g_1 \in \{1, \ldots, g-1\}$. By Lemma 4.1.15 we get $\Phi_j(F_a(A, \theta, \omega), F_b(A, \theta, \omega)) = 0$, hence $\mathcal{G}(A, \theta, \omega) = 0$. $\square$

3.8 On p-Adic Approximation

In this section we show how to p-adically approximate elements of $\mathbb{I}^r_{g,\mathrm{ord}}$ by quotients of elements in $\mathbb{I}^r_g$. We begin by introducing some new notation and concepts.

Let

$$W_{\mathrm{even}}(r) := \{w \in W(r) | \deg(w) \in 2\mathbb{Z}\}$$

and let $\mathbb{D} \subset \mathbb{I}_g$ be a multiplicative set such that the following two conditions hold.

(1) We have $\mathbb{D} \subset \cup_{r=1}^{\infty} \cup_{w \in W_{\mathrm{even}}(r)} \mathbb{I}^r_g(w) \backslash p\mathbb{I}^r_g(w)$.

(2) We have $\mathbb{D}^{\phi} \subset \mathbb{D}$.

Definition 3.8.1 An element $f \in \mathbb{D}^{-1}\mathbb{I}_g$ is said to have weight $w \in W$ if there exist $r \in \mathbb{N}$, $v \in W_{\mathrm{even}}(r)$, $F \in \mathbb{I}^r_g(v + w)$, and $G \in \mathbb{D} \cap \mathbb{I}^r_g(v)$, such that $f = G^{-1} F$.

We denote by $(\mathbb{D}^{-1}\mathbb{I}_g)(w)$ the R-submodule of $\mathbb{D}^{-1}\mathbb{I}_g$ of elements of weight w.

Definition 3.8.2 An element $f \in \mathbb{D}^{-1}\mathbb{I}_{g,\mathrm{ord}}$ is said to have weight $w \in W$ if there exist $r \in \mathbb{N}$, $v \in W_{\mathrm{even}}(r)$, $F \in \mathbb{I}^r_{g,\mathrm{ord}}(v+w)$, and $G \in \mathbb{D} \cap \mathbb{I}^r_{g,\mathrm{ord}}(v)$, such that $f = G^{-1}F$.

We denote by $(\mathbb{D}^{-1}\mathbb{I}_{g,\mathrm{ord}})(w)$ the R-submodule of $\mathbb{D}^{-1}\mathbb{I}_{g,\mathrm{ord}}$ of elements of weight w. Note that if a nonzero element of $\mathbb{D}^{-1}\mathbb{I}_g$ or $\mathbb{D}^{-1}\mathbb{I}_{g,\mathrm{ord}}$ has a weight w, then w is uniquely determined.

Definition 3.8.3 Fix a pair $(w', w'') \in \mathcal{W} \times \mathcal{W}$. We say that a quotient δ-modular function $f \in \mathbb{D}^{-1}M_{gg}$ has weight (w', w'') if there exists $r \in \mathbb{N}$ and a triple

$$(w, F, G) \in W_{\mathrm{even}}(r) \times M^r_{gg} \times (\mathbb{I}^r_g(w) \backslash p\mathbb{I}^r_g(w))$$

such that the following two conditions hold:

(i) we have $f = G^{-1}F$;

(ii) for each $S \in \mathrm{Ob}(\widehat{\mathbf{FilRing}_g})$, for all $(A, \theta, \omega) \in \mathbf{M}_g(S^0)$ and all $\lambda \in \mathbf{GL}_g(S^0)$ we have

$$F(A, \theta, \lambda\omega, S) = \det(\lambda)^{-w} \cdot \chi_{w'}(\lambda) \cdot F(A, \theta, \omega, S) \cdot \chi_{w''}(\lambda)^{\mathrm{t}}.$$

Note that if conditions (i) and (ii) hold and $(\tilde{w}, \tilde{F}, \tilde{G})$ is another triple in the set $W_{\mathrm{even}}(r) \times M^r_{gg} \times (\mathbb{I}^r_g(w) \backslash p\mathbb{I}^r_g(w))$ such that condition (i) holds for $(\tilde{F}, \tilde{G})$, then condition (ii) holds for $(\tilde{w}, \tilde{F})$.

We denote the R-submodule of $\mathbb{D}^{-1}M_{gg}$ of all elements of weight (w', w'') by

$$(\mathbb{D}^{-1}M_{gg})(w', w'').$$

Similar definitions can be made for $\mathbb{D}^{-1}M_{gg,\mathrm{ord}}$ in place of $\mathbb{D}^{-1}M_{gg}$. As in the case of Definition 2.3.12, for $g \geq 2$ a nonzero element of $\mathbb{D}^{-1}M_{gg}$ or $\mathbb{D}^{-1}M_{gg,\mathrm{ord}}$ could theoretically have two different weights (w', w'') and we do not know if the sum $w' + w''$ is uniquely determined.

If $f \in (\mathbb{D}^{-1}M_{gg})(w', w'') \cap (\mathbb{D}^{-1}M_{gg})^{\times}$, then $f^{-1} \in (\mathbb{D}^{-1}M_{gg})(-w'', -w')$.

Definition 3.8.4 Let $n \in \mathbb{N}$. Let $\{f_1, \ldots, f_n\}$ be a set of elements in $(\mathbb{D}^{-1}M_{gg,\mathrm{ord}})^{\times}$ that have weights in $\mathcal{W} \times \mathcal{W}$ and let $(w', w'') \in \mathcal{W} \times \mathcal{W}$. By a *coherent product* of elements in $\{f_1, \ldots, f_n\}$ of weight (w_1, w_2) we mean a product of the form

$$f_1^{\bullet} f_2^{\bullet} \cdots f_N^{\bullet} \in (\mathbb{D}^{-1}M_{gg,\mathrm{ord}})(w', w'') \cap (\mathbb{D}^{-1}M_{gg,\mathrm{ord}})^{\times}$$

for which the following two conditions hold:

(i) for each $i \in \{1, \ldots, N\}$ there exist $j \in \{1, \ldots, n\}$ and an integer $s \geq 0$ with the property that $f_i^{\bullet}$ equals either $f_j^{\phi^s}$ or $(f_j^{\phi^s})^{\mathrm{t}}$;

(ii) for each $i \in \{1, \ldots, N\}$ there exists $(w_i', w_i'') \in \mathcal{W} \times \mathcal{W}$ such that the form $f_i^{\bullet}$ has weight (w_i', w_i'') and moreover $w_1' = w'$, $w_N'' = w''$, and for all $i \in \{1, \ldots, N-1\}$ we have $w_i'' + w_{i+1}' = 0$.

Note that if $f_1, \ldots, f_n \in (\mathbb{D}^{-1}M_{gg})^{\times}$, then, in the notation above, we have:

$$f_1^{\bullet} f_2^{\bullet} \cdots f_N^{\bullet} \in (\mathbb{D}^{-1}M_{gg})(w', w'') \cap (\mathbb{D}^{-1}M_{gg})^{\times}.$$

Definition 3.8.5 Let $n \in \mathbb{N}$. Let $\{f_1, \ldots, f_n\}$ be a set of elements in $(\mathbb{D}^{-1}M_{gg,\mathrm{ord}})^{\times}$ that have weights in $\mathcal{W} \times \mathcal{W}$ and let $(w', w'') \in \mathcal{W} \times \mathcal{W}$. Let π belong to the p-adic completion $\widehat{\mathbb{D}^{-1}M_{gg,\mathrm{ord}}}$.

(a) We say π is a *coherent combination* of elements in $\{f_1, \ldots, f_n\}$ of weight (w', w'') if we have a series representation $\pi = \sum_{i=0}^{\infty} a_i \pi_i$ where the sequence $(a_i)_{i \geq 0}$ in R converges to 0 in the p-adic topology and each $\pi_i \in \mathbb{D}^{-1}M_{gg,\mathrm{ord}}(w', w'')$ is a coherent product of elements in $\{f_1, \ldots, f_n\}$ of weight (w', w'').

(b) We say that π is *regular* if in a series representation of it as in part (a) we can choose $a_0 \notin \mathfrak{m}_R$ and $a_i \in \mathfrak{m}_R$ for all $i \geq 1$.

(c) We say that π is *finite* if in a series representation of it as in part (a) we can choose $a_i = 0$ for all but finitely many i.

If $f_1, \ldots, f_n$ belong to $(\mathbb{D}^{-1}M_{gg})^{\times}$, then π comes from an element in $\widehat{\mathbb{D}^{-1}M_{gg}}$; if, in addition, π is finite, then π comes from an element in $\mathbb{D}^{-1}M_{gg}$.

In the notation above we have the following result.

Lemma 3.8.6 *We assume $n \geq 2$. Let π_1 be a coherent product of elements in the set*

$$\Sigma := \{f_2, \ldots, f_n, f_2^{-1}, \ldots, f_n^{-1}\},$$

of weight (w', w''), and let π_2 be a coherent combination of elements in the set $\Sigma \cup \{f_1\}$, of the same weight (w', w''). Then

$$(\pi_1 + p \cdot \pi_2)^{-1} = \pi_1^{-1} + p \cdot \pi_3$$

where π_3 is a coherent combination of elements in the set $\Sigma \cup \{f_1\}$, of the same weight $(-w'', -w')$ as π_1^{-1}. In particular, $(\pi_1 + p \cdot \pi_2)^{-1}$ is regular.

Proof Based on the identity

$$(\pi_1 + p \cdot \pi_2)^{-1} = \sum_{i=0}^{\infty} (-p)^i (\pi_1^{-1}\pi_2)^i \pi_1^{-1},$$

we can take $a_i := (-p)^i$ for $i \in \mathbb{N} \cup \{0\}$ and the lemma follows. □

For the remaining part of this section we set $f^r := f^r_{g,\mathrm{crys}}$ for $r \in \mathbb{N}$ and assume, in addition to the conditions (1) and (2) above, that the multiplicative set $\mathbb{D}$ also satisfies the third condition:

(3) We have $\det(f^1), \det(f^2) \in \mathbb{D}$.

For instance, as $\mathbb{D}$ we could take the minimal set

$$\mathbb{D}_{\min} := \{\det(f^1)^{w'} \det(f^2)^{w''} | w', w'' \in W_+\} \tag{3.91}$$

(contained in the image of the R-algebra monomorphism $\diamondsuit : \mathbb{H}_g^{\text{torus}} \to \mathbb{I}_{g,\text{ord}}$, see Theorem 3.6.29(h)) or the maximal set

$$\mathbb{D}_g := \cup_{r=1}^{\infty} \cup_{w \in W_{\text{even}}(r)} (\mathbb{I}_g^r(w) \backslash p\mathbb{I}_g^r(w)).$$

The key fact in what follows is that we have the equality in $\mathbb{I}_{g,\text{ord}}^r(w)$:

$$f^2 = (f^1)^{\phi} \cdot f^{\partial} + p \cdot (((f^{\partial})^{\phi})^{-1})^{\text{t}} \cdot f^1. \tag{3.92}$$

Indeed, by Theorems 3.6.16(a) and 3.6.17(a), the Serre–Tate expansions of the left and right hand sides of Eq. (3.92) coincide. On the other hand the two sides have the same weight. So the equality in Eq. (3.92) holds by the Serre–Tate expansion principle (Proposition 3.5.1(b)). In particular, by setting $f_{\partial} := (f^{\partial})^{-1}$, in $\mathbb{D}^{-1} M_{gg,\text{ord}}$ we get an equality

$$f^{\partial} = ((f^1)^{\phi})^{-1} \cdot f^2 - p \cdot ((f^1)^{\phi})^{-1} \cdot ((f_{\partial})^{\phi}))^{\text{t}} \cdot f^1 \tag{3.93}$$

where $((f^1)^{\phi})^{-1} \cdot f^2$ is a coherent product of elements in the set

$$\Pi := \{f^1, (f^1)^{-1}, f^2, (f^2)^{-1}\} \tag{3.94}$$

of weight $(-\phi, 1)$ and $((f^1)^{\phi})^{-1} \cdot ((f_{\partial})^{\phi}))^{\text{t}} \cdot f^1$ is a coherent product of elements in the set

$$\Pi' := \{f_{\partial}, f^1, (f^1)^{-1}, f^2, (f^2)^{-1}\} = \Pi \cup \{f_{\partial}\} \tag{3.95}$$

of the same weight. In particular, by Lemma 3.8.6, in $\mathbb{D}^{-1} M_{gg,\text{ord}}$ we have

$$f_{\partial} = (f^2)^{-1}(f^1)^{\varphi} + p\pi_1, \tag{3.96}$$

where $(f^2)^{-1}(f^1)^{\phi}$ is a coherent product of elements in the set Π and π_1 is a coherent combination of elements in the set Π', both of them having weight $(-1, \phi)$.

Lemma 3.8.7 *For every integer $n \geq 2$ there exist a finite regular coherent combination $\eta_n \in \mathbb{D}^{-1} M_{gg}$ of elements in Π above (see Display (3.94)) and a coherent combination $\pi_n \in \widehat{\mathbb{D}^{-1} M_{gg,ord}}$ of elements in Π' above (see Display (3.95)), both of weight $(-1, \phi)$, such that in $\widehat{\mathbb{D}^{-1} M_{gg,ord}}$ we have an identity*

$$f_{\partial} = \eta_n + p^n \pi_n. \tag{3.97}$$

Proof We proceed by induction on n. The base of the induction for $n = 2$ holds due to the Eq. (3.96). Assuming that the Eq. (3.97) holds for some $n \geq 2$, the passage from n to $n+1$ is achieved by substituting all occurrences of f_∂ in π_n in that equation by the right-hand side of Eq. (3.96) and regrouping the terms to get the desired expression. □

Corollary 3.8.8 *For every integer $n \geq 2$ there exist a finite regular coherent combination $\eta_n^\partial \in \mathbb{D}^{-1}M_{gg}$ of elements in Π above (see Display (3.94)) and a coherent combination $\pi_n^\partial \in \widehat{\mathbb{D}^{-1}M_{gg,ord}}$ of elements in Π' above (see Display (3.95)), both of weight $(-\phi, 1)$, such that in $\widehat{\mathbb{D}^{-1}M_{gg,ord}}$ we have an identity*

$$f^\partial = \eta_n^\partial + p^n \pi_n^\partial.$$

Proof Replace f_∂ in the right-hand side of Eq. (3.93) by the right-hand side of Eq. (3.97). □

Recall the forms $f^{\langle a \rangle}$ from Display (3.11) indexed by $a \in \mathbb{N}$; then Corollary 3.8.8 implies the following consequence.

Corollary 3.8.9 *For all integers $a \geq 1$ and $n \geq 2$ there exist a finite regular coherent combination $\eta_n^{(a)} \in \mathbb{D}^{-1}M_{gg}$ of elements in Π above (see Display (3.94)) and a coherent combination $\pi_n^{(a)} \in \widehat{\mathbb{D}^{-1}M_{gg,ord}}$ of elements in Π' above (see Display (3.95)), both of weight $(1, 1)$, such that in $\widehat{\mathbb{D}^{-1}M_{gg,ord}}$ we have an identity*

$$f^{\langle a \rangle} = \eta_n^{(a)} + p^n \pi_n^{(a)}.$$

Recall that we have R-algebra monomorphisms

$$\mathbb{H}_{g,\mathrm{tot}}^{\mathrm{torus}} \xrightarrow{\diamond} \mathbb{I}_{g,\mathrm{ord}} \to \mathbb{D}^{-1}\mathbb{I}_{g,\mathrm{ord}} \to \mathbb{D}^{-1}M_{g,\mathrm{ord}} \to \widehat{\mathbb{D}^{-1}M_{g,\mathrm{ord}}}$$

which, for $s \in \frac{1}{2}\mathbb{Z}$, induce R-linear maps

$$\mathbb{H}_g(s) \to \mathbb{I}_{g,\mathrm{ord}}(-2s) \to (\mathbb{D}^{-1}\mathbb{I}_{g,\mathrm{ord}})(-2s) \to \mathbb{D}^{-1}\mathbb{I}_{g,\mathrm{ord}},$$

where

$$\mathbb{H}_g(s) := \bigcup_{r \geq 1} \mathbb{H}_g^r(s), \quad \mathbb{I}_{g,\mathrm{ord}}(-2s) := \bigcup_{r \geq 1} \mathbb{I}_{g,\mathrm{ord}}^r(-2s).$$

Also, we have R-algebra homomorphisms

$$\mathbb{D}^{-1}\mathbb{I}_g \to \mathbb{D}^{-1}\mathbb{I}_{g,\mathrm{ord}} \to \mathbb{D}^{-1}M_{g,\mathrm{ord}} \to \widehat{\mathbb{D}^{-1}M_{g,\mathrm{ord}}},$$

that induce for each $w \in W$ the R-linear maps

$$(\mathbb{D}^{-1}\mathbb{I}_g)(w) \to (\mathbb{D}^{-1}\mathbb{I}_{g,\mathrm{ord}})(w) \to \widehat{\mathbb{D}^{-1}M_{g,\mathrm{ord}}}.$$

Lemma 3.8.10 *For every $w_0 \in W$ the reductions of the homomorphism*

$$(\mathbb{D}^{-1}\mathbb{I}_{g,ord})(w_0) \to \widehat{\mathbb{D}^{-1}M_{g,ord}}$$

modulo powers of p are injective.

Proof Let $n \in \mathbb{N}$ and let $G^{-1}F \in (\mathbb{D}^{-1}\mathbb{I}_{g,\mathrm{ord}})(w_0)$ where $G \in \mathbb{D} \cap \mathbb{I}_g(w)$ and $F \in \mathbb{I}_{g,\mathrm{ord}}(w + w_0)$ with $w \in W$ of even degree. We assume that the image of $G^{-1}F$ in $\widehat{\mathbb{D}^{-1}M_{g,\mathrm{ord}}}$ belongs to the submodule $p^n\widehat{\mathbb{D}^{-1}M_{g,\mathrm{ord}}}$. Then the image of $G^{-1}F$ in $\mathbb{D}^{-1}M_{g,\mathrm{ord}}$ belongs to $p^n\mathbb{D}^{-1}M_{g,\mathrm{ord}}$. So there exist $G_1 \in \mathbb{D} \cap \mathbb{I}_g(w_1)$, $H \in \mathbb{D} \cap \mathbb{I}_g(v)$ for some $v, w_1 \in W$ of even degree, and $F_1 \in M_{g,\mathrm{ord}}$ such that the image of HG_1F in $M_{g,\mathrm{ord}}$ equals the image of p^nHF_1G. By the flatness over R condition in the definition of the objects of $\widehat{\mathbf{FilRing}_\delta}$ (see Definition 2.1.8), as HG_1F is Hecke covariant of weight $v + w + w_1 + w_0$, we get that HF_1G is also Hecke covariant of the same weight. By Corollary 3.5.3(c) we conclude that F_1 is Hecke covariant of weight $w_1 + w_0$. So we have $G_1^{-1}F_1 \in (\mathbb{D}^{-1}\mathbb{I}_{g,\mathrm{ord}})(w_0)$. As HG_1F and p^nHGF_1 coincide in $M_{g,\mathrm{ord}}$, they coincide in $\mathbb{I}_{g,\mathrm{ord}}(v + w + w_1 + w_0)$. So $G^{-1}F = p^nG_1^{-1}F_1 \in p^n(\mathbb{D}^{-1}\mathbb{I}_{g,\mathrm{ord}})(w_0)$. □

The proof of the following elementary lemma is left as an exercise.

Lemma 3.8.11 *Let $C \subset B$ be an extension of flat $\mathbb{Z}_{(p)}$-algebras whose reduction modulo p, $C/pC \to B/pB$, is injective. If $b \in B$ is such that there exists $N \in \mathbb{N}$ with the property that p^Nb belongs to the p-adic closure of C in B, then b itself belongs to the p-adic closure of C in B.*

Here by the 'p-adic closure of C in B' we mean the closure of C in B where the latter is equipped with its p-adic topology.

Lemma 3.8.12 *The following properties hold.*

(a) *The R-algebra homomorphism*

$$\mathbb{D}^{-1}\mathbb{I}_g \to \mathbb{D}^{-1}\mathbb{I}_{g,ord}$$

is injective and its reduction modulo p is injective.

(b) *The R-algebras $\mathbb{D}^{-1}\mathbb{I}_g$ and $\mathbb{D}^{-1}\mathbb{I}_{g,ord}$ are p-adically separated integral domains having p as a prime element.*

Proof Part (a) follows from the Serre–Tate expansion principle (Proposition 3.5.1(b)) via localization which is an exact functor.

Part (b) follows directly from Corollary 3.5.2(b). □

In view of Lemma 3.8.12(a) we speak about inclusions $\mathbb{D}^{-1}\mathbb{I}_g \subset \mathbb{D}^{-1}\mathbb{I}_{g,\mathrm{ord}}$ and $\widehat{\mathbb{D}^{-1}\mathbb{I}_g} \subset \widehat{\mathbb{D}^{-1}\mathbb{I}_{g,\mathrm{ord}}}$. Here are our main approximation results.

Theorem 3.8.13 *Recall that the multiplicative set $\mathbb{D}$ satisfies the conditions (1) to (3) of this section. Then the following properties hold.*

(a) *For every rational number $s \in \frac{1}{2}\mathbb{Z}$ with $gs \in \mathbb{Z}$, the image of the R-linear map $\mathbb{H}_g(s) \to (\mathbb{D}^{-1}\mathbb{I}_{g,ord})(-2s)$ is contained in the p-adic closure of the image of the R-linear map*

$$(\mathbb{D}^{-1}\mathbb{I}_g)(-2s) \to (\mathbb{D}^{-1}\mathbb{I}_{g,ord})(-2s).$$

(b) *For every $w \in W$, the image of $\det(f^\partial)^w \in \mathbb{I}_{g,ord}((\phi-1)w)$ in the localization $(\mathbb{D}^{-1}\mathbb{I}_{g,ord})((\phi-1)w)$ is contained in the p-adic closure of the image of*

$$(\mathbb{D}^{-1}\mathbb{I}_g)((\phi-1)w) \to (\mathbb{D}^{-1}\mathbb{I}_{g,ord})((\phi-1)w).$$

(c) *We have*

$$\widehat{\mathbb{D}^{-1}\mathbb{I}_g} = \widehat{\mathbb{D}^{-1}\mathbb{I}_{g,ord}}.$$

Proof We start by proving part (a). Part (b) can be proved by an entirely similar argument.

In view of Lemma 3.8.10 and Theorem 3.6.20(b) it is enough to show that the image in $\widehat{\mathbb{D}^{-1}M_{g,\mathrm{ord}}}$ of each element of $\mathbb{I}^r_{g,\mathrm{ord}}(-2s)$ of the form $F(f^{\langle 1\rangle},\ldots,f^{\langle r\rangle})$, where $F \in \mathbb{H}^{r-1}_g(s)$, is a p-adic limit of images of elements in $(\mathbb{D}^{-1}\mathbb{I}_g)(-2s)$.

Fix such an F and an integer $n \geq 1$. We seek an element $F[n] \in (\mathbb{D}^{-1}\mathbb{I}_g)(-2s)$ whose image in the ring $\widehat{\mathbb{D}^{-1}M_{g,\mathrm{ord}}}$ differs from the image of $F(f^{\langle 1\rangle},\ldots,f^{\langle r\rangle})$ by an element in $p^n\widehat{\mathbb{D}^{-1}M_{g,\mathrm{ord}}}$.

By Corollary 3.8.9 there exist finite regular coherent combinations

$$\eta_n^{(1)},\ldots,\eta_n^{(r-1)} \in \mathbb{D}^{-1}M_{gg} = \mathbf{Mat}_g(\mathbb{D}^{-1}M_g)$$

of elements in Π above (see Display (3.94)) of weight $(1,1)$ such that

$$F(f^{\langle 1\rangle},\ldots,f^{\langle r-1\rangle}) \equiv F(\eta_n^{(1)},\ldots,\eta_n^{(r-1)}) \mod p^n \tag{3.98}$$

in the ring $\widehat{\mathbb{D}^{-1}M_{g,\mathrm{ord}}}$. The elements $\eta_n^{(a)}$ can be written as

$$\eta_n^{(a)} = G^{-1}\Phi_{n,a} \tag{3.99}$$

where $G \in \mathbb{D}_{\min}$ (see Eq. (3.91)) is of some weight $w \in W$ of even degree and $\Phi_{n,a} \in M_{gg}$ are such that they satisfy the following two conditions.

(i) For each triple $(S,(A,\theta,\omega),\lambda) \in \mathrm{Ob}(\mathbf{FilRing}_\delta) \times \mathbf{M}_g(S^0) \times \mathbf{GL}_g(S^0)$ we have:

$$\Phi_{n,a}(A,\theta,\lambda\omega,S) = \det(\lambda)^{-w}\cdot\lambda\cdot\Phi_{n,a}(A,\theta,\omega,S)\cdot\lambda^{\mathrm{t}}. \tag{3.100}$$

(ii) For each $S \in \mathrm{Ob}(\widehat{\mathbf{FilRing}_\delta})$ and every triple $(x_1, x_2, u) \in \mathbf{M}'_g(S^0)$, where $x_1 = (A_1, \theta_1, \omega_1)$, $x_2 = (A_2, \theta_2, \omega_2)$, $uu^t = d \in \mathbb{N}\backslash p\mathbb{N}$, and $u^*\omega_2 = \omega_1$, the following equality holds:

$$\Phi_{n,a}(A_2, \theta_2, \omega_2, S) = d^{-1+\frac{g\cdot\deg(w)}{2}} \cdot \Phi_{n,a}(A_1, \theta_1, \omega_1, S). \tag{3.101}$$

Condition (i) follows from the fact that f^1 and f^2 have weights $(\phi, 1)$ and $(\phi^2, 1)$, respectively. Condition (ii) follows from the Hecke covariance of f^1 and f^2 (see Eq. (2.23)). As F is homogeneous of degree gs, by Eq. (3.99) we get that

$$F(\eta_n^{(1)}, \ldots, \eta_n^{(r-1)}) = (G^{gs})^{-1}\Phi, \quad \Phi := F(\Phi_{n,1}, \ldots, \Phi_{n,r-1}) \in M_g. \tag{3.102}$$

Using Eq. (3.100) we get that for every triple $(S, (A, \theta, \omega), \lambda)$ as in condition (i) above we have

$$\Phi(A, \theta, \lambda\omega, S) = \det(\lambda)^{2s-gsw} \cdot \Phi(A, \theta, \omega, S).$$

Thus Φ has weight $gsw - 2s$. Also, by Eq. (3.101), for every S and (x_1, x_2, u) as in condition (ii) above we have

$$\Phi(A_2, \theta_2, \omega_2, S) = d^{gs(-1+\frac{g\cdot\deg(w)}{2})}\Phi(A_1, \theta_1, \omega_1, S).$$

By Eq. (2.23) we get that $\Phi \in \mathbb{I}_g^{r'}(gsw - 2s)$ for some $r' \in \mathbb{N}$. (Note that a priori one cannot take $r' = r$.) Thus we have $(G^{gs})^{-1}\Phi \in (\mathbb{D}^{-1}\mathbb{I}_g)(-2s)$. Then the element $F[n] := F(\eta_n^{(1)}, \ldots, \eta_n^{(r-1)})$ has the desired properties in view of Eqs. (3.98) and (3.102).

By parts (a) and (b), the image $\mathrm{Im}(\mathbb{H}_{g,\mathrm{tot}}^{\mathrm{torus}} \to \mathbb{D}^{-1}\mathbb{I}_{g,\mathrm{ord}})$ (see Theorem 3.6.29(a)) is contained in the p-adic closure of $\mathbb{D}^{-1}\mathbb{I}_g$ in $\mathbb{D}^{-1}\mathbb{I}_{g,\mathrm{ord}}$. By Lemmas 3.8.11 and 3.8.12(a), we get that every element of $\mathbb{D}^{-1}\mathbb{I}_{g,\mathrm{ord}}$ is contained in the p-adic closure of $\mathbb{D}^{-1}\mathbb{I}_g$ in $\mathbb{D}^{-1}\mathbb{I}_{g,\mathrm{ord}}$. Now, by Lemma 3.8.12(a) again, each Cauchy sequence in $\mathbb{D}^{-1}\mathbb{I}_{g,\mathrm{ord}}$ which is contained in $\mathbb{D}^{-1}\mathbb{I}_g$ is as well a Cauchy sequence in $\mathbb{D}^{-1}\mathbb{I}_g$. We conclude that every element of $\widehat{\mathbb{D}^{-1}\mathbb{I}_{g,\mathrm{ord}}}$ that belongs to $\mathbb{D}^{-1}\mathbb{I}_{g,\mathrm{ord}}$ is contained in $\widehat{\mathbb{D}^{-1}\mathbb{I}_g}$. This implies that part (c) holds. □

3.9 Flows on Affine Ordinary Loci

In this section we show that the form $f^1_{g,\mathrm{crys}}$ induces δ-flows in the sense recalled below on affine open subschemes of $\mathcal{X}_{\mathrm{ord}}$ whose reductions modulo p are nonempty; the main result (see Theorem 3.9.12) will be used to prove in the next section a Reciprocity Theorem for CL_R points for every $g \geq 1$.

Definition 3.9.1 Let X be an affine smooth scheme over $\mathrm{Spec}(R)$ and let $J^1(X)$ be its first p-jet space. Following [Bu17], Ch. 3, Sect. 3.6, Def. 3.122, an ideal $\mathfrak{a}$ of $\mathcal{O}(J^1(X))$ is called a *δ-flow* on X if the composite projection

$$Z^1(\mathfrak{a}) := \mathrm{Spf}(\mathcal{O}(J^1(X))/\mathfrak{a}) \subset J^1(X) \to \widehat{X}$$

is an isomorphism.

By the universal property of p-jet spaces, giving a δ-flow on X is equivalent to giving a p-derivation on $\mathcal{O}(\widehat{X})$. (The notion of *δ-flow* was suggested by the analogy between p-derivations and vector fields which was in the background of [Bu95a, Bu05].)

Remark 3.9.2 If $\mathfrak{a}$ is a δ-flow on X, then for all integers $n \geq 1$ the projection

$$Z^n(\mathfrak{a}) := \mathrm{Spf}(\mathcal{O}(J^n(X))/(\mathfrak{a}, \delta\mathfrak{a}, \dots, \delta^{n-1}\mathfrak{a})) \subset J^n(X) \to \widehat{X}$$

is an isomorphism. To check this we show that the R-algebra homomorphism $\mathcal{O}(\widehat{X}) \to \mathcal{O}(Z^n(\mathfrak{a}))$ is an isomorphism. To check the injectivity we note that, by the universal property of p-jet spaces, the p-derivation on $\mathcal{O}(\widehat{X})$ defines a right inverse (retraction) $\mathcal{O}(J^n(X)) \to \mathcal{O}(\widehat{X})$ of the R-algebra homomorphism $\mathcal{O}(\widehat{X}) \to \mathcal{O}(J^n(X))$; but clearly this right inverse factors through an R-algebra homomorphism $\mathcal{O}(Z^n(\mathfrak{a})) \to \mathcal{O}(\widehat{X})$ which is a right inverse for $\mathcal{O}(\widehat{X}) \to \mathcal{O}(Z^n(\mathfrak{a}))$. Hence $\mathcal{O}(\widehat{X}) \to \mathcal{O}(Z^n(\mathfrak{a}))$ is injective. We check that the R-algebra homomorphism $\mathcal{O}(\widehat{X}) \to \mathcal{O}(Z^n(\mathfrak{a}))$ is surjective by induction on $n \in \mathbb{N}$. The case $n = 1$ is our δ-flow hypothesis. Now assume the statement for some $n \geq 1$ and take an element u of $\mathcal{O}(Z^{n+1}(\mathfrak{a}))$. So u is a p-adic limit of elements of the form v and δw with $v, w \in \mathcal{O}(Z^n(\mathfrak{a}))$. By the induction hypothesis v and w come from elements of $\mathcal{O}(\widehat{X})$. Hence δw comes from an element of $\mathcal{O}(Z^1(\mathfrak{a}))$. By case $n = 1$, δw comes then from an element of $\mathcal{O}(\widehat{X})$. Thus u comes from an element of $\mathcal{O}(\widehat{X})$. This ends the induction.

Our next purpose is to show that Hecke covariant Siegel δ-modular forms of weight $(\phi, 1)$ naturally define δ-flows on $\mathcal{X}_{\mathrm{ord}} \subset \mathcal{X}$. We first introduce a few extra concepts on δ-flows.

Let X be a smooth affine scheme over $\mathrm{Spec}(R)$ of relative dimension d. Assume $\bar{X}$ is irreducible. Let $f = (f_1, \dots, f_d) \in \mathcal{O}(J^1(X))^d$. Assume furthermore there exist étale coordinates $\tau = (\tau_1, \dots, \tau_d) \in \mathcal{O}(X)^d$, i.e., their corresponding morphism of R-schemes $\tau : X \to \mathbb{A}^d_R$ is étale.

For a d-tuple of indeterminates $\tau' := (\tau'_1, \dots, \tau'_d)$, let

$$\mathcal{O}(X)[\tau'] := \mathcal{O}(X)[\tau'_1, \dots, \tau'_n]$$

be the polynomial $\mathcal{O}(X)$-algebra in the entries of τ'; we also view τ' as an element of $\mathcal{O}(X)[\tau']^d$. By [Bu95a], Prop. 1.4 we have an $\mathcal{O}(X)$-algebra isomorphism

$$\widehat{\mathcal{O}(X)[\tau']} \simeq \mathcal{O}(J^1(X))$$

that maps τ_i' to $\delta\tau_i$ for $i \in \{1, \dots, d\}$.

Definition 3.9.3 We say f is *quasi-linear with respect to* τ if there exist

$$(L_{ij})_{1 \le i,j \le d} \in \mathbf{GL}_d(\mathcal{O}(\widehat{X})), \quad (E_1, \dots, E_d) \in \mathcal{O}(\widehat{X})^d, \quad (J_1, \dots, J_d) \in \mathcal{O}(J^1(X))^d$$

such that for all $i \in \{1, \dots, d\}$, inside $\mathcal{O}(J^1(X))$ we have

$$f_i = \sum_{j=1}^{d} L_{ij}\tau_j' + E_i + pJ_i. \tag{3.103}$$

Next consider a point $P \in X(R)$, $P : \mathrm{Spec}(R) \to X$, and an isomorphism between the completion of X along the image of P (i.e., the completion of the local ring of X at its k-valued closed point defined by P) and the formal power series ring $R[[\underline{z}]] := R[[z_1, \dots, z_d]]$ where $\underline{z} := (z_1, \dots, z_d)$ is a d-tuple of indeterminates to be called *formal coordinates* at P. For simplicity we view this isomorphism as an identification.

We also assume that the image of P under the map $\tau(R) : X(R) \to \mathbb{A}^d(R)$ is $0 \in \mathbb{A}^d(R)$. Let $R[[\underline{z}]][\underline{z}'] := R[[\underline{z}]][z_1', \dots, z_d']$ where $\underline{z}' := (z_1', \dots, z_d')$ is a d-tuple of new indeterminates. We have an R-algebra isomorphism (which we view also as an identification) $R[[\tau]] := R[[\tau_1, \dots, \tau_d]] \simeq R[[\underline{z}]]$ induced by the composite R-algebra homomorphism $R[\tau_1, \dots, \tau_d] \to \mathcal{O}(X) \to R[[\underline{z}]]$; let $S_i \in R[[\tau]]$ be the image of z_i for $i \in \{1, \dots, d\}$. Let $R[[\tau]][\tau'] := R[[\tau]][\tau_1', \dots, \tau_d']$ and let $\delta : R[[\tau]] \to \widehat{R[[\tau]][\tau']}$ be the unique p-derivation relative to the inclusion such that $\delta\tau_i = \tau_i'$ for $i \in \{1, \dots, d\}$. So we have a natural R-algebra composite homomorphism

$$\mathcal{O}(J^1(X)) \to \widehat{\mathcal{O}(X)[\tau']} \to \widehat{R[[\tau]][\tau']} \to \widehat{R[[\underline{z}]][\underline{z}']} \tag{3.104}$$

where the first and the third homomorphisms are isomorphisms, the inverse of the last isomorphism mapping z_i' to δS_i for $i \in \{1, \dots, d\}$; it is injective as $J^1(X)$ is irreducible and smooth over k.

Definition 3.9.4 We say f is *formally quasi-linear at* P *with respect to* $\underline{z}$ if there exist

$$(u_{ij})_{a \le i,j \le d} \in \mathbf{GL}_d(R[[\underline{z}]]), \quad (V_1, \dots, V_d) \in R[[\underline{z}]]^d, \quad (W_1, \dots, W_d) \in \widehat{R[[\underline{z}]][\underline{z}']}^d$$

such that for all $i \in \{1, \dots, d\}$ we have the following equality in the ring $\widehat{R[[\underline{z}]][\underline{z}']}$:

$$f_i = \sum_{j=1}^{d} u_{ij}\tau_j' + V_i + pW_i.$$

Our terminology comes from the analogy with classical differential equations where *quasi-linearity* means 'linearity in the derivatives'.

Clearly, if f is quasi-linear with respect to τ, it is also formally quasi-linear at P with respect to $\underline{z}$. We have the following partial converse of this which is an extension of [BPII], Lem. 4.66.

Lemma 3.9.5 *Let X, P, f, τ, $\underline{z}$ be as above. Assume f is formally quasi-linear at P with respect to $\underline{z}$. Then there exists an affine open subscheme Y of X that contains P and such that the image of f in $\mathcal{O}(J^1(Y))^d$ is quasi-linear with respect to the image of τ in $\mathcal{O}(Y)^d$.*

Proof We view the R-algebra monomorphism $\mathcal{O}(J^1(X)) \to \widehat{R[[\underline{z}]][\underline{z}']}$ as an inclusion (see Diagram (3.104)). For $\alpha = (\alpha_1, \ldots, \alpha_d) \in \mathbb{Z}^d_{\geq 0}$ and $b = (b_1, \ldots, b_d) \in B^d$ with B a ring, we write $b^\alpha := b_1^{\alpha_1} \cdots b_d^{\alpha_d} \in B$. For $i \in \{1, \ldots, d\}$ we write

$$S_i := S_i(\tau) = z_i = \sum_{\alpha \in \mathbb{Z}^d_{\geq 0}} c_{\alpha i} \tau^\alpha \in R[[\tau]],$$

where each $c_{\alpha i} \in R$. Let $\tau^p := (\tau_1^p, \ldots, \tau_d^p) \in \mathcal{O}(X)^d$ and consider the d-tuple $\tau^p + p\tau' \in \mathcal{O}(X)[\tau']^d$. Then

$$z_i' = \frac{1}{p}\left[\sum_{\alpha \in \mathbb{Z}^d_{\geq 0}} c^\phi_{\alpha i}(\tau^p + p\tau')^\alpha - \left(\sum_{\alpha \in \mathbb{Z}^d_{\geq 0}} c_{\alpha i}\tau^\alpha\right)^p\right] = \sum_{j=1}^d (\partial S_i/\partial \tau_j)^p \tau_j' + a_i + pb_i$$

for some $a_i \in R[[\tau]]$ and $b_i \in \widehat{R[[\tau]][\tau']}$. Denoting by $\bar{f}_i$ the reduction modulo p of f_i, we get the identity

$$\overline{f}_i = \sum_{j=1}^d \overline{l}_{ij}\tau_j' + \overline{a}_i \tag{3.105}$$

in $k[[\tau]][\tau']$, where $\overline{a}_i \in k[[\tau]]$ is the reduction modulo p of a_i and for the reduction

$$(\overline{u}_{ij})_{1 \leq i,j \leq d} \in \mathbf{GL}_d(k[[\underline{z}]]) = \mathbf{GL}_d(k[[\tau]])$$

of $(u_{ij})_{1 \leq i,j \leq d}$ modulo p, we denote

$$\overline{l}_{ij} := \sum_{m=1}^d \overline{u}_{im} \cdot (\overline{\partial S_m/\partial \tau_j})^p \in k[[\tau]].$$

Here $\overline{\partial S_m/\partial \tau_j} \in k[[\tau]]$ is the reduction modulo p of $\partial S_m/\partial \tau_j \in R[[\tau]]$. On the other hand we have

$$\overline{f}_i \in \mathcal{O}(\overline{X})[\tau'] \subset k[[\tau]][\tau'].$$

As $\bar{l}_{ij},\ \bar{a}_i \in k[[\tau]]$, from Eq. (3.105) we get that

$$\bar{l}_{ij},\ \bar{a}_i \in \mathcal{O}(\overline{X}).$$

Choose $L_{ij}, E_i \in \mathcal{O}(X)$ that lift $\bar{l}_{ij}, \overline{a_i}$ respectively. Then Eq. (3.103) holds for some $J_i \in \mathcal{O}(J^1(X))$ as $\overline{\mathcal{O}(J^1(X))} \to k[[\tau]][\tau']$ is a k-algebra monomorphism. As the matrices

$$(u_{ij}(0))_{1\le i,j\le d},\ ((\partial S_i/\partial \tau_j)(0))_{1\le i,j\le d} \in \mathbf{Mat}_d(R)$$

are invertible so is $(L_{ij}(P))_{1\le i,j\le d} \in \mathbf{Mat}_d(R)$. We conclude by taking Y to be the nonzero locus of the determinant of the matrix $(L_{ij})_{1\le i,j\le d} \in \mathbf{Mat}_d(\mathcal{O}(X))$. □

Remark 3.9.6 With X, f, τ as in the part preceding Lemma 3.9.5, we consider additional étale coordinates $\tilde{\tau} = (\tilde{\tau}_1, \ldots, \tilde{\tau}_d) \in \mathcal{O}(X)^d$. If f is quasi-linear with respect to $\tilde{\tau}$, then it is also quasi-linear with respect to τ. To check this, let $P \in X(R)$. Compositing τ and $\tilde{\tau}$ with translations of $\mathbb{A}^d$ we may assume that the maps $X(R) \to \mathbb{A}^d(R) = R^d$ induced by τ and $\tilde{\tau}$ map P into 0. Then we take $[\underline{z}]$ such that we have $R[[\underline{z}]] = R[[\tilde{\tau}]] := R[[\tilde{\tau}_1, \ldots, \tilde{\tau}_d]]$. As f is quasi-linear with respect to $\tilde{\tau}$, it is formally quasi-linear at P with respect to $\underline{z}$ at P. Based on the proof of Lemma 3.9.5 we have $\bar{Y} = \bar{X}$, as the matrix $(\bar{l}_{ij})_{1\le i,j\le n} \in \mathbf{Mat}_d(\mathcal{O}(\bar{X}))$ is invertible.

The link between quasi-linearity and δ-flows is given by the following lemma.

Lemma 3.9.7 *Let X, f, τ be as in the part preceding Lemma 3.9.5. Assume f is quasi-linear with respect to τ. Then the ideal $\mathfrak{a}$ of $\mathcal{O}(J^1(X))$ generated by the entries of f is a δ-flow on X.*

Proof Recall that $Z^1(\mathfrak{a}) = \mathrm{Spf}(\mathcal{O}(J^1(X)/\mathfrak{a}))$ and that $\overline{\bigstar}$ means the reduction modulo p of $\bigstar$. It is enough to check that the projection

$$\overline{Z^1(\mathfrak{a})} \subset \overline{J^1(X)} \to \overline{X}$$

is an isomorphism; but this is clear from the definition of quasi-linearity. □

We denote by $(A_{\mathcal{X}}, \theta_{\mathcal{X}})$ the universal principally polarized abelian scheme over $\mathcal{X} = \mathcal{A}_{g,1,N,R}$. For a scheme X over $\mathcal{X}$, let $(A_X, \theta_X) := X \times_{\mathcal{X}} (A_{\mathcal{X}}, \theta_{\mathcal{X}})$. Until Sect. 3.10 we let $d := \frac{g(g+1)}{2}$ be the relative dimension of $\mathcal{X}$ over R.

For each ring B and matrix $M = (m_{ij})_{1\le i,j\le g} \in \mathbf{Mat}_g(B)$, we define the d-tuple

$$\overrightarrow{M^{\mathrm{upp}}} := (m_{11}, m_{12}, \ldots, m_{1g}, m_{22}, \ldots, m_{2g}, m_{33}, \ldots, m_{gg}) \in B^d$$

whose entries m_{ij} with $1 \le i \le j \le g$ are listed in the lexicographic order.

Lemma 3.9.8 *There exists an affine open subscheme $\mathcal{Y} \subset \mathcal{X}$ which admits étale coordinates τ, for which $\overline{\mathcal{Y}}$ is nonempty, for which the abelian scheme $A_{\mathcal{Y}}$ has a column basis $\omega_{\mathcal{Y}}$ of 1-forms on $A_{\mathcal{Y}}/\mathcal{Y}$ and such that for the matrix*

$$f_{\mathcal{Y}} := f^1_{g,crys}(A_{\mathcal{Y}}, \theta_{\mathcal{Y}}, \omega_{\mathcal{Y}}, \mathcal{O}^{\infty}(\mathcal{Y})) \in \boldsymbol{Mat}_g(\mathcal{O}(J^1(\mathcal{Y})))$$

the following properties hold.

(a) *The d-tuple $\overrightarrow{f^{upp}_{\mathcal{Y}}} \in \mathcal{O}(J^1(\mathcal{Y}))^d$ is quasi-linear with respect to τ.*

(b) *The ideal $\mathfrak{a}$ generated by the entries of the matrix $f_{\mathcal{Y}}$ is equal to the ideal $\mathfrak{a}^{upp}$ generated by the entries of the d-tuple $\overrightarrow{f^{upp}_{\mathcal{Y}}}$.*[2]

(c) *The ideal $\mathfrak{a}$ is a δ-flow on $\mathcal{Y}$.*

Proof We recall that (A_0, θ_0, e) is as in Display (3.3). Equipping (A_0, θ_0) with a symplectic similitude level-N structure, we get a k-valued point P_0 of $\mathcal{X}$. Let $P \in \mathcal{X}(R)$ be the R-valued point which is the canonical lift of P_0, i.e., is defined by the canonical lift A_P of A_0.

We take $X \subset \mathcal{X}$ to be an affine open subscheme that admits étale coordinates τ and contains P. We denote $T_{11}, T_{12}, \ldots, T_{1g}, T_{22}, \ldots, T_{2g}, T_{33}, \ldots, T_{gg}$ by $z_1, \ldots, z_d$ and for $1 \leq j < i \leq g$ we will use as before $T_{ij} := T_{ji}$; so $T^{\mathrm{t}} = T$ and $R[[\underline{z}]] = R[[T]] = S^0_{\mathrm{for}}$ is as before. Let $\omega_{S^0_{\mathrm{for}}}$ be the column basis of 1-forms on the corresponding abelian scheme $A_{S^0_{\mathrm{for}}}/\mathrm{Spec}(S^0_{\mathrm{for}})$ that is attached to e. The restriction ω_P of $\omega_{S^0_{\mathrm{for}}}$ to A_P is obtained from $\omega_{S^0_{\mathrm{for}}}$ by setting $T = 0$. Replacing X by a smaller open subscheme we can assume there exists a column basis ω_X of 1-forms on A_X/X whose restriction to A_P equals ω_P. So we have

$$\omega_X = \lambda \cdot \omega_{S^0_{\mathrm{for}}}, \quad \lambda = \lambda(T) \in \mathbf{GL}_g(S^0_{\mathrm{for}}), \quad \lambda(0) = 1_g.$$

Write $\lambda = (\lambda_{ij})_{1 \leq i,j \leq g}$, $\lambda_{ij} = \lambda_{ij}(T)$. Replacing f by a $R^{\times}$-multiple of it and using the homogeneity property of $f = f^1_{g,\mathrm{crys}}$ and Theorem 3.6.16(a), we get that

$$f_X := f(A_X, \theta_X, \omega_X) = \lambda \cdot f(A_{S^0_{\mathrm{for}}}, \theta_{S^0_{\mathrm{for}}}, \omega_{S^0_{\mathrm{for}}}) \cdot (\lambda^{\mathrm{t}})^{\phi} = \lambda \cdot \Psi \cdot (\lambda^{\mathrm{t}})^{\phi}.$$

We have

$$\overrightarrow{f^{\mathrm{upp}}_X} = (f_{X,11}, f_{X,12}, \ldots, f_{X,1g}, f_{X,22}, \ldots, f_{X,2g}, f_{X,33}, \ldots, f_{X,gg}),$$

where for $1 \leq i, j \leq g$ we have (recall that Ψ is symmetric)

[2] It is also equal to the ideal generated by the entries of the symmetric matrix $f^{\langle 1 \rangle}_{g,\mathrm{crys}}(A_{\mathcal{Y}}, \theta_{\mathcal{Y}}, \omega_{\mathcal{Y}})$, provided $\mathcal{Y}$ modulo p is contained in the ordinary locus.

$$f_{X,ij} = \sum_{m,s=1}^{g} \lambda_{im}\Psi_{ms}\lambda_{sj}^{\phi}$$

$$= \sum_{1\leq m<s\leq g}(\lambda_{im}\lambda_{sj}^{\phi} + \lambda_{is}\lambda_{mj}^{\phi})\Psi_{ms} + \sum_{s=1}^{g}\lambda_{is}\lambda_{sj}^{\phi}\Psi_{ss}.$$

The d-tuple $\overrightarrow{\Psi^{\text{upp}}}$ is easily seen to be formally quasi-linear at P with respect to $\underline{z}$. Recall the ring $S^1_{\text{for}} := \widehat{S^0_{\text{for}}[T']}$ in Sect. 2.3.5 and write

$$\overrightarrow{f_X^{\text{upp}}} = Q \cdot \overrightarrow{\Psi^{\text{upp}}}, \quad Q = (q_{ij,ms})_{1\leq i\leq j\leq g, 1\leq m\leq s\leq g} \in \mathbf{Mat}_d(S^1_{\text{for}}),$$

where a d-tuple is identified with a matrix of size $d \times 1$ and

$$q_{ij,ms} := \lambda_{im}\lambda_{sj}^{\phi} + \lambda_{is}\lambda_{mj}^{\phi} \quad \text{for } m < s,$$

$$q_{ij,ss} := \lambda_{is}\lambda_{sj}^{\phi}.$$

Claim 3.9.9 *We have $Q = Q_0 + pQ_1$, $Q_0 \in \boldsymbol{GL}_d(S^0_{for})$ and $Q_1 \in \boldsymbol{Mat}_d(S^1_{for})$.*

Indeed the reduction $\overline{q}_{ij,ms} \in \overline{S^1_{\text{for}}} = k[[T]][T']$ of $q_{ij,ms}$ modulo p is actually in $\overline{S^0_{\text{for}}} = k[[T]]$, so we can write $\overline{q}_{ij,ms} = \overline{q}_{ij,ms}(T)$ and, using the Kronecker delta, we have

$$\overline{q}_{ij,ms}(0) = \delta_{im}\delta_{js} + \delta_{is}\delta_{jm} \quad \text{for } m < s,$$

and

$$\overline{q}_{ij,ss}(0) = \delta_{is}\delta_{js}.$$

So $\overline{q}_{ij,ms}(0) = 0$ for $(i, j) \neq (m, s)$ and $\overline{q}_{ms,ms} = 1$; in other words, the reduction modulo p of the matrix Q has entries in $\overline{S^0_{\text{for}}}$ and its evaluation at 0 is the identity matrix. This proves Claim 3.9.9.

By Claim 3.9.9 and the fact that $\overrightarrow{\Psi^{\text{upp}}}$ is formally quasi-linear at P with respect to $\underline{z}$ it follows that $\overrightarrow{f_X^{\text{upp}}}$ is formally quasi-linear at P with respect to $\underline{z}$. By Lemma 3.9.5 there exists an affine open subscheme $\mathcal{Y} \subset X$ as in the part (a) of the lemma, and we let $f_{\mathcal{Y}}$ and $\overrightarrow{f_{\mathcal{Y}}^{\text{upp}}}$ be the analogs of f_X and $\overrightarrow{f_X^{\text{upp}}}$ (respectively).

We prove part (b) of the lemma. Clearly $\mathfrak{a}^{\text{upp}} \subset \mathfrak{a}$. Thus we have closed embeddings

$$Z^1(\mathfrak{a}) = \text{Spf}(\mathcal{O}(J^1(\mathcal{Y}))/\mathfrak{a}) \subset Z^1(\mathfrak{a}^{\text{upp}}) = \text{Spf}(\mathcal{O}(J^1(\mathcal{Y}))/\mathfrak{a}^{\text{upp}}) \subset J^1(\mathcal{Y})$$

between formal schemes. By part (a), the projection

$$\overline{Z^1(\mathfrak{a}^{\text{upp}})} \subset \overline{J^1(\mathcal{Y})} \to \overline{\mathcal{Y}}$$

is an isomorphism. To prove that $\mathfrak{a} = \mathfrak{a}^{\text{upp}}$ it is enough to show that the closed embedding $Z^1(\mathfrak{a}) \subset Z^1(\mathfrak{a}^{\text{upp}})$ is an isomorphism, hence that the closed embedding $\overline{Z^1(\mathfrak{a})} \subset \overline{Z^1(\mathfrak{a}^{\text{upp}})}$ is an isomorphism. As $\overline{\mathcal{Y}}$ is reduced it is enough to show that the image of the composite homomorphism

$$\overline{Z^1(\mathfrak{a})} \subset \overline{Z^1(\mathfrak{a}^{\text{upp}})} \to \overline{\mathcal{Y}} \tag{3.106}$$

contains a Zariski open sense subset. Now the ordinary locus $\overline{\mathcal{Y}}_{\text{ord}}$ of $\overline{\mathcal{Y}}$ is open and dense. Let $P'_0 \in \overline{\mathcal{Y}}_{\text{ord}}(k)$ be a k-valued point and let $P' \in \mathcal{Y}(R)$ be the R-valued point that defines the canonical lift of the abelian variety over Spec(k) that corresponds to P'_0. Let $J^1(P') \in J^1(\mathcal{Y})(R)$ be the first jet of P'. Then all the components $f_{\mathcal{Y},ij}$ of $\overrightarrow{f_{\mathcal{Y}}^{\text{upp}}}$ vanish at $J^1(P')$ hence $\overline{J^1(P')} \in \overline{Z^1(\mathfrak{a})}(k)$, and thus $\overline{J^1(P')}$ maps to P'_0. This shows that the image of the composite homomorphism (3.106) contains a dense open subscheme which ends our proof of part (b).

Part (c) follows from parts (a) and (b) plus Lemma 3.9.7. □

Lemma 3.9.10 *Let $\pi : X' \to X$ be an étale morphism between affine smooth schemes over Spec(R) and let $\mathfrak{a} \subset \mathcal{O}(J^1(X))$. Let $\mathfrak{a}' := \mathfrak{a} \cdot \mathcal{O}(J^1(X'))$. If $\mathfrak{a}$ is a δ-flow on X then $\mathfrak{a}'$ is a δ-flow on X'. The converse also holds provided π is surjective.*

Proof As π is étale, based on [Bu95a], Prop. 1.4, we have an isomorphism

$$J^1(X') \simeq J^1(X) \times_{\widehat{X}} \widehat{X}'.$$

Hence $Z^1(\mathfrak{a}') \simeq Z^1(\mathfrak{a}) \times_{\widehat{X}} \widehat{X}'$. Clearly, if $Z^1(\mathfrak{a}) \to \widehat{X}$ is an isomorphism, then $Z^1(\mathfrak{a}') \to \widehat{X}'$ is an isomorphism; the converse also holds if π is surjective because then π is faithfully flat. □

Corollary 3.9.11 *Let X_1, X_2, Y be affine smooth schemes over Spec(R) and let*

$$X_1 \xleftarrow{\pi_1} Y \xrightarrow{\pi_2} X_2$$

be étale morphism with π_1 surjective. For $m \in \mathbb{N}$ let $g^{(1)} \in \mathcal{O}(J^1(X_1))^m$ and $g^{(2)} \in \mathcal{O}(J^1(X_2))^m$ be m-tuples of elements (viewed as $m \times 1$ matrices) such that we have

$$\pi_1^* g^{(1)} = M_{12} \cdot \pi_2^* g^{(2)}$$

for some matrix $M_{12} \in \boldsymbol{GL}_m(\mathcal{O}(J^1(Y)))$. Let $\mathfrak{a}_1$ and $\mathfrak{a}_2$ be the ideals in $\mathcal{O}(J^1(X_1))$ and $\mathcal{O}(J^1(X_2))$ generated by the entries of $g^{(1)}$ and $g^{(2)}$ (respectively). If $\mathfrak{a}_2$ is a δ-flow on X_2, then $\mathfrak{a}_1$ is a δ-flow on X_1.

Proof This follows directly from Lemma 3.9.10 applied to π_2 and π_1. □

If X is an affine open subscheme of $\mathcal{X}$ such that there exists a column basis ω_X of 1-forms on A_X/X, then the ideal $\mathfrak{a}_X$ of $\mathcal{O}(J^1(X))$ generated by the entries of

$$f_X := f^1_{g,\mathrm{crys}}(A_X, \theta_X, \omega_X, \mathcal{O}^\infty(X)) \in \mathbf{Mat}_g(\mathcal{O}(J^1(X)))$$

does not depend on the choice of the basis ω_X.

We now take X to be an arbitrary affine open subscheme of $\mathcal{X}$. For each affine cover $X = \cup X_\alpha$ with every X_α such that there exists a column basis ω_{X_α} of 1-forms on A_{X_α}/X_α, the ideals $\mathfrak{a}_{X_\alpha}$ glue together to give a well-defined ideal $\mathfrak{a}_X$ in $\mathcal{O}(J^1(X))$ that does not depend on the affine cover.

Theorem 3.9.12 *Let X be an affine open subscheme of $\mathcal{X}$ whose reduction modulo p is nonempty and contained in the ordinary locus $\overline{\mathcal{X}}_{ord}$ of $\overline{\mathcal{X}}$. Then the ideal $\mathfrak{a}_X$ is a δ-flow on X.*

Proof Let $\mathcal{Y} \subset X$ and $\omega_{\mathcal{Y}}$ be as in Lemma 3.9.8 and its proof. It suffices to show that X can be covered with affine open subschemes X_α such that $\mathfrak{a}_{X_\alpha}$ is a δ-flow on X_α. So we can assume that there exists a column basis ω_X of 1-forms on A_X/X and it suffices to show that each point $P_0 \in X(k)$ is contained in an affine open subscheme X_0 of X such that $\mathfrak{a}_{X_0}$ is a δ-flow on X_0.

Let ℓ be a prime that does not divide pN. By the density of the ℓ-power polarized isogeny classes of ordinary points (see [Cha], Thm. 2) there exists an isogeny

$$u_0 : A_{P_0} \to A_{P_0'},$$

where (A_{P_0}, θ_{P_0}) is the principally polarized abelian variety over $\mathrm{Spec}(k)$ that corresponds to P_0 and where $(A_{P_0'}, \theta_{P_0'})$ is the principally polarized abelian variety over $\mathrm{Spec}(k)$ that corresponds to some point $P_0' \in \mathcal{Y}(k)$, such that u_0 preserves the principal polarizations up to multiplication by ℓ^{s_0} for some $s_0 \in \mathbb{N} \cup \{0\}$ (i.e., the pullback of $\theta_{P_0'}$ by u_0 equals θ_{P_0} times ℓ^{s_0}). Thus $\deg(u_{\mathbb{C}}) = \ell^{2gs_0}$ is prime to p.

Let

$$\mathcal{X} \overset{\pi_1}{\leftarrow} \mathcal{X}' \overset{\pi_2}{\rightarrow} \mathcal{X}$$

be the Hecke correspondence that parameterizes pairs of principally polarized abelian schemes of relative dimension g with symplectic similitude level-N structures together with an isogeny of degree ℓ^{2gs_0} which preserves the principal polarizations up to multiplication by ℓ^{s_0} and let

$$u_{\mathcal{X}'} : \pi_1^* A_{\mathcal{X}} \to \pi_2^* A_{\mathcal{X}}$$

be the corresponding universal isogeny of abelian schemes over $\mathcal{X}'$ (compatible with the principal polarizations).

It is well-known that $\mathcal{X}'$ is smooth and that the two projections π_1, π_2 are finite étale.[3] We recall the argument for this.

We fix a perfect alternating form $\psi_{2g} : \mathbb{Z}_\ell^{2g} \times \mathbb{Z}_\ell^{2g} \to \mathbb{Z}_\ell$. Working in the finite étale topology of $\mathcal{X}$, we can assume that there exists a level-ℓ^{s_0} structure on $A_{\mathcal{X}}$. But

[3] For the case of the stack of polarized isogenies of degree a power of ℓ (with $\ell \neq p$ a prime) over the stack $\mathcal{A}_{g,1,\mathbb{Z}_{(p)}}$ see [FC], Ch. 7, Sect. 3, p. 252.

in this case $\mathcal{X}'$ is the disjoint union of copies of $\mathcal{X}$ indexed by $\mathbb{Z}_\ell$-lattices $\mathcal{L}$ of $\mathbb{Q}_\ell^{2g}$ which satisfy the following property.

(i) *We have* $\ell^{s_0}\mathbb{Z}_\ell^{2g} \subset \mathcal{L} \subset \ell^{-s_0}\mathbb{Z}_\ell^{2g}$, s_0 *is the smallest whole number with this property and we get a perfect alternating form* $\ell^m \psi_{2g} : \mathcal{L} \times \mathcal{L} \to \mathbb{Z}_\ell$ *for a suitable* $m \in \mathbb{Z}$.

Moreover, the two projections π_1 and π_2 induce isomorphisms when restricted to each $\mathcal{X}$ copy of $\mathcal{X}'$.

We view $u_0 : A_{P_0} \to A_{P_0'}$ as a k-valued point of $\mathcal{X}'$. Setting

$$X' := \pi_1^{-1}(X) \cap \pi_2^{-1}(\mathcal{Y}) \subset \mathcal{X}'$$

we have that u_0 belongs to $X'(k)$. As π_1 is flat, $\pi_1(X')$ is open in X. Hence there exists an affine open subscheme X_0 of $\pi_1(X')$ that contains P_0. Setting $X_0' := \pi_1^{-1}(X_0)$ we have a diagram

$$X_0 \overset{\Pi_1}{\leftarrow} X_0' \overset{\Pi_2}{\rightarrow} \mathcal{Y}$$

with Π_1, Π_2 induced by π_1, π_2, respectively, and an induced isogeny

$$u_{X_0'} : \Pi_1^* A_{X_0} \to \Pi_2^* A_{\mathcal{Y}}$$

of abelian schemes over X_0'. Consider the matrix

$$f_{X_0} := f^1_{g,\mathrm{crys}}(A_{X_0}, \theta_{X_0}, \omega_{X_0}, \mathcal{O}^\infty(X_0)) \in \mathbf{Mat}_g(\mathcal{O}(J^1(X_0)))$$

where θ_{X_0} is induced by θ_X. By Hecke covariance we have

$$\Pi_1^* f_{X_0} = \beta_1 \cdot \Pi_2^* f_{\mathcal{Y}} \cdot \beta_2$$

for some $\beta_1, \beta_2 \in \mathbf{GL}_g(\mathcal{O}(J^1(X_0')))$. We apply Corollary 3.9.11 by taking the quadruple (X_1, X_2, Y, m) to be equal to $(X_0, \mathcal{Y}, X_0', g^2)$ and with

$$Q_{12} \in \mathbf{GL}_{g^2}(\mathcal{O}(J^1(X_0')))$$

as the matrix representation (with respect to some standard bases) of the $\mathcal{O}(J^1(X_0'))$-linear automorphism defined by the rule $\gamma \mapsto \beta_1 \cdot \gamma \cdot \beta_2$, $\gamma \in \mathbf{Mat}_g(\mathcal{O}(J^1(X_0')))$; as the ideal $\mathfrak{a}_{\mathcal{Y}}$ generated by the entries of $f_{\mathcal{Y}}$ is a δ-flow on $\mathcal{Y}$ (see Lemma 3.9.8(c)), it follows that the ideal $\mathfrak{a}_{X_0}$ generated by the entries of f_{X_0} is also a δ-flow on X_0. □

3.10 Reciprocity for Canonical Lifts

The purpose of this section is to derive an application of our flow theorem (see Theorem 3.9.12) which extends to arbitrary genus g and all primes p the 'Reciprocity Theorem for CL points' in [BPII], Thm. 3.5.

Definition 3.10.1 Consider a finite normal extension R' of $\mathbb{Z}_p$ contained in R and A' an abelian scheme over R'. Let p^ν be the cardinality of the field R'/pR' and let A be the abelian scheme over R obtained from A' via base change. We say that A' is *nondegenerate* if the following two conditions, thought as nondegeneration axioms, hold.

(NAX1) The characteristic polynomial of the p^ν-power Frobenius endomorphism of $\overline{A'}$ has simple roots only.

(NAX2) The Serre–Tate matrix $q(A) := (q_A(e_i, \check{e}_j))_{1\le i,j\le d}$ of A with respect to some (equivalently every) bases $e_1, \dots, e_d$ and $\check{e}_1, \dots, \check{e}_d$ of $T_p(A_0)$ and $T_p(\check{A}_0)$ respectively satisfies the condition

$$\det\left(p^{-1}(q(A) - \mathbb{E}_d)\right) \not\equiv 0 \mod p$$

(recall that $\mathbb{E}_d$ is the $d \times d$ matrix all of whose entries are 1).

For A and A' as above we denote by $A(\mathbb{Z}_p^{\text{ur}})$ the set $A'(\mathbb{Z}_p^{\text{ur}})$ of $\mathbb{Z}_p^{\text{ur}}$-points of A' where $\mathbb{Z}_p^{\text{ur}}$ is the maximal unramified extension extension of $\mathbb{Z}_p$ in R; this makes sense as $R' \subset \mathbb{Z}_p^{\text{ur}}$. We identify $R = \widehat{\mathbb{Z}_p^{\text{ur}}}$. Note that the set $A'(\mathbb{Z}_p^{\text{ur}})$ is independent of the choice of the model A'/R' of A/R. This is so as the pullback functor from the category of abelian schemes over $\text{Spec}(\mathbb{Z}_p^{\text{ur}})$ to the category of abelian schemes over $\text{Spec}(R)$ is full faithful.

Recall that $\mathcal{X}_{\text{ord}}$ denotes the ordinary locus of $\mathcal{X} := \mathcal{A}_{g,1,N,R}$ with $N \ge 3$. Let $\text{CL}_R \subset \mathcal{X}_{\text{ord}}(R)$ be the set of canonical lift points (see Remark 2.4.11). Also, for an abelian group Γ we denote by Γ_{tors} its torsion subgroup.

The main result of this section is the following one.

Theorem 3.10.2 *Assume we are given morphisms of schemes over R*

$$\Pi : Y \to X, \quad \Phi : Y \to A$$

with $X \subset \mathcal{X}_{ord}$ affine, Y affine, Π étale, and A an abelian scheme of relative dimension d. Assume that X, Y, A, Π, Φ descend to corresponding objects X', Y', A', Π', Φ' defined over a finite normal extension R' of $\mathbb{Z}_p$ contained in R and assume A' is nondegenerate. Then there exists a morphism of p-adic formal schemes

$$\Phi^+ : \widehat{Y} \to \widehat{\mathbb{G}_a^d}$$

such that for all integers $n \ge 1$, all points $P_1, \dots, P_n \in \Pi^{-1}(CL_R) \subset Y(R)$ and all integers $m_1, \dots, m_n \in \mathbb{Z}$, the following equivalence holds

$$\sum_{i=1}^{n} m_i \Phi(P_i) \in A(R)_{tors} \quad \Longleftrightarrow \quad \sum_{i=1}^{n} m_i \Phi^+(P_i) = 0 \in R^d.$$

Above, the three maps $\Pi(R) : Y(R) \to X(R)$, $\Phi(R) : Y(R) \to A(R)$, and $\Phi^+(R) : Y(R) = \widehat{Y}(R) \to \widehat{\mathbb{G}_a^d}(R) = R^d$ were denoted simply by Π, Φ, and Φ^+

(respectively); here $\widehat{Y}(R)$ denotes, as usual, the set of Spf(R)-points of $\widehat{Y}$. A map such as Φ^+ above was called in [BPII] a *reciprocity map*; we refer to [BPII], Rmk. 3.7 for comments on this terminology. In particular we get the following consequence.

Corollary 3.10.3 *In the situation of Theorem 3.10.2 assume in addition that there exists $Q \in \Pi^{-1}(CL_R)$ such that $\Phi(Q) \notin A(R)_{tors}$. Then the image of the map*

$$\Pi^{-1}(CL_R) \cap \Phi^{-1}(A(R)_{tors}) \to Y(R) \to Y(k) \tag{3.107}$$

is not Zariski dense in $\overline{Y}$. (Here the first arrow in Diagram (3.107) is the inclusion and $Y(k) = \overline{Y}(k)$.)

Proof Taking $n = 1$ and $m_1 = 1$ in Theorem 3.10.2 we get that for every valued point $P \in \Pi^{-1}(\mathrm{CL}_R) \subset Y(R)$ we have

$$\Phi(P) \in A(R)_{\mathrm{tors}} \iff \Phi^+(P) = 0. \tag{3.108}$$

By the existence of Q there exists an index $j \in \{1, \ldots, d\}$ such that the jth component $\Phi_j^+ \in \mathcal{O}(\widehat{Y})$ of Φ^+ is nonzero. Write $\Phi_j^+ = p^N \Phi_{j,0}^+$ with $N \geq 0$ and $\Phi_{j,0}^+ \in \mathcal{O}(\widehat{Y}) \backslash p\mathcal{O}(\widehat{Y})$. Hence the image $\overline{\Phi_{j,0}^+}$ of $\Phi_{j,0}^+$ in $\mathcal{O}(\overline{Y})$ is nonzero. By the equivalence in Display (3.108) the image of the composite of maps in Diagram (3.107) is contained in the zero locus of $\overline{\Phi_{j,0}^+} \in \mathcal{O}(\overline{Y})$. □

In order to prove Theorem 3.10.2 recall from [Bu95a], Introd. that a *δ-character* of order $\leq r$ of an abelian scheme A over R is a δ-function $\psi : A(R) \to R = \mathbb{G}_a(R)$ of order $\leq r$ in the sense of Definition 2.1.10 which is also a group homomorphism. We need the following result from [Bu97], Cor. 1.7.

Theorem 3.10.4 *Consider a finite normal extension R' of $\mathbb{Z}_p$ contained in R, a nondegenerate abelian scheme A' over R' of relative dimension d, and the abelian scheme A over R obtained from A' via base change. Then there exist δ-characters $\psi_1, \ldots, \psi_d : A(R) \to R$ of order ≤ 2 such that the map*

$$\psi := (\psi_1, \ldots, \psi_d) : A(R) \to R^d = \mathbb{G}_a^d(R)$$

is surjective and satisfies

$$A(\mathbb{Z}_p^{ur}) \cap Ker(\psi) = A(\mathbb{Z}_p^{ur})_{tors}.$$

The theorem above can be viewed as an arithmetic analog of Manin's Theorem of the Kernel [Man], Sect. 5, Thm. 2.

Proof of Theorem 3.10.2. Let $\psi_1, \ldots, \psi_d$ be as in Theorem 3.10.4. For $j \in \{1, \ldots, d\}$ we consider the δ-function $\psi_j \circ \Phi$ on $Y(R)$, identified with an element of $\mathcal{O}(J^2(Y))$. On the other hand, by Theorem 3.9.12 the ideal $\mathfrak{a}_X \subset \mathcal{O}(J^1(X))$

is a δ-flow on X. By Lemma 3.9.10 the ideal $\mathfrak{a}_X \cdot \mathcal{O}(J^1(Y))$ (generated by the image of $\mathfrak{a}_X$ via the homomorphism $\Pi^* : \mathcal{O}(J^1(X)) \to \mathcal{O}(J^1(Y))$ induced by Π) is a δ-flow on Y. By Remark 3.9.2 the algebra homomorphism

$$\mathcal{O}(\widehat{Y}) \to \mathcal{O}(J^2(Y))/(\mathfrak{a}_X, \delta\mathfrak{a}_X)$$

is an isomorphism which, for simplicity, we view as an identity. Thus for every $j \in \{1, \ldots, d\}$ we have an equality of functions $Y(R) \to R$,

$$\psi_j \circ \Phi = \Phi_j^+ + \sum_{m=1}^{M} a_{j,m} \cdot (\varphi_{j,m} \circ \Pi) + \sum_{m=1}^{M} b_{j,m} \cdot (\delta \circ \varphi_{j,m} \circ \Pi), \tag{3.109}$$

with $\Phi_j^+ \in \mathcal{O}(\widehat{Y})$, $M \in \mathbb{N}$, where for $m \in \{1, \ldots, M\}$ $a_{j,m}, b_{j,m} \in \mathcal{O}(J^2(Y))$ are identified with functions $Y(R) \to R$ and $\varphi_{j,m} \in \mathfrak{a}_X$ is identified with a function $X(R) \to R$. These identifications are obtained by viewing the injective R-algebra homomorphism of Diagram (2.3) as an inclusion. Also the dots in Eq. (3.109) denote usual multiplication of R-valued functions. Consider the morphism of p-adic formal schemes

$$\Phi^+ := (\Phi_1^+, \ldots, \Phi_d^+) : \widehat{Y} \to \widehat{\mathbb{G}_a^d}.$$

The functions $\varphi_{j,m} \circ \Pi$ vanish on every point $P \in \Pi^{-1}(\mathrm{CL}_R)$. So for each such P we have

$$\psi_j(\Phi(P)) = \Phi_j^+(P). \tag{3.110}$$

Assume now that $P_1, \ldots, P_n \in \Pi^{-1}(\mathrm{CL}_R)$ and $m_1, \ldots, m_n \in \mathbb{Z}$.

Claim 3.10.5 *We have $P_1, \ldots, P_n \in Y'(\mathbb{Z}_p^{ur})$.*

Indeed as Π is étale it is sufficient to show that for each $j \in \{1, \ldots, n\}$, the image $\Pi(P_j) \in X(R)$ in fact belongs to $X'(\mathbb{Z}_p^{\mathrm{ur}})$. Recall now the Shimura–Taniyama theorem that abelian varieties with complex multiplication over an algebraically closed field of characteristic 0 are defined over number fields (for instance, see [CCO], Thm. 1.7.2.1). Note that all polarizations and symplectic similitude level-N structures on such abelian varieties are then also defined over (possibly larger) number fields. We deduce that $\Pi(P_j) \in X'(F)$ for some finite extension F of the fraction field of R' contained in an algebraic closure of K. As $F \cap R \subset \mathbb{Z}_p^{\mathrm{ur}}$ our claim follows.

By Claim 3.10.5, we have $\Phi(P_1), \ldots, \Phi(P_n) \in A(\mathbb{Z}_p^{\mathrm{ur}})$. By Theorem 3.10.4, and by the fact that ψ_j are group homomorphisms, the condition

$$\sum_{i=1}^{n} m_i \Phi(P_i) \in A(R)_{\mathrm{tors}}$$

is equivalent to the system of equation

$$\sum_{i=1}^{n} m_i \psi_j(\Phi(P_i)) = 0 \quad \text{for} \quad j \in \{1, \dots, d\}. \tag{3.111}$$

By Eq. (3.110) we get that System of Eqs. (3.111) is equivalent to the equation

$$\sum_{i=1}^{n} m_i \Phi^+(P_i) = 0.$$

□

3.11 Some Computations of Cohomology Groups

In this section we present a few basic computations of cohomology groups of some simple ss-δ-spaces that appeared naturally in our monograph. We will not pursue here more general cases.

Proposition 3.11.1 *For* $\mathbb{P}^1_\delta := Proj_\delta(\mathbb{S}_1, (z_0, z_1))$ *we have*

$$H^0(\mathbb{P}^1_\delta) = R \quad and \quad \phi\text{-}rank_R(H^1(\mathbb{P}^1_\delta)) = \infty.$$

In what follows we use the forms $f^{\langle r \rangle}$ in Display (2.62).

Proposition 3.11.2 *For* $Z_{\delta,n} := Proj_\delta(\mathbb{I}_{1,ord}, (f^{\langle 1 \rangle}, \dots, f^{\langle n \rangle}))$, $n \geq 2$ *we have*

$$H^0(Z_{\delta,n}) = \mathbb{I}_{1,ord,\langle f^{\langle n \rangle}\rangle} \quad and \quad H^d(Z_{\delta,n}) = 0 \quad for \quad d \geq 1.$$

Theorem 3.11.3 *For* $\lambda = (\lambda_{ij})_{1 \leq i,j \leq 2} \in \boldsymbol{GL}_2(R)$ *set* $\begin{pmatrix} f_1^\lambda \\ f_2^\lambda \end{pmatrix} := \lambda \cdot \begin{pmatrix} f^{\langle 1 \rangle} \\ f^{\langle 2 \rangle} \end{pmatrix}$ *and*

$$Z^\lambda_{\delta,2} := Proj_\delta(\mathbb{I}_{1,ord}, (f_1^\lambda, f_2^\lambda)).$$

If all entries λ_{ij} *of* λ *are nonzero, then we have*

$$H^0(Z^\lambda_{\delta,2}) = R \quad and \quad \phi\text{-}rank_R(H^1(Z^\lambda_{\delta,2})) = \infty.$$

To state a partial generalization of Theorem 3.11.3 we first recall the maps $\sigma, \tau : \mathbb{H}_g(s) \to \mathbb{H}_g(s)$ in Displays (3.52) and (3.53) and $\diamond : \mathbb{H}_g(s) \to \mathbb{I}_{g,\mathrm{ord}}(-2s)$ in Proposition 3.6.19(a).

Definition 3.11.4 The elements of the sets

$$\mathbb{H}_g(1) \backslash \sigma(\mathbb{H}_g(1)) \quad \text{and} \quad \diamond(\mathbb{H}_g(1) \backslash \sigma(\mathbb{H}_g(1)))$$

are called primitive.

Proposition 3.11.5 *Assume $f = (f_0, \dots, f_n)$ is a linear system of weight -2 in $\mathbb{I}_{g,ord}$ such that f_0 and f_1 are primitive and R-linearly independent. Then we have*

$$H^0(Proj_\delta(\mathbb{I}_{g,ord}, f)) = R.$$

For $\mathbb{I}_g$ we have the following property.

Proposition 3.11.6 *There exists a linear system (φ_1, φ_2) of the following weight $-1 - \phi - \phi^2 - \phi^3$ in $\mathbb{I}_g$ such that*

$$H^0(Proj_\delta(\mathbb{I}_g, (\varphi_1, \varphi_2))) = R.$$

To prove Proposition 3.11.2 we first need the following lemma.

Lemma 3.11.7 *Let B be an integral δ-graded ring. For $w \in W$ and $n \in \mathbb{N}$ let $f_0, \dots, f_n \in B(w) \backslash pB(w)$. If there exist units $g_1, \dots, g_n \in B^\times$ such that we have $f_i = g_i (f_0)^{\phi^i}$ for all $i \in \{1, \dots, n\}$, then for $Z_\delta := Proj_\delta(B, (f_0, \dots, f_n))$ we have*

$$H^0(Z_\delta) = B_{\langle f_n \rangle} \text{ and } H^d(Z_\delta) = 0 \text{ for } d \geq 1.$$

Proof The hypotheses imply that for all integers $d \geq 1$ and $0 \leq i_0 < \dots < i_d \leq n$ we have

$$B_{\langle f_{i_0} \cdots f_{i_d} \rangle} = B_{\langle f_{i_0} \rangle} \supset \dots \supset B_{\langle f_{i_d} \rangle},$$

and this implies that $H^0(Z_\delta) = B_{\langle f_n \rangle}$. Now let $(a_{i_0 \cdots i_d})_{0 \leq i_0 < \dots < i_d \leq n}$ be a cocycle in

$$B_d := \prod_{0 \leq i_0 < \dots < i_\bullet \leq d} B_{\langle f_{i_0} \cdots f_{i_\bullet} \rangle}.$$

For integers $0 \leq i_0 < \dots < i_{d-1} \leq n$ define

$$b_{i_0 \cdots i_{d-1}} := a_{i_0 \cdots i_{d-1} n} \text{ if } i_{d-1} < n,$$

$$b_{i_0 \cdots i_{d-1}} := 0 \text{ if } i_{d-1} = n.$$

One easily checks, using the cocycle condition for $(a_{i_0 \cdots i_d})_{0 \leq i_0 < \dots < i_d \leq n}$, that the latter is the image by $\partial_{d-1} : B_{d-1} \to B_d$ of $(b_{i_0 \cdots i_{d-1}})_{0 \leq i_0 < \dots < i_{d-1} \leq n} \in B_{d-1}$ hence $H^d(Z_\delta) = 0$. □

Proof of Proposition 3.11.2. This follows from Lemma 3.11.7 applied to the $n+3$-tuple $(B, n, f_0, \dots, f_n)$ equal to $(\mathbb{I}_{1,\text{ord}}, n-1, f^{\langle 1 \rangle}, \dots, f^{\langle n \rangle})$. This is possible as $n \geq 2$ and for $i \in \{2, \dots, n\}$ we have $f^{\langle i \rangle} = (f^\partial)^{\frac{\phi^i + \phi^{i-1} - 2}{\phi - 1}} (f^{\langle 0 \rangle})^{\phi^{i-1}}$ by Eq. (2.62) and $f^\partial = f^\partial_{1,\text{crys}}$ and hence also $(f^\partial)^{\frac{\phi^i + \phi^{i-1} - 2}{\phi - 1}}$ is a unit of $\mathbb{I}_{1,\text{ord}}$ by Theorem 2.4.5(b) which follows from Theorem 2.4.5(a) proved in Theorem 3.6.17(a) and (b). □

The following lemma will be used in the proofs of Proposition 3.11.1 and Theorem 3.11.3.

Lemma 3.11.8 *Let B be an integral δ-graded ring with $B(0) = R$ and with p as a prime element. Let $w \in W$ and let $(f_i)_{i\in\{0,\dots,n\}}$ be a family of elements of $B(w)\backslash pB(w)$. Assume that the principal ideals generated in $B \otimes_R K$ by each of the members of the family $(f_i^{\phi^j})_{(i,j)\in\{0,1\}\times(\mathbb{N}\cup\{0\})}$ are prime and distinct. Then, for $Z_\delta := Proj_\delta(B, (f_0, \dots, f_n))$ we have $H^0(Z_\delta) = R$.*

Proof Every element in $H^0(Z_\delta)$ can be written as

$$\frac{F_0}{f_0^{w_0}} = \frac{F_1}{f_1^{w_1}}$$

with $w_i \in W_+$ and $F_i \in B(ww_i)$ for $i \in \{0, 1\}$. Clearing out the denominators we get that $f_0^{w_0}$ divides F_0 in $B \otimes_R K$ hence in B (because f_0 is not divisible in B by the prime element p). So

$$\frac{F_0}{f_0^{w_0}} \in B(0) = R.$$

□

For the proof of Proposition 3.11.1 we will need the following extra lemma.

Lemma 3.11.9 *Let $(x_1), (y_1), (x_2), (y_2)$ be distinct principal ideals of an integral domain B such that $y_1x_2 \notin (x_1, y_2)$ and let $f_1, f_2 \in B$. Then for all nonnegative integers n_1, m_1, n_2, m_2 we have*

$$\frac{y_1x_2}{x_1y_2} \neq \frac{f_1}{x_1^{n_1}y_1^{m_1}} - \frac{f_2}{x_2^{n_2}y_2^{m_2}}.$$

Proof We show that the assumption that there exist $n_1, m_1, n_2, m_2 \in \mathbb{N} \cup \{0\}$ such that $\frac{y_1x_2}{x_1y_2} = \frac{f_1}{x_1^{n_1}y_1^{m_1}} - \frac{f_2}{x_2^{n_2}y_2^{m_2}}$ leads to a contradiction. We can assume the sum $n_1 + m_1 + n_2 + m_2$ is minimal. Clearing out the denominators and using the fact that the principal prime ideals $(x_1), (y_1), (x_2), (y_2)$ are distinct, we get that $n_1 = 1$, $m_1 = 0, n_2 = 0, m_2 = 1$, hence $y_1x_2 = y_2f_1 - x_1f_2 \in (x_1, y_2)$, a contradiction. □

We view Lemma 3.11.9 as an abstraction of the fact that if B is a field and $P, Q \in \mathbb{P}^1_B(B)$ are two distinct points, then we have $H^1(X, \mathcal{O}_X) \neq 0$ for

$$X := (\mathbb{P}^1_B \times \mathbb{P}^1_B)\backslash\{(P, Q), (Q, P)\}.$$

Proof of Proposition 3.11.1. Its identity $H^0(\mathbb{P}^1_\delta) = R$ follows from Lemma 3.11.8 applied to $B = \mathbb{S}_1$. For $H^1(\mathbb{P}^1_\delta)$ it is enough to check that there exists no nontrivial R-linear combination

$$\sum_{b,a\in\mathbb{Z},b>a\geq 0} \lambda_{cb} \frac{z_1^{\phi^a} z_0^{\phi^b}}{z_0^{\phi^a} z_1^{\phi^b}} \in \mathbb{S}_{1,\langle z_0 z_1\rangle}, \quad \lambda_{ab} \in R, \tag{3.112}$$

which can be written as a difference

$$\frac{F_0}{z_0^{w_0}} - \frac{F_1}{z_1^{w_1}}, \quad F_0 \in \mathbb{S}_1(w_0),\ F_1 \in \mathbb{S}_1(w_1),\ w_0, w_1 \in W_+. \tag{3.113}$$

To check this, we assume there exists an equality

$$\sum_{b,a\in\mathbb{Z},b>a\geq 0} \lambda_{ab} \frac{z_1^{\phi^a} z_0^{\phi^b}}{z_0^{\phi^a} z_1^{\phi^b}} = \frac{F_0}{z_0^{w_0}} - \frac{F_1}{z_1^{w_1}}$$

between the elements appearing in Displays (3.112) and (3.113) such that we have $\lambda_{a_0 b_0} \neq 0$ for some integers $b_0 > a_0 \geq 0$. We view this equality as an identity between rational functions in the indeterminates $z_i^{\phi^c}$ and substitute $z_i^{\phi^c}$ by 1 for $c \in (\mathbb{N} \cup \{0\})\backslash\{a_0, b_0\}$ and $i \in \{0, 1\}$ in this identity. Under this substitution, if (a, b) is a pair of integers with $b > a \geq 0$ and $(a, b) \neq (a_0, b_0)$, then the term $\lambda_{ab} \frac{z_1^{\phi^a} z_0^{\phi^b}}{z_0^{\phi^a} z_1^{\phi^b}}$ becomes $\lambda_{a_0 b} \frac{z_1^{\phi^{a_0}}}{z_0^{\phi^{a_0}}}$ if $a = a_0$, becomes $\lambda_{ab_0} \frac{z_0^{\phi^{b_0}}}{z_1^{\phi^{b_0}}}$ if $b = b_0$, and becomes λ_{ab} if $a \neq a_0$ and $b \neq b_0$. Moreover, the difference $\frac{F_0}{z_0^{w_0}} - \frac{F_1}{z_1^{w_1}}$ becomes a difference $\frac{F_0'}{z_0^{w_0'}} - \frac{F_1'}{z_1^{w_1'}}$ in which

$$F_0', F_1' \in R[z_0^{\phi^{a_0}}, z_0^{\phi^{b_0}}, z_1^{\phi^{a_0}}, z_1^{\phi^{b_0}}]$$

and w_0', w_1' are linear combinations of ϕ^{a_0} and ϕ^{b_0} with nonnegative integers coefficients . We get an equality

$$\frac{z_1^{\phi^{a_0}} z_0^{\phi^{b_0}}}{z_0^{\phi^{a_0}} z_1^{\phi^{b_0}}} = \frac{G_0}{z_0^{v_0}} - \frac{G_1}{z_1^{v_1}}$$

where $v_0 := w_0' + \phi^{a_0}$, $v_1 := w_1' + \phi^{b_0}$,

$$G_0 := \lambda_{a_0 b_0}^{-1} z_0^{v_0} \left(\frac{F_0'}{z_0^{w_0'}} - \sum_{b\neq b_0} \lambda_{a_0 b} \frac{z_1^{\phi^{a_0}}}{z_0^{\phi^{a_0}}} - \sum_{a\neq a_0, b\neq b_0} \lambda_{ab} \right) \in K[z_0^{\phi^{a_0}}, z_0^{\phi^{b_0}}, z_1^{\phi^{a_0}}, z_1^{\phi^{b_0}}],$$

$$G_1 := \lambda_{a_0 b_0}^{-1} z_1^{v_1} \left(\frac{F_1'}{z_1^{w_1'}} + \sum_{a\neq a_0} \lambda_{ab_0} \frac{z_0^{\phi^{b_0}}}{z_1^{\phi^{b_0}}} \right) \in K[z_0^{\phi^{a_0}}, z_0^{\phi^{b_0}}, z_1^{\phi^{a_0}}, z_1^{\phi^{b_0}}].$$

We get a contradiction based on Lemma 3.11.9 applied to

$$(B, x_1, y_1, x_2, y_2) := (\mathbb{S}_1, z_0^{\phi^{a_0}}, z_0^{\phi^{b_0}}, z_1^{\phi^{a_0}}, z_1^{\phi^{b_0}}).$$

□

Proof of Theorem 3.11.3. By Theorem 3.6.29 (a), we have

$$\mathbb{I}_{1,\mathrm{ord}} \otimes_R K = K[f^{\langle 1\rangle}, f^{\langle 2\rangle}, f^{\langle 3\rangle}, \dots, f^{\partial}, (f^{\partial})^{-1}, (f^{\partial})^{\phi}, (f^{\partial})^{-\phi}, (f^{\partial})^{\phi^2}, (f^{\partial})^{-\phi^2} \dots],$$

where the ordinary Siegel δ-modular forms $f^{\langle 1\rangle}, f^{\langle 2\rangle}, f^{\langle 3\rangle}, \dots, f^{\partial}, (f^{\partial})^{\phi}, (f^{\partial})^{\phi^2}, \dots$ are K-algebraically independent. For all integers $a \geq 0$ and $i \geq 1$ we have

$$(f^{\langle i\rangle})^{\phi^a} = (f^{\partial})^{-2\frac{\phi^a-1}{\phi-1}} \cdot f^{\langle i+a\rangle}.$$

Setting $f_i := f_i^{\lambda}$ for $i \in \{1, 2\}$ and

$$u_{ia} := \lambda_{i1}^{\phi^a} f^{\langle a+1\rangle} + \lambda_{i2}^{\phi^a} f^{\langle a+2\rangle}$$

we have

$$f_i^{\phi^a} = (f^{\partial})^{-2\frac{\phi^a-1}{\phi-1}} \cdot u_{ia}.$$

Claim 3.11.10 *The classes in $H^1(Z_\delta^\lambda)$ of the elements*

$$\eta_{ab} := \frac{f_2^{\phi^a} f_1^{\phi^b}}{f_1^{\phi^a} f_2^{\phi^b}} = \frac{u_{2a}u_{1b}}{u_{1a}u_{2b}} \in \mathbb{I}_{1,ord,\langle f_1 f_2\rangle}, \quad 0 \leq a < b,$$

are R-linearly independent; in other words, no nontrivial R-linear combination of the elements η_{ab} can be written as a difference of the type

$$\frac{F_1}{f_1^{w_1}} - \frac{F_2}{f_2^{w_2}} \tag{3.114}$$

with $w_i \in W$ and $F_i \in \mathbb{I}_{1,ord}(-2w_i)$ for $i \in \{1, 2\}$.

Claim 3.11.10 implies the equality ϕ-$\mathrm{rank}_R(H^1(Z_{\delta,2}^\lambda)) = \infty$ of Theorem 3.11.3.
To prove Claim 3.11.10 assume there exists a nontrivial R-linear combination of the elements η_{ab} which are differences as in Display (3.114) and seek a contradiction. Fix a pair (a_0, b_0) of integers with $0 \leq a_0 < b_0$ such that the coefficient of $\eta_{a_0b_0}$ in this linear combination is nonzero. We take

$$B := K[f^{\langle n\rangle} | n \in \mathbb{N}]$$

and let $\Sigma \subset B$ be the multiplicative monoid generated by all u_{ic} with $i \in \{1, 2\}$ and integers $c \geq 0$, $c \notin \{a_0, b_0\}$. Setting $(f^{\partial})^{\phi^j} = 1$ in F_1, F_2 for all $j \in \mathbb{N} \cup \{0\}$ we get (exactly as in the proof of Proposition 3.11.1) an equality in the ring of fractions $\Sigma^{-1}B$ of the form

$$\frac{u_{2a_0}u_{1b_0}}{u_{1a_0}u_{2b_0}} = \frac{H_1}{u_{1a_0}^{n_1}u_{1b_0}^{m_1}} - \frac{H_2}{u_{2a_0}^{n_2}u_{2b_0}^{m_2}}$$

where $n_1, m_1, n_2, m_2 \in \mathbb{N} \cup \{0\}$ and $H_1, H_2 \in \Sigma^{-1}B$. By Lemma 3.11.9 we get a contradiction if we prove the next claim.

Claim 3.11.11 *For $0 \leq a < b$ the following properties hold.*

(a) *The principal ideals $(u_{1a}), (u_{1b}), (u_{2a}), (u_{2b})$ of $\Sigma^{-1}B$ are prime and distinct.*

(b) *We have $u_{1b}u_{2a} \notin (u_{1a}, u_{2b}) \cdot \Sigma^{-1}B$.*

It is an easy exercise to show that Claim 3.11.11 follows from the following last claim.

Claim 3.11.12 *For $\mathbb{O} := \{1, 2\} \times \mathbb{Z}_{\geq 0}$ the following properties hold.*

(a) *The principal ideals (u_{ij}) of B indexed by $(i, j) \in \mathbb{O}$ are prime and pairwise distinct.*

(b) *For every two distinct pairs $(i_1, j_1), (i_2, j_2) \in \mathbb{O}$ the ideal $(u_{i_1 j_1}, u_{i_2 j_2})B$ is prime in B.*

(c) *For every three distinct pairs $(i_1, j_1), (i_2, j_2), (i_3, j_3) \in \mathbb{O}$ we have the property that $u_{i_3 j_3} \notin (u_{i_1 j_1}, u_{i_2 j_2})B$.*

Indeed Part (a) in Claim 3.11.12 clearly implies Part (a) in Claim 3.11.11. Furthermore Parts (b) and (c) in Claim 3.11.12 imply Part (b) in Claim 3.11.11; indeed, if $u_{1b}u_{2a} \in (u_{1a}, u_{2b}) \cdot \Sigma^{-1}B$ then a product of us times $u_{1b}u_{2a}$ belongs to ideal (u_{1a}, u_{2b}) in B. As the latter is a prime ideal by Part (b) in Claim 3.11.12 we get a contradiction by Part (c) in Claim 3.11.12.

To check Claim 3.11.12 we view $B = K[f^{\langle n\rangle} | n \in \mathbb{N}]$ as a graded K-algebra by giving $f^{\langle n\rangle}$ degree 1 for all $n \geq 1$. Parts (a) and (b) of Claim 3.11.12 are clear because u_{ij} are pairwise K-linearly independent linear forms in B; here we are using that all entries of the matrix λ are nonzero.

Now using the grading on B we see that part (c) of Claim 3.11.12 is equivalent to showing that for all three distinct pairs $(i_1, j_1), (i_2, j_2), (i_3, j_3) \in \mathbb{O}$ the linear forms $u_{i_1 j_1}, u_{i_2 j_2}, u_{i_3 j_3}$ are K-linearly independent. Let Γ be the $3 \times \infty$ matrix whose (k, l)-entries are the coefficients of $f^{\langle l\rangle}$ in $u_{i_k j_k}$. We have to show that Γ has a nonzero 3×3 minor. We may assume $j_1 \leq j_2 \leq j_3$. As the case $j_1 = j_2 = j_3$ does not occur (as $i_1, i_2, i_3 \in \{1, 2\}$) are not distinct), the 3×3 minor corresponding to the columns l_1, l_2, l_3 is nonzero, where, depending on the three disjoint possible cases, the triple (l_1, l_2, l_3) is as follows:

Case 1: If $j_1 < j_2 < j_3$, then $(l_1, l_2, l_3) := (j_1 + 1, j_2 + 1, j_3 + 1)$.

Case 2: If $j_1 = j_2 < j_3$, then $(l_1, l_2, l_3) := (j_1 + 1, j_2 + 2, j_3 + 2)$.

Case 3: If $j_1 < j_2 = j_3$, then $(l_1, l_2, l_3) := (j_1 + 1, j_2 + 1, j_3 + 2)$.

In all three cases we used, again, the fact that all entries of the matrix λ are nonzero. At this point Claim 3.11.12 is proved. So the first identity

$H^0(Z^{\lambda}_{\delta,2}) = R$ in Theorem 3.11.3 is proved. The second identity ϕ-$\text{rank}_R(H^1(Z^{\lambda}_{\delta,2})) = \infty$ in Theorem 3.11.3 follows from Lemma 3.11.8 and Claim 3.11.12(a). □

Proof of Proposition 3.11.5. Let $f_i := (F_i)^{\diamond}$, $i \in \{0, 1\}$, $F_i \in \mathbb{H}_g(1)$ primitive. Let $B := \mathbb{I}_{g,\text{ord}}$. By Lemma 3.11.8 it is enough to show that the principal ideals generated in $B \otimes_R K$ by each of the members of the family $(f_i^{\phi^j})_{(i,j)\in\{0,1\}\times(\mathbb{N}\cup\{0\})}$ are prime and distinct. By Eq. (3.54) it is enough to show that the principal ideals

$$\mathfrak{a}_{ij} = ((F_i^{\tau^j\sigma^j})^{\diamond})$$

generated in $B \otimes_R K$ by the members of Family

$$((F_i^{\tau^j\sigma^j})^{\diamond})_{(i,j)\in\{0,1\}\times(\mathbb{N}\cup\{0\})} \tag{3.115}$$

are prime and distinct. By Corollary 3.6.31(e) and (f) the ideals $\mathfrak{a}_{ij}$ are prime and it is enough to show that the members of Family (3.115) are pairwise R-linear independent. Assume this is not the case. As the $\mathbb{Z}_p$-algebra monomorphism $\sigma \circ \tau$ fixes no polynomial of degree ≥ 1 and as $\diamond$ is a $\mathbb{Z}_p$-algebra monomorphism, there exist two indices $j_0, j_1 \in \mathbb{N} \cup \{0\}$ with $j_1 \geq j_0$ and $\lambda \in K$ such that

$$F_{i_0}^{\tau^{j_0}\sigma^{j_0}} = \lambda \cdot F_{i_1}^{\tau^{j_1}\sigma^{j_1}} = (\phi^{-j_1}(\lambda) \cdot F_{i_1})^{\tau^{j_1}\sigma^{j_1}}$$

where $\{i_0, i_1\} = \{0, 1\}$. So we have

$$F_{i_0} = (\phi^{-j_1}(\lambda) \cdot F_{i_1})^{\tau^j\sigma^j}$$

where $j := j_1 - j_0 \geq 0$. As F_0 and F_1 are primitive we get $j = 0$. But then F_0 and F_1 are K-linearly dependent, a contradiction. □

In order to prove Proposition 3.11.6 we need the following lemma in which we use the forms $f^{[a]}$ introduced in Display (3.12).

Lemma 3.11.13 *The forms* $\det(f^{[1]}), \det(f^{[2]}) \in \mathbb{I}_{g,ord}(-2)$ *are primitive and R-linearly independent.*

Proof The linear independence follows from the Serre–Tate expansion principle, see Corollary 3.6.18(b). To prove that $\det(f^{[1]})$ and $\det(f^{[2]})$ are primitive, recall (cf. Display (3.44)) the definition of the partial polarizations $\Theta_0, \ldots, \Theta_g \in \mathbb{H}_g(1)$ of the invariant polynomial det via the identity

$$\det(y_0T + y_1T') = \sum_{i=0}^{g} \Theta_i(T, T')y_0^{g-i}y_1^i$$

and (see Corollary 3.6.18(d)) the formula $f^{[2]} = pf^{\langle 1\rangle} + f^{\langle 2\rangle}$. Consider the polynomial $\Theta^{[2]} := \sum_{i=0}^{g} p^{g-i}\Theta_i(T, T')$. As $\det(f^{[1]}) = \Theta_0(f^{\langle 1\rangle}, 0) = (\Theta_0(T, 0))^{\diamond}$, $\det(f^{[1]})$ is primitive. On the other hand we have:

$$\det(f^{[2]}) = \det(pf^{\langle 1\rangle} + f^{\langle 2\rangle}) = \sum_{i=0}^{g} \Theta(f^{\langle 1\rangle}, f^{\langle 2\rangle})p^{g-i} = (\Theta^{[2]})^{\diamond}.$$

So we are left to check that $\Theta^{[2]}$ is primitive. But the coefficient of the monomial $T_{11} \cdots T_{g-1,g-1}T'_{gg}$ in $\Theta_i(T, T')$ is 1 if $i = 1$ and 0 if $i \in \{0, 2, \ldots, g\}$. So $\Theta^{[2]} \notin R[T']$ which ends our proof. □

Proof of Proposition 3.11.6. For $\varphi_1 := \det(f^1)^{1+\phi^2}$, $\varphi_2 := \det(f^2)^{1+\phi}$ we have

$$R \subset H^0(Z_\delta, (\varphi_1, \varphi_2)) = \mathbb{I}_{g,\langle\varphi_1\rangle} \cap \mathbb{I}_{g,\langle\varphi_2\rangle} \subset \mathbb{I}_{g,\mathrm{ord},\langle\varphi_1\rangle} \cap \mathbb{I}_{g,\mathrm{ord},\langle\varphi_2\rangle}$$

$$\subset \mathbb{I}_{g,\mathrm{ord},\langle\det(f^1)\rangle} \cap \mathbb{I}_{g,\mathrm{ord},\langle\det(f^2)\rangle} \subset \mathbb{I}_{g,\mathrm{ord},\langle\det(f^{[1]})\rangle} \cap \mathbb{I}_{g,\mathrm{ord},\langle\det(f^{[2]})\rangle} = R,$$

where the last equality follows from Proposition 3.11.5 and Lemma 3.11.13. □

Chapter 4
Background and Required Results in GIT

We recall that this chapter is completely self-contained and contains results in geometric invariant theory (GIT). The results are often used and cited in the prior chapters, this chapter being a prerequisite for them. In fact, the four chapters form a continuous and integrated sequence, with the first three chapters being a main motivation for the last one while the last one is an indispensable and powerful tool for the first three ones. As such, a few references to prior chapters are also made in what follows as comments or remarks.

The aim of the chapter is to study the structure of algebras of invariants of the group $\mathbf{SL}_g$ acting on tuples of quadratic forms and on tuples of endomorphisms in arbitrary (mixed) characteristics. Though this theory has classical roots, unless explicitly stated, all results in this chapter are new.

We start, in Sect. 4.1, by giving formulas (see Sect. 4.1.1) for the (Krull) dimensions of algebras of invariants of $\mathbf{SL}_g$ acting on tuples of quadratic forms (see Theorem 4.1.1) and on tuples of endomorphisms (see Theorem 4.1.2); these formulas may be classical but we could not find a reference for them. We continue by giving an estimate (see Theorem 4.1.4) for the dimensions of the images of certain basic homomorphisms between the two types of algebras of invariants; this estimate plays a key role in our applications to Chaps. 2 and 3. The basic tool in the proofs of the above three theorems carried on respectively in Sects. 4.1.3–4.1.5 is a study of Hilbert–Mumford stability in the corresponding contexts (see Sect. 4.1.2).

Section 4.2 offers a detailed study of the invariants of $\mathbf{SL}_g$ acting on tuples of quadratic forms for arbitrary g. In particular we introduce some special 'Θ and Υ invariants' (see Sect. 4.2.6) and we compare the algebras of all invariants with the algebras generated by the Θ invariants (see, for instance, Theorems 4.2.8 and 4.2.13 and Corollary 4.2.23). We deduce information on the Hilbert series of the algebras of invariants (see Corollaries 4.2.14 and 4.2.15), the smallest number of homogeneous generators (see Proposition 4.2.10(a)), and transcendence bases (see Theorem 4.2.17 and Corollary 4.2.19(b)).

A. Buium and A. Vasiu, *Delta-Invariant Theory for Hecke Correspondences*,
Lecture Notes in Mathematics 2396, https://doi.org/10.1007/978-3-032-20571-1_4

Section 4.3 gives finer results in the case $g = 2$ on Hilbert series (see Proposition 4.3.4 and Corollary 4.3.8), the smallest number of homogeneous generators (see Proposition 4.3.3), and relations (see Corollary 4.3.6).

In Sect. 4.4 we prove a general result about the σ-finite generation of algebras of invariants for arbitrary reductive groups.

Throughout this chapter we adopt some general notation as follows.

Let $B^\times$ be the group of units of a unital commutative ring B.

Let $\mathbb{K}$ be an arbitrary algebraically closed field. Let $p := \mathrm{char}(\mathbb{K})$; so either $p = 0$ or $p > 0$ is a prime number. Product of schemes over $\mathrm{Spec}(\mathbb{K})$ are denoted by $\times$. All semisimple groups over $\mathrm{Spec}(\mathbb{K})$ are assumed connected.

For a finite dimensional vector space V over $\mathbb{K}$, let

$$\mathbf{V} := \mathrm{Spec}(\mathrm{Sym}(V^*)),$$

where $\mathcal{O}(\mathbf{V}) := \mathrm{Sym}(V^*)$ is the symmetric algebra of the dual V^* of V. Thus, for a $\mathbb{K}$-algebra $\mathbb{B}$ we have $\mathbf{V}(\mathbb{B}) = \mathbb{B} \otimes_{\mathbb{K}} V$, and in particular $\mathbf{V}(\mathbb{K}) = V$.

Let $r \in \mathbb{N} \cup \{0\}$. Let V^{r+1} be the direct sum of $r + 1$-copies of V.

Let $g \in \mathbb{N}$. Let $\mathbf{GL}_g := \mathbf{GL}_{g,\mathbb{K}}$ be the reductive group over $\mathrm{Spec}(\mathbb{K})$ of $g \times g$ invertible matrices. Let $\mathbf{SL}_g := \mathbf{SL}_{g,\mathbb{K}}$ be the semisimple group over $\mathrm{Spec}(\mathbb{K})$ which is the kernel of the determinant homomorphism $\det : \mathbf{GL}_g \to \mathbb{G}_{m,\mathbb{K}}$. Let $\mathbf{SO}_g$ be the closed subgroup scheme of $\mathbf{SL}_g$ of orthogonal matrices of determinant 1. If $p = 2$, then $\mathbf{SO}_g$ is smooth if and only if $g = 1$. If $p \neq 2$, then $\mathbf{SO}_1$ is trivial, $\mathbf{SO}_2 \simeq \mathbb{G}_{m,\mathbb{K}}$, and $\mathbf{SO}_g$ is semisimple for $g \geq 3$. Let $\mathbf{GT}_g$ be the standard maximal torus of $\mathbf{GL}_g$ of invertible diagonal matrices. Thus $\mathbf{T}_g := \mathbf{GT}_g \cap \mathbf{SL}_g$ is the standard maximal torus of $\mathbf{SL}_g$. Let 1_g be the identity element of $\mathbf{GL}_g(\mathbb{K})$ and of its subgroups.

Let G be a reductive group over $\mathrm{Spec}(\mathbb{K})$. Let G^0 be the connected component of its identity element and let $Z(G)$ be its center. Each G-module V is a finite dimensional left G-module given by a representation $\rho_V : G \to \mathbf{GL}_V$. Thus V^* is also a G-module given by the (composite homomorphism) representation $\rho_{V^*} : G \xrightarrow{\mathrm{inv}} G \xrightarrow{\rho_V} \mathbf{GL}_V \simeq \mathbf{GL}_{V^*}$, where $G \xrightarrow{\mathrm{inv}} G$ is the involutory inverse antiisomorphism and $\mathbf{GL}_V \simeq \mathbf{GL}_{V^*}$ is the canonical antiisomorphism. For a triple $(h, v, w) \in G(\mathbb{K}) \times V \times V^*$, we have $(h \cdot w)(v) = w(h^{-1} \cdot v) \in \mathbb{K}$.

4.1 Stable Points and Dimensions of Quotients

Let Mat_g be the $\mathbb{K}$-vector space of $g \times g$ matrices with entries in $\mathbb{K}$, and let

$$\mathrm{V}_g := \{M \in \mathrm{Mat}_g | M^{\mathrm{t}} = M\} \subset \mathrm{Mat}_g$$

be its subspace of symmetric matrices, where the upper right index t stands for transpose. Let Z_g be the subspace of V_g of all diagonal matrices. For $\lambda_1, \ldots, \lambda_g \in \mathbb{K}$ let

$$\mathrm{Diag}(\lambda_1, \ldots, \lambda_g) \in \mathrm{Z}_g$$

be such that for $i \in \{1, \ldots, g\}$ its ii entry is λ_i. Let

$$\mathrm{L}_g := \mathbb{K}1_g \times \mathrm{Z}_g \subset \mathrm{Z}_g^2 \subset \mathrm{V}_g^2;$$

we have $\dim_{\mathbb{K}}(\mathrm{L}_g) = g + 1$. We recall that $M \in \mathrm{Mat}_g$ is called *alternate* if $M^{\mathrm{t}} = -M$ and, in case $p = 2$, all its diagonal entries are 0. As such, if $p = 2$, $Q \in \mathrm{V}_g$ is called *nonalternate* if at least one of its diagonal entries is nonzero.

Let $\mathrm{Tr} : \mathrm{Mat}_g \to \mathbb{K}$ be the trace $\mathbb{K}$-linear map.

By our convention above we have

$$\mathbf{Z}_g = \mathrm{Spec}(\mathrm{Sym}(\mathrm{Z}_g^*)) \subset \mathbf{V}_g = \mathrm{Spec}(\mathrm{Sym}(\mathrm{V}_g^*)) \subset \mathbf{Mat}_g = \mathrm{Spec}(\mathrm{Sym}(\mathrm{Mat}_g^*)).$$

In[1] what follows by $\mathbf{SL}_g$-module we mean a left $\mathbf{SL}_g$-module.

We view Mat_g as an $\mathbf{SL}_g$-module via the classical congruence action

$$\mathbf{SL}_g \times \mathbf{Mat}_g \xrightarrow{\mathrm{cong}} \mathbf{Mat}_g$$

given on $\mathbb{K}$-valued points by the rule: $(\Lambda, M) \mapsto \Lambda M \Lambda^{\mathrm{t}}$. Then V_g is an $\mathbf{SL}_g$-submodule of Mat_g and thus V_g^{r+1} and its dual $(\mathrm{V}_g^{r+1})^* = (\mathrm{V}_g^*)^{r+1}$ naturally become $\mathbf{SL}_g$-modules.

We also consider the classical conjugation action of $\mathbf{SL}_g$ on Mat_g (when not clear from context, we specify which one of the two structures of Mat_g as an $\mathbf{SL}_g$-module is used). We view Mat_g as a left $\mathbf{SL}_g$-module via the action

$$\mathbf{SL}_g \times \mathbf{Mat}_g \xrightarrow{\mathrm{conj}} \mathbf{Mat}_g$$

given on $\mathbb{K}$-valued points by the rule: $(\Lambda, M) \mapsto \Lambda M \Lambda^{-1}$. Then Mat_g^{r+1} naturally becomes an $\mathbf{SL}_g$-module isomorphic to its dual $(\mathrm{Mat}_g^*)^{r+1}$.

4.1.1 Main Results on Dimensions of Quotients

Our first goal is to prove the following formula which is likely to be classical but for which we could not find a reference.

Theorem 4.1.1 *For all integers $r \geq 1$ we have*

$$\dim(Sym((V_g^{r+1})^*)^{\mathbf{SL}_g}) = (r + 1) \cdot \frac{g(g + 1)}{2} - g^2 + 1.$$

[1] In Chaps. 1–3, $\mathbf{Mat}_g$ is a ring scheme over $\mathrm{Spec}(\mathbb{Z})$, but in this section, $\mathbf{Mat}_g$ is a shorter notation for $\mathbf{Mat}_{g,\mathbb{K}}$ viewed just as a vector group variety over $\mathrm{Spec}(\mathbb{K})$.

We prove the following formula which, also, is likely to be classical.

Theorem 4.1.2 *For all integers $r \geq 1$, for the conjugation action we have*

$$\dim(Sym((Mat_g^{r+1})^*)^{SL_g}) = r \cdot g^2 + 1.$$

Let $n \in \mathbb{N}$. We consider now the morphism

$$\pi_n : \mathbf{V}_g^{n+1} \to \mathbf{Mat}_g^n \tag{4.1}$$

given on $\mathbb{K}$-valued points by the map

$$\pi_n := \pi_n(\mathbb{K}) : \mathrm{V}_g^{n+1} \to \mathrm{Mat}_g^n, \quad \pi_n(Q_0, \ldots, Q_n) = (Q_0 Q_1^*, Q_1 Q_2^*, \ldots, Q_{n-1} Q_n^*), \tag{4.2}$$

where we recall that $M^* \in \mathrm{Mat}_g$ denotes the adjugate of $M \in \mathrm{Mat}_g$; we have $M^* M = M M^* = \det(M) 1_g$. The morphism of Display (4.2) is $\mathbf{SL}_g$-equivariant (with $\mathbf{SL}_g$ acting on V_g by congruence and on Mat_g by conjugation) and hence induces a $\mathbb{K}$-algebra homomorphism

$$\mathrm{Sym}((\mathrm{Mat}_g^n)^*)^{\mathbf{SL}_g} \to \mathrm{Sym}((\mathrm{V}_g^{n+1})^*)^{\mathbf{SL}_g}$$

whose image we denote by $\widetilde{\mathrm{Sym}((\mathrm{Mat}_g^n)^*)^{\mathbf{SL}_g}}$.

Remark 4.1.3 With q a prime, let K be the field of fractions of the ring of q-typical Witt vectors with coefficients in an algebraic closure of $\mathbb{F}_q$. In Chaps. 2 and 3, the analogs of $\mathbb{K}$-algebras

$$\mathrm{Sym}((\mathrm{Mat}_g^n)^*)^{\mathbf{SL}_g} \text{ and } \widetilde{\mathrm{Sym}((\mathrm{Mat}_g^n)^*)^{\mathbf{SL}_g}}$$

for $p = 0$, in which the field $\mathbb{K}$ is replaced by the field K, were denoted by

$$\mathbb{H}_{g,\mathrm{conj},K}^n \text{ and } \widetilde{\mathbb{H}_{g,\mathrm{conj},K}^n}$$

(respectively, and the prime q was also denoted differently, by p)

We prove the following formula used in Theorem 3.7.1.

Theorem 4.1.4 *For all integers $n \geq 2$ we have*

$$\dim(\widetilde{Sym((Mat_g^n)^*)^{SL_g}}) = (n-1)\frac{g(g+1)}{2} + g.$$

For the proofs of Theorems 4.1.1, 4.1.2 and 4.1.4 we need to first recall from [MFK] some basic facts of geometric invariant theory.

4.1.2 Stable Points

Throughout this subsection we fix a G-module V. We recall that a point $v \in V = \mathbf{V}(\mathbb{K})$ is called *stable* if its G-orbit $G \cdot v$ is closed and its stabilizer $\mathrm{Stab}_G(v)$ is finite (see [MFK], App. to Ch. I, Sect. B, Def.). Recall that the $\mathbb{K}$-algebra of invariant homogeneous polynomials $\mathcal{O}(\mathbf{V})^G$ is finitely generated (see [MFK], App. to Ch. I, Sect. A, Thm. A.1.0) and normal; if in addition G is semisimple, $\mathcal{O}(\mathbf{V})^G$ is a unique factorization domain.

Let $\mathbf{V}/G := \mathrm{Spec}(\mathcal{O}(\mathbf{V})^G)$ and let

$$\varpi : \mathbf{V} \to \mathbf{V}/G$$

be the morphism of affine $\mathbb{K}$-schemes which is the spectrum of the injective $\mathbb{K}$-algebra homomorphism that is the inclusion $\mathcal{O}(\mathbf{V})^G \subset \mathcal{O}(\mathbf{V})$.[2]

If $p = 0$, then ϖ is submersive and a universal categorical quotient by [MFK], Ch. I, Sect. 2, Thm. 1.1. If $p > 0$, then ϖ is submersive and (only) a uniform categorical quotient by [MFK], App. to Ch. I, Sect. C, Thm. A.1.1. In what follows, whenever for a statement independent on p two references to [MFK] are required, one for $p = 0$ and one for $p > 0$, they are both quoted but without mentioning that the first one is for $p = 0$ and the second one is for $p > 0$.

Let $V^s \subset V$ be the set of stable points. Then V^s is the set of $\mathbb{K}$-valued points of a G-invariant and open (possibly empty) subscheme $\mathbf{V}^s$ of $\mathbf{V}$ which is also $\mathbb{K}^\times$-invariant (see [MFK], App. to Ch. I, Sect. B, Def.). If $\mathbf{X}$ is a locally closed subscheme of $\mathbf{V}$, let $\mathbf{X}^s := \mathbf{X} \cap \mathbf{V}^s$.

Lemma 4.1.5 *The following properties hold.*

(a) *For each $v \in V^s = \mathbf{V}^s(\mathbb{K})$, the reduced fiber $(\varpi^{-1}(\varpi(v)))_{red}$ is $G \cdot v$ and is a closed subscheme of $\mathbf{V}$ of dimension* $\dim(G) - \dim(Stab_G(v)) = \dim(G)$.

(b) *The image $\varpi(\mathbf{V}^s)$ is an open subvariety of $\mathbf{V}/G$.*

Proof Clearly, $\dim(G) - \dim(\mathrm{Stab}_G(v)) = \dim(G)$ and $(\varpi^{-1}(\varpi(v)))_{\mathrm{red}}$ is closed in $\mathbf{V}$. As ϖ is a categorical quotient, we have $G \cdot v \subset \varpi^{-1}(\varpi(v))$. Thus to prove part (a) it suffices to show that the assumption that $(\varpi^{-1}(\varpi(v)))_{\mathrm{red}} \backslash G \cdot v$ has a $\mathbb{K}$-valued point v' leads to a contradiction.

The Zariski closure $\overline{G \cdot v'}$ of $G \cdot v'$ in $\mathbf{V}$ is contained in $(\varpi^{-1}(\varpi(v)))_{\mathrm{red}}$. If $\overline{G \cdot v'} \cap G \cdot v \neq \emptyset$, then $\dim(G \cdot v') > \dim(G \cdot v) = \dim(G)$, a contradiction. So $\overline{G \cdot v'} \cap G \cdot v = \emptyset$. From this and [MFK], Ch. I, Sect. 2, Cor. 1.2 and App. to Ch. I, Sect. C, Cor. A.1.3 we get that there exists a global function on $\mathbf{V}/G$ which is 0 on $\overline{G \cdot v} = G \cdot v$ and is 1 on $\overline{G \cdot v'}$. This contradicts the fact that $\overline{G \cdot v'} \cup G \cdot v$ maps to $\varpi(v) \in (\mathbf{V}/G)(\mathbb{K})$. So part (a) holds.

From the first part we get that $\varpi^{-1}(\varpi(\mathbf{V}^s)) = \mathbf{V}^s$. So $\varpi^{-1}(\varpi(\mathbf{V}^s))$ is open in $\mathbf{V}$. As ϖ is submersive, we get that $\varpi(\mathbf{V}^s)$ is open in $\mathbf{V}/G$. □

---

[2] In the literature one also uses the notation $\mathbf{V}//G$ in place of $\mathbf{V}/G$; for simplicity we stick here to the notation $\mathbf{V}/G$.

For uniform or universal geometric quotients we refer to [MFK], Ch. 0, Defs. 0.6 and 07.

Lemma 4.1.6 *Let* $\mathbf{W}$ *be a nonempty open subvariety of* $\varpi(\mathbf{V}^s)$. *Then the following properties hold.*

(a) *The morphism* $\varpi^{-1}(\mathbf{W}) \to \mathbf{W}$ *induced by* ϖ *is a uniform geometric quotient.*

(b) *We have* $\dim(\mathbf{V}/G) = \dim_{\mathbb{K}}(V) - \dim(G)$.

(c) *If* $p = 0$, *then the morphism* $\varpi^{-1}(\mathbf{W}) \to \mathbf{W}$ *is a universal geometric quotient.*

Proof For parts (a) and (c) we can assume that $\mathbf{W}$ is affine and this case is a particular case of [MFK], Ch. I, Sect. 2, Amplific. 1.3 as the geometric fibers of $\varpi^{-1}(\mathbf{W}) \to \mathbf{W}$ are closed (see Lemma 4.1.5); see also [MFK], Ch. I, Prop. 1.9 or Thm. 1.10 for the 'uniform' part.

Part (b) follows from the dimension formula for the generic fibers of the dominant morphism $\varpi^{-1}(\mathbf{W}) \to \mathbf{W}$ (see [E], Part II, Cor. 14.6), these fibers having dimension $\dim(G)$ by Lemma 4.1.5(a). □

Definition 4.1.7 A G-invariant homogeneous polynomial $F \in \mathcal{O}(\mathbf{V})$ is called a *G-separating polynomial on* V (or on $\mathbf{V}$) if $F \neq 0$ and for all $v \in V = \mathbf{V}(\mathbb{K})$ with $F(v) \neq 0$ the stabilizer $\mathrm{Stab}_G(v)$ is finite.

We have the following interpretation of stable points in terms of G-separating polynomials on V which is a slight improvement of a particular case of [MFK], Ch. 1, Amplific. 1.11.

Proposition 4.1.8 *The following properties hold.*

(a) *The stable locus* V^s *is nonempty if and only if there exists a* G*-separating polynomial* F *on* V. *More precisely, for a point* $w \in V$ *the following two statements are equivalent.*

(a.i) *We have* $w \in V^s$.

(a.ii) *There exist a* G*-separating polynomial* F *on* V *such that* $F(w) \neq 0$.

(b) *If* F *is a* G*-separating polynomial on* V *and* $D(F) = (\mathbf{V}/G)\backslash Z(F)$ *is its principal open of* $\mathbf{V}/G$ *(so* $Z(F)$ *is the zero locus of* F *in* $\mathbf{V}/G$*), then we have an open embedding* $\varpi^{-1}(D(F)) \subset \mathbf{V}^s$.

(c) *With* F *as in part (b), the morphism* $\varpi^{-1}(D(F)) \to D(F)$ *is a uniform geometric quotient.*

(d) *If* V^s *is nonempty, then we have* $\dim(\mathbf{V}/G) = \dim_{\mathbb{K}}(V) - \dim(G)$.

(e) *If* $p = 0$ *and* F *is as in part (b), then the morphism* $\varpi^{-1}(D(F)) \to D(F)$ *is a universal geometric quotient.*

Proof For part (a), we first prove that $(a.ii) \Rightarrow (a.i)$. Let $v \in \varpi^{-1}(D(F))(\mathbb{K})$ and we consider $v' \in V = \mathbf{V}(\mathbb{K})$ in the Zariski closure of the orbit $G \cdot v$. Then $F(v') = F(v) \neq 0$ and thus $\mathrm{Stab}_G(v')$ is finite; therefore the orbits $G \cdot v$ and $G \cdot v'$ have the same dimension equal to $\dim(G)$. This forces $v' \in G \cdot v$, i.e., the orbit $G \cdot v$ is closed and we get that $v \in V^s$. Thus $\varpi^{-1}(D(F)) \subset \mathbf{V}^s$, hence $w \in V^s$, i.e., $(a.ii) \Rightarrow (a.i)$.

Next we prove that $(a.i) \Rightarrow (a.ii)$. If $w \in V^s$, then the closed G-invariant subsets $G \cdot w$ and $\mathbf{V}\setminus\mathbf{V}^s$ of $\mathbf{V}$ are disjoint. From this and [MFK], Ch. I, Sect. 2, Cor. 1.2 and App. to Ch. I, Sect. C, Cor. A.1.3 we get that there exists a global function F of $\mathbf{V}/G$ which is 1 on $G \cdot w$ and is 0 on $\mathbf{V}\setminus\mathbf{V}^s$. Clearly $F \neq 0$. As $\mathbf{V}\setminus\mathbf{V}^s$ is $\mathbb{K}^\times$-invariant, we can assume that F is homogeneous. If $v \in V$ is such that $F(v) \neq 0$, then $v \in V^s$ and thus F is a G-separating polynomial on V with $F(w) \neq 0$, i.e., $(a.i) \Rightarrow (a.ii)$.

For part (b), the inclusion $\varpi^{-1}(D(F)) \subset \mathbf{V}^s$ follows from part (a).

Parts (c)–(e) follow respectively from this inclusion and Lemma 4.1.6(a)–(c). □

We have the following trivial functorial property which is also a particular case of [MFK], Ch. 1, Prop. 1.18 and App. to Ch. I, end of Sect. C.

Lemma 4.1.9 *Let $f : V \to V'$ be a surjective morphism of G-modules. Let $F \in \mathcal{O}(\mathbf{V}')$ be a G-separating polynomial on V'. Then $F \circ f \in \mathcal{O}(\mathbf{V})$ is a G-separating polynomial on V (equivalently, we have $f^{-1}((V')^s) \subset V^s$).*

Lemma 4.1.10 *We assume that V is a faithful G-module such that under the closed embedding homomorphism $\rho_V : G \to \boldsymbol{GL}_V$ we have $\rho_V(G^0) \subset \boldsymbol{SL}_V$. Then there exists $m \in \{1, \ldots, \dim_{\mathbb{K}}(V)\}$ such that $(V^m)^s$ is nonempty.*

Proof As the sets $(V^m)^s$ do not change if G is replaced by G^0, we can assume that $G = G^0 \subset \mathbf{SL}_V$. Let $d := \dim_{\mathbb{K}}(V)$. We identify $V = \mathbb{K}^d$, $\mathbf{SL}_V = \mathbf{SL}_d$ and $V^d = \mathrm{Mat}_d$ as $\mathbf{SL}_d$-modules acted upon via left multiplications.

Let $\Lambda \in V^d = \mathrm{Mat}_d$ be such that it has determinant 1. The orbit $\mathbf{SL}_d \cdot \Lambda$ is the closed subset of $\mathbf{Mat}_d$ formed by all matrices of determinant 1 and the stabilizers $\mathrm{Stab}_{\mathbf{SL}_d}(\Lambda)$ and $\mathrm{Stab}_G(\Lambda)$ are trivial. Thus the orbit $G \cdot \Lambda$ inside $\mathbf{SL}_d \cdot \Lambda$ can be identified with the closed subgroup G of $\mathbf{SL}_d$ and thus it is closed. So $\Lambda \in (V^d)^s$. □

4.1.3 Proof of Theorem 4.1.1

To prove Theorem 4.1.1 we can assume that $g \geq 2$ and we first review a few classical facts for $g \geq 2$ (which for $g = 1$ are trivial).

Lemma 4.1.11 *Let $Q_0 \in V_g$ be invertible. If $p = 2$, we assume Q_0 is nonalternate. Then for each element $\nu \in \mathbb{K}$ such that $\nu^g = \det(Q_0)$, Q_0 is $\boldsymbol{SL}_g$-equivalent to $\nu 1_g$ (i.e., $Q_0 \in \boldsymbol{SL}_g \cdot \nu 1_g$).*

Proof As Q_0 is nonalternate (no matter what p is), it is congruent to a diagonal matrix (see [A], Thm. 6). As Q_0 is invertible and $\mathbb{K}$ is algebraically closed, we get that Q_0 is congruent to 1_g. Thus Q_0 is $\mathbf{SL}_g$-equivalent to $\lambda 1_g$ for some $\lambda \in \mathbb{K}$ such that $\lambda^g = \det(Q_0)$. We can assume $\nu \neq \lambda$. As $\lambda^g = \nu^g$, $\lambda 1_g$ and $\nu 1_g$ belong to the same $\mathbf{T}_g$-orbit but not to the same $\mathbf{SO}_g$-orbit and not necessarily to the same $Z(\mathbf{SL}_g)$-orbit: concretely, if $a := \sqrt[2g]{\frac{\nu}{\lambda}} \in \mathbb{K}^\times$, then $a^g \in \{-1, 1\}$ and for $\Lambda \in \{a 1_g, \mathrm{Diag}(-a, a, \dots, a)\} \cap \mathbf{T}_g(\mathbb{K})$ we have $\Lambda \lambda 1_g \Lambda^t = \nu 1_g$. □

Proposition 4.1.12 *Let the pair* $(Q_0, Q_1) \in V_g^2$ *be such that* Q_0 *is invertible and* $Q_0^{-1} Q_1$ *has* g *distinct eigenvalues. If* $p = 2$ *we also assume that* $Tr(Q_0^{-1} Q_1) \neq 0$. *Then* (Q_0, Q_1) *is* $\mathbf{SL}_g$*-equivalent to a pair* $(\lambda 1_g, D_1) \in L_g$ *with* $\lambda \in \mathbb{K}^\times$.

Proof We first show that the assumption that $p = 2$ and Q_0 is alternate leads to a contradiction. As the rank of an alternate matrix is even and as Q_0 is invertible, this assumption implies that g is even and from [A], Thm. 4 we get that Q_0 is congruent to a block diagonal matrix $\mathcal{J}_g := \mathrm{Diag}(\mathcal{J}, \dots, \mathcal{J})$ whose $g/2$ diagonal blocks are the elementary matrix $\mathcal{J} := \begin{pmatrix} 0 & 1 \\ 1 & 0 \end{pmatrix}$ of type I. Thus Q_0 is $\mathbf{SL}_g$-equivalent to $\nu \mathcal{J}_g$ with $\nu \in \mathbb{K}^\times$ such that $\nu^g = \det(Q_0)$ and as such we can assume that $Q_0 = \nu \mathcal{J}_g$. For $a, b, c \in \mathbb{K}$ we have $\mathrm{Tr}\left(\mathcal{J} \begin{pmatrix} a & b \\ b & c \end{pmatrix}\right) = 2b = 0$. From this, as Q_1 is symmetric and g is even, decomposing $Q_1 = (q_{ij})_{1 \le i, j \le g/2}$ into blocks $q_{ij} = q_{ji}^t \in \mathrm{Mat}_2$ we compute $\mathrm{Tr}(Q_0^{-1} Q_1) = \nu^{-1} \sum_{i=1}^{g/2} \mathrm{Tr}(\mathcal{J} q_{ii}) = 0$, contradiction.[3]

Thus, if $p = 2$, then Q_0 is nonalternate, and from Lemma 4.1.11 we get that we can assume that $Q_0 = \lambda 1_g$ for some $\lambda \in \mathbb{K}^\times$. Thus, as $Q_0^{-1} Q_1$ has g distinct eigenvalues, the same holds for Q_1. We write $Q_1 = \Lambda_2 D_2 \Lambda_2^{-1}$, where $\Lambda_2 \in \mathbf{GL}_g(\mathbb{K})$ and $D_2 \in \mathrm{Z}_g$ has g distinct eigenvalues. As $Q_1 \in \mathrm{V}_g$, from the identity $Q_1^t = Q_1$ we easily get that $\Lambda_2^t \Lambda_2$ commutes with D_2 and hence, as D_2 has g distinct eigenvalues, $\Lambda_2^t \Lambda_2 \in \mathrm{Z}_g$ is diagonal and we write $\Lambda_2^t \Lambda_2 = \Lambda_1^t \Lambda_1$ with $\Lambda_1 \in \mathbf{GL}_g(\mathbb{K}) \cap \mathrm{Z}_g$. For $\Lambda := \Lambda_1 \Lambda_2^{-1} \in \mathbf{SO}_g(\mathbb{K})$ we compute

$$D_1 := \Lambda Q_1 \Lambda^t = \Lambda Q_1 \Lambda^{-1} = \Lambda_1 \Lambda_2^{-1} Q_1 \Lambda_2 \Lambda_1^{-1} = \Lambda_1 D_2 \Lambda_1^{-1} \in \mathrm{Z}_g.$$

Thus the pair (Q_0, Q_1) is $\mathbf{SL}_g$-equivalent to $\Lambda \cdot (Q_0, Q_1) = (\lambda 1_g, D_1)$. □

We recall that if

$$\Theta_0, \dots, \Theta_g \in \mathcal{O}(\mathbf{V}_g^2)^{\mathbf{SL}_g} = \mathcal{O}(\mathbf{V}_g^2 / \mathbf{SL}_g)$$

are the partial polarizations of the invariant polynomial det defined by the identity

$$\det(y_0 Q_0 + y_1 Q_1) = \sum_{i=0}^{g} y_0^{g-i} y_1^i \Theta_i(Q_0, Q_1), \tag{4.3}$$

[3] As $(\Lambda Q_0 \Lambda^t)^{-1} \Lambda Q_1 \Lambda^t = (\Lambda^t)^{-1} Q_0^{-1} Q_1 \Lambda^t$, the trace of $Q_0^{-1} Q_1$ only depends on the $\mathbf{SL}_g$-orbit of (Q_0, Q_1).

where y_0, y_1 are two indeterminates, then the Cayley's *tact-invariant* (see [Dol], Sect. 2.3.2, Thm. 2.3.12) is

$$J := \mathrm{Disc}_g(\Theta_0, \ldots, \Theta_g) \in \mathcal{O}(\mathbf{V}_g^2)^{\mathbf{SL}_g} = \mathcal{O}(\mathbf{V}_g^2/\mathbf{SL}_g),$$

where Disc_g is the discriminant of the generic binary form of degree g; so Disc_g is a homogeneous polynomial in $g+1$ indeterminates with coefficients in $\mathbb{K}$ and of degree $2g-2$ (by convention, for $g=1$ it is the constant polynomial 1).

Proof of Theorem 4.1.1. Based on Proposition 4.1.8(d), to prove Theorem 4.1.1, it suffices to prove the following claim.

Claim 4.1.13 *If* $r \geq 1$, *then* $(V_g^{r+1})^s$ *is nonempty.*

To prove this claim we can assume that $r=1$ by Lemma 4.1.9. If $p \neq 2$, let $F_1 := 1$. If $p=2$, let $F_1 := \Theta_1$. Based on Proposition 4.1.8(a), to end the proof it suffices to prove that

$$F_0 := \Theta_0 \cdot F_1 \cdot J \in \mathcal{O}(\mathbf{V}_g^2)^{\mathbf{SL}_g} = \mathcal{O}(\mathbf{V}_g^2/\mathbf{SL}_g)$$

is an $\mathbf{SL}_g$-separating polynomial on V_g^2.

To check this, let $(Q_0, Q_1) \in \mathrm{V}_g^2$ be such that $F_0(Q_0, Q_1) \neq 0$, i.e., $\det(Q_0) \neq 0$ and $J(Q_0, Q_1) \neq 0$ and, in case $p=2$, $\Theta_1(Q_0, Q_1) \neq 0$. We have to show that $\mathrm{Stab}_{\mathbf{SL}_g}((Q_0, Q_1))$ is finite. If $p=2$, then $\mathrm{Tr}(Q_0^{-1}Q_1) = \det(Q_0)^{-1}\Theta_1(Q_0, Q_1) \neq 0$. The pair (Q_0, Q_1) is $\mathbf{SL}_g$-equivalent to a pair $(\lambda 1_g, D_1) \in \mathrm{L}_g \subset \mathrm{V}_g^2$ with $\lambda \in \mathbb{K}^\times$ (see Proposition 4.1.12).

As $J(\lambda 1_g, D_1) = J(Q_0, Q_1) \neq 0$ and

$$\det(y_0 1_g + y_1 \lambda^{-1} D_1) = \lambda^{-g} \sum_{i=0}^{g} y_0^{g-i} y_1^i \Theta_i(\lambda 1_g, D_1), \tag{4.4}$$

we get that $\lambda^{-1}D_1$ has distinct eigenvalues. Thus D_1 has distinct diagonal entries and hence its centralizer in $\mathbf{SL}_g$ is $\mathbf{T}_g$. Therefore we have $\mathrm{Stab}_{\mathbf{SO}_g}(D_1) = \mathbf{SO}_g \cap \mathbf{T}_g$. As $\mathrm{Stab}_{\mathbf{SL}_g}((Q_0, Q_1))$ is $\mathbf{SL}_g(\mathbb{K})$-conjugate to

$$\mathrm{Stab}_{\mathbf{SL}_g}((\lambda 1_g, D_1)) = \mathrm{Stab}_{\mathrm{Stab}_{\mathbf{SL}_g}(\lambda 1_g)}(D_1) = \mathrm{Stab}_{\mathbf{SO}_g}(D_1) = \mathbf{SO}_g \cap \mathbf{T}_g \simeq \boldsymbol{\mu}_{2,\mathbb{K}}^{g-1},$$

we get that $\mathrm{Stab}_{\mathbf{SL}_g}((Q_0, Q_1))$ is a finite group scheme over $\mathrm{Spec}(\mathbb{K})$. Hence F_0 is an $\mathbf{SL}_g$-separating polynomial on V_g^2. Thus the claim and Theorem 4.1.1 hold. □

4.1.4 Proof of Theorem 4.1.2

To prove Theorem 4.1.2 we need a few preparations. Let $e_1 := (1\,0\,0\,\cdots 0)^{\mathrm{t}}$, $\dots, e_g := (0\,0\,0\,\cdots 1)^{\mathrm{t}}$ be the standard ordered basis of $\mathbb{K}^g$. For $M \in \mathrm{Mat}_g$ and $q \in \{1, \dots, g-1\}$ let $M^{\wedge q} \in \mathrm{Mat}_{\binom{g}{q}}$ denote the matrix of the q-th wedge power of the endomorphism of $\mathbb{K}^g$ defined by M with respect to the standard basis $\{e_{i_1} \wedge \cdots \wedge e_{i_q} | 1 \le i_1 < \cdots < i_q \le g\}$ ordered lexicographically.

For $D := \mathrm{Diag}(\lambda_1, \dots, \lambda_g)$, $D^{\wedge q} \in \mathrm{Z}_{\binom{g}{q}}$ has diagonal entries $\lambda_{i_1} \cdots \lambda_{i_q}$s with indices $1 \le i_1 < \cdots < i_q \le g$. For $M_0, M_1 \in \mathrm{Mat}_g$ we have $(M_0 M_1)^{\wedge q} = M_0^{\wedge q} M_1^{\wedge q}$ and we set

$$\Phi_q(M_0, M_1) := \det(M_0^{\wedge q} M_1^{\wedge q} - M_1^{\wedge q} M_0^{\wedge q}).$$

We recall that $\mathbf{SL}_g$ acts on Mat_g^2 by conjugation.

Lemma 4.1.14 *For every $q \in \{1, \dots, g-1\}$, $\Phi_q(M_0, M_1)$ is a nonzero $\mathbf{SL}_g$-invariant homogeneous polynomial in the entries of M_0, M_1.*

Proof The only nonobvious thing is that $\Phi_q \neq 0$. To check that $\Phi_q \neq 0$, it suffices to find $\Lambda_0, \Lambda_1 \in \mathbf{GL}_g(\mathbb{K})$ such that

$$\det(\Lambda_0^{\wedge q} - (\Lambda_1 \Lambda_0 \Lambda_1^{-1})^{\wedge q}) \neq 0.$$

We take $\Lambda_0 \in \mathrm{Z}_g$ such that $\Lambda_0^{\wedge q}$ has distinct nonzero eigenvalues and Λ_1 to be the permutation matrix attached to the g-cycle $(1 \cdots g)$. With these choices, $(\Lambda_1 \Lambda_0 \Lambda_1^{-1})^{\wedge q}$ becomes a diagonal matrix obtained from $\Lambda_0^{\wedge q}$ by permuting its diagonal entries via a permutation without fixed points, hence $\Lambda_0^{\wedge q} - (\Lambda_1 \Lambda_0 \Lambda_1^{-1})^{\wedge q} \in \mathrm{Z}_{\binom{g}{q}}$ has nonzero entries on the diagonal and thus is invertible. To give a concrete example of such a Λ_0, let $N \ge 2^g$ be an integer not divisible by p, let ζ be a primitive N-th root of unity in $\mathbb{K}$, and choose

$$\Lambda_0 := \mathrm{Diag}(\zeta^{2^0}, \zeta^{2^1}, \zeta^{2^2}, \dots, \zeta^{2^{g-1}}) \in \mathrm{Z}_g. \qquad \square$$

Lemma 4.1.15 *Let $q \in \{1, \dots, g-1\}$ and let $M_0, M_1 \in Mat_g$ be such that they have a common invariant subspace W of dimension q. Then $\Phi_q(M_0, M_1) = 0$.*

Proof As each nonzero vector w_q of the line $\bigwedge^q(W)$ of $\bigwedge^q(\mathbb{K}^g) = \mathbb{K}^{\binom{g}{q}}$ is an eigenvector of both $M_0^{\wedge q}$ and $M_1^{\wedge q}$, we have

$$(M_0^{\wedge q} M_1^{\wedge q} - M_1^{\wedge q} M_0^{\wedge q}) w_q = 0.$$

This implies that $\Phi_q(M_0, M_1) = 0$. $\square$

Proof of Theorem 4.1.2. The $\mathbf{SL}_g$-invariant homogeneous polynomial

$$\Phi(M_0, M_1) := \prod_{q=1}^{g-1} \Phi_q(M_0, M_1) \in \mathcal{O}(\mathbf{Mat}_g^2)$$

is nonzero (see Lemma 4.1.14). Similar to the proof of Theorem 4.1.1, by Proposition 4.1.8(a), (b) and (d) and Lemma 4.1.9, to prove Theorem 4.1.2 it suffices to prove that Φ is an $\mathbf{SL}_g$-separating polynomial on Mat_g^2. Let $(M_0, M_1) \in \mathrm{Mat}_g^2$ with $\Phi(M_0, M_1) \neq 0$. We know that M_0 and M_1 do not have a common nontrivial invariant subspace, see Lemma 4.1.15. If $\Lambda \in \mathrm{Stab}_{\mathbf{SL}_g}((M_0, M_1))(\mathbb{K})$, then each eigenspace of Λ, being M_0-invariant and M_1-invariant, must be $\mathbb{K}^g$. This implies that

$$\mathrm{Stab}_{\mathbf{SL}_g}((M_0, M_1))(\mathbb{K}) = Z(\mathbf{SL}_g)(\mathbb{K}) \simeq \boldsymbol{\mu}_g(\mathbb{K})$$

is finite. Thus Φ is indeed an $\mathbf{SL}_g$-separating polynomial on Mat_g^2. □

4.1.5 Proof of Theorem 4.1.4

As there exists an $\mathbf{SL}_g$-separating polynomial on $\mathbf{Mat}_g^2$, $(\mathbf{Mat}_g^2)^{\mathrm{s}}$ is a nonempty open subscheme of $\mathbf{Mat}_g^2$ by Proposition 4.1.8(b). But to prove Theorem 4.1.4 we need that in fact the closed subscheme $\mathbf{V}_g^2$ and the two open subschemes $\mathbf{GL}_g^2$ and $(\mathbf{Mat}_g^2)^{\mathrm{s}}$ of $\mathbf{Mat}_g^2$ have a nonempty intersection.

Lemma 4.1.16 *We have* $(\mathbf{GL}_g^2)^s \cap (\mathbf{Z}_g \times \mathbf{V}_g) \neq \emptyset$.

Proof If $p = 0$ let $\mathbb{K}_0$ be the algebraic closure of $\mathbb{Q}$ in $\mathbb{K}$ and let $\mathbb{K}_1 := \mathbb{C}$. If $p > 0$ let $\mathbb{K}_0$ be the algebraic closure of $\mathbb{F}_p$ in $\mathbb{K}$ and let $\mathbb{K}_1$ be the completion of an algebraic closure of $\mathbb{K}_0((x))$ with respect to its standard topology (here x is an indeterminate). The lemma holds for $\mathbb{K}$ if and only if it holds for $\mathbb{K}_0$ and if and only if it holds for $\mathbb{K}_1$. Thus we can assume that $\mathbb{K} = \mathbb{K}_1$ is complete with respect to an absolute value. The set of $\mathbb{K}$-valued points of a variety over $\mathbb{K}$ has a natural topology induced from that on $\mathbb{K}$ to be referred as the metric topology.

For $\nu_1, \ldots, \nu_g \in \mathbb{K}^\times$ distinct let $D_0 := \mathrm{Diag}(\nu_1, \ldots, \nu_g) \in \mathbf{GL}_g(\mathbb{K})$. Also, let $\mathbb{E}_g \in \mathrm{Mat}_g$ be the matrix whose g^2 entries are all 1. For $\lambda \in \mathbb{K}$ the condition that $\lambda\mathbb{E}_g + D_0$ is an invertible matrix with g distinct eigenvalues is expressed by λ not being a zero of a polynomial $\chi(x) \in \mathbb{K}[x]$. As $\chi(0) \neq 0$, there exists $a \in \mathbb{K}^\times$ such that $\chi(a) \neq 0$. This implies that the matrix $Q_1 := \mathbb{E}_g + a^{-1}D_0$ is invertible and has g distinct eigenvalues.

Let $\overline{\mathbf{o}}$ be the closure in $\mathbf{Mat}_g^2$ of the $\mathbf{SL}_g$-orbit $\mathbf{o}$ of (D_0, Q_1). We note that $(D_0, Q_1) \in \mathbf{GL}_g^2(\mathbb{K}) \cap (\mathrm{Z}_g \times \mathrm{V}_g)$. As $\mathrm{Stab}_{\mathbf{SL}_g}(D_0) = \mathbf{T}_g$ we compute

$$\mathrm{Stab}_{\mathbf{SL}_g}((D_0, Q_1)) = \mathrm{Stab}_{\mathbf{T}_g}(Q_1) = \mathrm{Stab}_{\mathbf{T}_g}(\mathbb{E}_g) = Z(\mathbf{SL}_g) \simeq \boldsymbol{\mu}_{g,\mathbb{K}}.$$

Thus $(D_0, Q_1) \in (\mathrm{Mat}_g^2)^{\mathrm{s}}$ if and only if $\overline{\mathbf{o}} = \mathbf{o}$. Hence to prove the lemma it suffices to check that $\overline{\mathbf{o}} = \mathbf{o}$. Let $(Q_0, Q_1') \in \overline{\mathbf{o}}(\mathbb{K})$. We check that there exists $\Lambda \in \mathbf{SL}_g(\mathbb{K})$

such that $(\Lambda D_0 \Lambda^{-1}, \Lambda Q_1 \Lambda^{-1}) = (Q_0, Q_1')$. The characteristic polynomials of Q_0 and D_0 are the same and hence Q_0 is similar to D_0. Thus to prove the existence of Λ we can assume that $Q_0 = D_0$.

Let $(\Lambda_n)_{n\geq 1}$ be a sequence in $\mathbf{SL}_g(\mathbb{K})$ such that $(\Lambda_n D_0 \Lambda_n^{-1}, \Lambda_n Q_1 \Lambda_n^{-1})$ converges in the metric topology to (D_0, Q_1'). Its existence can be checked by considering an irreducible closed affine curve $C \subset \overline{\mathbf{o}}$ that passes through (D_0, Q_1') and such that the intersection $C \cap (\overline{\mathbf{o}} \backslash \mathbf{o})$ is finite and using the fact that the complement of finitely many points in $\tilde{C}(\mathbb{K})$, where $\tilde{C}$ is the normalization of C, is dense in $\tilde{C}(\mathbb{K})$ in the metric topology.

If $\mathbf{U}_g^+$ (resp. $\mathbf{U}_g^-$) is the unipotent subgroup of $\mathbf{SL}_g$ of upper (resp. lower) triangular matrices with all diagonal entries 1, then the product morphism

$$\mathbf{U}_g^- \times \mathbf{U}_g^+ \times \mathbf{T}_g \to \mathbf{SL}_g$$

is an open embedding (by the standard Bruhat decomposition, see [Bore], Ch. IV, Sect. 14, Sect. 14.12, Thm. (a)) and thus we can identify $\mathbf{U}_g^- \times \mathbf{U}_g^+$ with an open subscheme of the quotient scheme $\mathbf{Q}_q := \mathbf{SL}_g/\mathbf{T}_g$. As $\Lambda_n D_0 \Lambda_n^{-1}$ converges to D_0, $\Lambda_n \mathbf{T}]_g \in \mathbf{Q}_g(\mathbb{K})$ converges to $\mathbf{T}_g \in \mathbf{Q}_g(\mathbb{K})$. Thus there exists $N \in \mathbb{N}$ such that for each $n \geq N$ there exists $\Sigma_n \in (\mathbf{U}_g^- \times \mathbf{U}_g^+)(\mathbb{K}) \subset \mathbf{SL}_g(\mathbb{K})$ with the properties that we have $\Lambda_n \mathbf{T}_g = \Sigma_n \mathbf{T}_g$ and the sequence $(\Sigma_n)_{n\geq N}$ converges to 1_g.

For $n \geq N$, let $D_n := \Sigma_n^{-1}\Lambda_n$; we have $D_n \in \mathbf{T}_g(\mathbb{K})$. As $(\Sigma_n D_n D_0 D_n^{-1} \Sigma_n^{-1}, \Sigma_n D_n Q_1 D_n^{-1} \Sigma_n^{-1})_{n\geq N}$ converges to (D_0, Q_1') and both $(\Sigma_n)_{n\geq N}$ and $(\Sigma_n^{-1})_{n\geq N}$ converge to 1_g, the sequence $(D_n D_0 D_n^{-1}, D_n Q_1 D_n^{-1})_{n\geq N}$ is equal to the sequence $(D_0, D_n Q_1 D_n^{-1})_{n\geq N}$ and converges to (D_0, Q_1').

As $Z(\mathbf{GL}_g)$ centralizes Q_1, by translating the D_ns by invertible scalar matrices and by not requiring any more that they are in $\mathbf{T}_g(K)$, we can assume they are in $\mathbf{GT}_g(\mathbb{K})$ and their 11 entries are 1. Thus for $n \geq N$ there exists $D_n = \mathrm{Diag}(1, \lambda_{2,n}, \ldots, \lambda_{g,n}) \in \mathbf{GT}_g(\mathbb{K})$ such that $(D_n Q_1 D_n^{-1})_{n\geq N}$ converges to Q_1'. For an integer $1 < i \leq g$, the $i1$ entry of $D_n Q_1 D_n^{-1}$ is $\lambda_{i,n}$ and thus $(\lambda_{i,n})_{n\geq N}$ converges to the $i1$ entry λ_i of Q_1'. Similarly, $(\lambda_{i,n}^{-1})_{n\geq N}$ converges to the $1i$ entry of Q_1'. The last two sentences imply that $\lambda_i \neq 0$. We conclude that the sequence $(D_n)_{n\geq N}$ converges to the matrix $D := (1, \lambda_2, \ldots, \lambda_g) \in \mathbf{GT}_g(\mathbb{K})$ and therefore $DQ_1D^{-1} = Q_1'$. Defining $\Lambda := \sqrt[g]{\det(D^{-1})}D \in \mathbf{T}_g(\mathbb{K})$, we have $(\Lambda D_0 \Lambda^{-1}, \Lambda Q_1 \Lambda^{-1}) = (D_0, Q_0')$. This implies that $\overline{\mathbf{o}} = \mathbf{o}$. □

For the next two lemmas we recall the morphism π_n in Diagram (4.1).

Lemma 4.1.17 *We assume that $n \geq 2$. Then there exist invertible symmetric matrices*
$Q_0, \ldots, Q_n \in V_g$ such that we have $\pi_n(Q_0, \ldots, Q_n) \in (\mathbf{GL}_g^n)^s(\mathbb{K})$.

Proof By Lemma 4.1.16 we can find invertible symmetric matrices $Q_0, Q_1' \in \mathrm{V}_g$ such that $(Q_0, Q_1') \in (\mathrm{Mat}_g^2)^{\mathrm{s}}$. Let $Q_1 := 1_g$. Write $Q_1' = Q_2^*$, so $Q_2 \in \mathrm{V}_g$. By Proposition 4.1.8(a) there exists an $\mathbf{SL}_g$-separating homogeneous polynomial F on Mat_g^2 such that $F(Q_0, Q_1') = F(Q_0 Q_1^*, Q_1 Q_2^*) \neq 0$. The function

$\mathrm{Mat}_g^n \to \mathbb{K}$ defined by $(M_0, \ldots, M_{n-1}) \mapsto F(M_0, M_1)$ is a separating polynomial (see Lemma 4.1.9) and does not vanish at a point of the form $\pi_n(Q_0, Q_1, Q_2, \ldots, Q_n)$ with $Q_3, \ldots, Q_n \in \mathrm{V}_g$. Hence the latter points are stable by Proposition 4.1.8(a) and thus belong to $(\mathbf{GL}_g^n)^{\mathrm{s}}(\mathbb{K})$ provided $Q_3, \ldots, Q_n$ are also invertible. □

Lemma 4.1.18 *We assume that $n \geq 2$. There exists a Zariski dense open subscheme $\boldsymbol{W}_{g,n}$ of $\boldsymbol{GL}_g^{n+1} \cap \boldsymbol{V}_g^{n+1}$ such that for each point $(Q_0, \ldots, Q_n)$ in $\boldsymbol{W}_{g,n}(\mathbb{K})$ the fiber $\boldsymbol{\pi}_n^{-1}(\boldsymbol{\pi}_n(Q_1, \ldots, Q_{n+1}))$ has dimension* 1.

Proof Let $(Q_0, \ldots, Q_n) \in \mathrm{V}_g^{n+1}$ and $\lambda \in \mathbb{K}^\times$. Let $\lambda_n, \lambda_{n-1}, \ldots, \lambda_0 \in \mathbb{K}^\times$ be defined recursively by $\lambda_n := \lambda$ and $\lambda_i := \lambda_{i+1}^{1-g}$ for $i \in \{n-1, n-2, \ldots, 0\}$. Then we have

$$\pi_n(Q_0, \ldots, Q_n) = \pi_n(\lambda_0 Q_0, \ldots, \lambda_n Q_n);$$

hence $\boldsymbol{\pi}_n^{-1}(\boldsymbol{\pi}_n(Q_0, \ldots, Q_n))$ has dimension ≥ 1 if $(Q_0, \ldots, Q_n)$ is not the zero vector of V_g^{n+1} and it is easy to see that the same holds even if $(Q_0, \ldots, Q_n)$ is the zero vector of V_g^{n+1}.

We claim that there exists

$$(Q_0, \ldots, Q_n) \in \mathbf{GL}_g^{n+1}(\mathbb{K}) \cap \mathrm{V}_g^{n+1}$$

such that $\boldsymbol{\pi}_n^{-1}(\boldsymbol{\pi}_n(Q_0, \ldots, Q_n))$ has dimension ≤ 1 and thus has dimension 1; as the dimension of fibers of $\boldsymbol{\pi}_n$ is upper semicontinuous on the source (see [E], Part II, Thm. 14.8), this holds for all $\mathbb{K}$-valued points $(Q_0, \ldots, Q_n)$ of a Zariski dense open subscheme $\mathbf{W}_{g,n}$ of $\mathbf{GL}_g^{n+1} \cap \mathbf{V}_g^{n+1}$.

To check our claim, recall from Lemma 4.1.16 that we have a non-identity $(\mathbf{GL}_g^2)^{\mathrm{s}} \cap \mathbf{V}_g^2 \neq \emptyset$. Let $Q_0 := 1_g$, let $(Q_1, Q_2) \in (\mathbf{GL}_g^2)^{\mathrm{s}}(\mathbb{K}) \cap \mathrm{V}_g^2$, and, if $n \geq 3$, let $(Q_3, \ldots, Q_n) \in (\mathbf{GL}_g^{n-2})(\mathbb{K}) \cap \mathrm{V}_g^{n-2}$. Let $\widetilde{Q}_0, \ldots, \widetilde{Q}_n \in \mathrm{V}_g$ be such that

$$\pi_n(\widetilde{Q}_0, \ldots, \widetilde{Q}_n) = \pi_n(Q_0, \ldots, Q_n).$$

We have

$$\widetilde{Q}_i \widetilde{Q}_{i+1}^* = Q_i Q_{i+1}^* \quad \text{for all} \quad i \in \{0, \ldots, n-1\}. \tag{4.5}$$

It follows that the matrices $\widetilde{Q}_0, \ldots, \widetilde{Q}_{n-1}$ are invertible and thus there exist unique matrices $\Lambda_0, \ldots, \Lambda_n \in \mathbf{GL}_g(\mathbb{K})$ such that we have identities $\widetilde{Q}_0 = Q_0\Lambda_0 = \Lambda_0$, $\widetilde{Q}_1 = Q_1\Lambda_1, \ldots, \widetilde{Q}_n = Q_n\Lambda_n$. From Eq. (4.5) we get that

$$\Lambda_i \Lambda_{i+1}^* = 1_g \quad \text{for all} \quad i \in \{0, \ldots, n-1\}. \tag{4.6}$$

As $\widetilde{Q}_0 = \Lambda_0$ is symmetric, by induction on $i \in \{0, \ldots, n\}$ we get from Eq. (4.6) that $\Lambda_i \in \mathrm{V}_g$. As $\widetilde{Q}_1, \widetilde{Q}_2 \in \mathrm{V}_g$, we have $Q_1\Lambda_1 = (Q_1\Lambda_1)^{\mathrm{t}} = \Lambda_1 Q_1$ and $Q_2\Lambda_2 = (Q_2\Lambda_2)^{\mathrm{t}} = \Lambda_2 Q_2$. Thus Λ_1 commutes with Q_1 and Λ_2 commutes with Q_2. Hence Λ_2^{-1} commutes with Q_2 and thus so does Λ_1. Therefore Λ_1 is a $\mathbb{K}$-valued point of the one dimensional subvariety $Z(\mathbf{GL}_g)\mathrm{Stab}_{\mathbf{SL}_g}((Q_1, Q_2))$ of $\mathbf{GL}_g$ and thus, as $\Lambda_0, \Lambda_2, \dots, \Lambda_n$ are determined by Λ_1 up to a finite number of possibilities by Eq. (4.6) and the fact that $\Lambda \in \mathbf{GL}_g(K)$ is determined by its adjugate Λ^* up to a scalar multiple by a $(g-1)$-th root of unity in $\mathbb{K}$, our claim is proved. □

Remark 4.1.19 The morphism $\pi_1 : \mathbf{V}_g^2 \to \mathbf{Mat}_g$ is dominant and thus for a generic point $(Q_0, Q_1) \in \mathrm{V}_g^2$, $\pi_1^{-1}(\pi_1(Q_0, Q_1))$ has dimension g. Even more is true, namely: the morphism $\mathbf{V}_g^2 \to \mathbf{Mat}_g \times_{\mathrm{Spec}(\mathbb{K})} \mathbb{A}^1_{\mathbb{K}}$ defined on $\mathbb{K}$-valued points by the rule $(Q_0, Q_1) \mapsto (Q_0 Q_1^*, \det(Q_0))$ is dominant. To check this it suffices to show that for each pair $(\Lambda, \lambda) \in \mathbf{GL}_g(\mathbb{K}) \times \mathbb{K}^\times$ with Λ diagonalizable there exist $Q_0, Q_1 \in \mathrm{V}_g$ such that $Q_0 Q_1^* = \Lambda$ and $\det(Q_0) = \lambda$. We write $\Lambda = \Lambda_1 D \Lambda_1^{-1}$, where $D \in \mathrm{Z}_g$ is invertible and $\det(\Lambda_1) = 1$. Let $\nu := \sqrt[g]{\lambda}$. Then we can write $\Lambda = Q_0 Q_2$, where $Q_0 := \nu \cdot \Lambda_1 \Lambda_1^{\mathrm{t}}$, $Q_2 := \nu^{-1} \cdot (\Lambda_1^{-1})^{\mathrm{t}} D \Lambda_1^{-1}$. Clearly $Q_0, Q_2 \in \mathrm{V}_g$ are invertible and $\det(Q_0) = \nu^g = \lambda$. We can write $Q_2 = Q_1^*$, where $Q_1 \in \mathrm{V}_g$ (as it is a scalar multiple of Q_2^{-1}), so $Q_0 Q_1^* = \Lambda$.[4]

Proof of Theorem 4.1.4. Consider the morphisms

$$\mathbf{V}_g^{n+1} \xrightarrow{\pi_n} \mathbf{Mat}_g^n \xrightarrow{\varpi_n} \mathbf{Mat}_g^n/\mathbf{SL}_g,$$

the Zariski closure $\mathbf{X}_n$ of $\pi_n(\mathbf{V}_g^{n+1})$ in $\mathbf{Mat}_g^n$, and the Zariski closure $\mathbf{Y}_n$ of $\varpi_n(\mathbf{X}_n)$ (or of $(\varpi_n \circ \pi_n)(\mathbf{V}_g^{n+1})$) in $\mathbf{Mat}_g^n/\mathbf{SL}_g$. By Lemma 4.1.18 and the dimension formula for the generic fibers of the dominant morphism $\mathbf{V}_g^{n+1} \to \mathbf{X}_n$ (see [E], Part II, Cor. 14.6) we have

$$\dim(\mathbf{X}_n) = (n+1)\frac{g(g+1)}{2} - 1.$$

By Lemma 4.1.17, $\mathbf{X}_n \cap (\mathbf{Mat}_g^n)^{\mathrm{s}} \neq \emptyset$. As $\mathbf{X}_n$ is $\mathbf{SL}_g$-invariant and all fibers of $(\mathbf{Mat}_g^n)^{\mathrm{s}} \xrightarrow{\varpi_n} (\mathbf{Mat}_g^n)^{\mathrm{s}}/\mathbf{SL}_g$ are $\mathbf{SL}_g$-orbits of dimension g^2-1, based on the same dimension formula we get that

$$\dim(\mathbf{Y}_n) = \dim(\mathbf{X}_n) - (g^2-1) = (n+1)\frac{g(g+1)}{2} - g^2 = (n-1)\frac{g(g+1)}{2} + g.$$

This equality implies that Theorem 4.1.4 holds. □

4.1.6 On 'Main Theorems'

We recall the 'first main theorem' for the conjugation action of $\mathbf{SL}_g$ on $\mathbf{Mat}_g^{r+1}$ (conjectured by Artin, proved for $p = 0$ in [G], Thm. 17.4 and [P], Thm. 1.3, and for $p > 0$ in [Don92, Don93]; cf. also [DCP], Thm. 1.10). For a $\mathbb{K}$-algebra $\mathbb{B}$ and $M \in \mathbf{Mat}_g(\mathbb{B})$, the characteristic polynomial

[4] We have $\mathrm{Mat}_g = (\mathrm{V}_g \cap \mathbf{GL}_g(\mathbb{K})) \cdot \mathrm{V}_g$ (the proof of [Bos], Thm. 1 works for all square matrices with entries in a ring which are similar to a Jordan matrix).

$$\chi_M(x) := \det(x \cdot 1_g - M) = \sum_{j=0}^{g} (-1)^j c_j(M) x^{g-j} \in \mathbb{B}[x]$$

satisfies $c_0(M) = 1$, $c_1(M) = \mathrm{Tr}(M)$, and $c_g(M) = \det(M)$. We identify

$$\mathrm{Sym}((\mathrm{Mat}_g^{r+1})^*) = \mathbb{K}[X_0, \ldots, X_r],$$

where $\mathbb{K}[X_0, \ldots, X_r] := \mathbb{K}[X_{l,ij} | l \in \{0, \ldots, r\}$ and $i, j \in \{1, \ldots, g\}]$ with each $X_l = (X_{l,ij})_{1 \le i,j \le g}$ a $g \times g$ matrix with indeterminates as entries. Here we view every $X_{l,ij}$ as the element of $(\mathrm{Mat}_g^{r+1})^*$ that sends each $(M_0, \ldots, M_r) \in \mathrm{Mat}_g^{r+1}$ to the ij entry of M_l.

As the identifications $X_{l,ij} = X_{l,ij}(X_0, \ldots, X_r)$ make sense for all $1 \le i, j \le g$ and all $l \in \{0, \ldots, r\}$, for $F \in \mathrm{Sym}((\mathrm{Mat}_g^{r+1})^*)$ we can write

$$F = F(X_0, \ldots, X_r) = F(X_{0,11}, X_{0,12}, \ldots, X_{r,gg})$$

and we can similarly evaluate F at elements of $\mathrm{Mat}_g^{r+1} \otimes_{\mathbb{K}} \mathrm{Sym}((\mathrm{Mat}_g^{r+1})^*)$.

The action of $\Lambda \in \mathbf{SL}_g(\mathbb{K})$ on F is given by the rule

$$(\Lambda \cdot F)(X_0, \ldots, X_r) := F(\Lambda^{-1} X_0 \Lambda, \ldots, \Lambda^{-1} X_r \Lambda).$$

To check this we can assume $r = 0$ and $F = X_{0,ij}$. We need to check that $\Lambda \cdot X_{0,ij}$ is the ij entry of $\Lambda^{-1} X_0 \Lambda$. But this follows from the fact that for all $M \in \mathrm{Mat}_g$, $(\Lambda \cdot X_{0,ij})(M)$ is the ij entry of the matrix $\Lambda^{-1} M \Lambda$, confirm the very definition of the duals of $\mathbf{SL}_g$-modules.

The 'first main theorem' is the following one.

Theorem 4.1.20 *The $\mathbb{K}$-algebra $Sym((Mat_g^{r+1})^*)^{\mathbf{SL}_g}$ is generated by the following set of homogeneous polynomials:*

$$\{c_j(X_{l_1} \cdots X_{l_N}) | j \in \{1, \ldots, g\}, N \in \mathbb{N}, l_1, \ldots, l_N \in \{0, \ldots, r\}\}.$$

If $p \nmid g!$, then the $\mathbb{K}$-algebra $Sym((Mat_g^{r+1})^*)^{\mathbf{SL}_g}$ is generated by the subset defined by $j = 1$, i.e., by $\{c_1(X_{l_1} \cdots X_{l_N}) | N \in \mathbb{N}, l_1, \ldots, l_N \in \{0, \ldots, r\}\}$.[5]

[5] If $p = 0$, one can take $N \in \{1, \ldots, 2^g - 1\}$ (see [P], Thm. 3.4(a)). Also, there exists a 'second main theorem' (see [P], Thm. 4.5(a) for $p = 0$ and [DCP], Introd. for $p > 0$) which describes the relations among these generators but it is not used in the monograph.

Remark 4.1.21 We recall the following general result (see [MFK], App. to Ch. I, Sect. F, Thm.) If G is semisimple and acts on $\mathbf{X} = \mathrm{Spec}(\mathbb{B})$ with $\mathbb{B}$ a finitely generated $\mathbb{K}$-algebra which is a unique factorization domain, then the following two statements are equivalent.

(1) There exists a nonempty G-invariant open subscheme $\mathbf{Y}$ of $\mathbf{X}$ such that for each $x \in \mathbf{Y}(\mathbb{K})$, the orbit of x is closed in $\mathbf{X}$.

(2) There exists a nonempty G-invariant open subscheme $\mathbf{S}$ of $\mathbf{X}$ such that for each $x \in \mathbf{S}(\mathbb{K})$, the reduced stabilizer of x in G is reductive.

This result can be used to get less explicit proofs of Theorems 4.1.1 and 4.1.2. Our explicit proofs of Theorems 4.1.1 and 4.1.2 helped in proving Theorems 2.4.21 and 4.1.4 and are repeatedly used in what follows.

4.2 General Results on Quadratics for Arbitrary $g \geq 1$

We return to the geometric invariant theory of quadratic forms.

4.2.1 *The $\mathbf{SL}_g$-Module V_g*

Let $\varepsilon_g \in \{0, 1\}$ be the remainder of the division of g by 2. As we have $\mathrm{Ker}(\mathbf{SL}_g \to \mathbf{GL}_{V_g}) \simeq \boldsymbol{\mu}_{2,\mathbb{K}}^{1-\varepsilon_g}$, we use $\mathbb{W}_{\varepsilon_g}$, where

$$\mathbb{W}_0 := \frac{1}{2}\mathbb{Z}_{\geq 0} \quad \text{and} \quad \mathbb{W}_1 := \mathbb{Z}_{\geq 0}.$$

Let $\tau_g : \mathbf{SL}_g \simeq \mathbf{SL}_g$ be the involutory automorphism which on $\mathbb{K}$-valued points is given by the rule: $\Lambda \mapsto (\Lambda^{\mathrm{t}})^{-1}$. If $g \geq 3$, then τ_g is the 'standard' outer automorphism of $\mathbf{SL}_g$. For a finite dimensional $\mathbf{SL}_g$-module V given by a representation $\rho_V : \mathbf{SL}_g \to \mathbf{GL}_V$, let $\tau_g^*(V)$ be the $\mathbf{SL}_g$-module given by the representation $\rho_V \circ \tau_g$; thus we have $\tau_g^*(V) = V$ as $\mathbb{K}$-vector spaces. If $g = 2$, then τ_2 is an inner automorphism of $\mathbf{SL}_2$ and therefore $\tau_2^*(V)$ and V are isomorphic $\mathbf{SL}_2$-modules.

We consider the standard g-dimensional $\mathbf{SL}_g$-module $\mathrm{W}_g := \mathbb{K}^g$, where the action $\mathbf{SL}_g \times \mathbf{W}_g \to \mathbf{W}_g$ is given on $\mathbb{K}$-valued points by the rule: $(\Lambda, v) \mapsto \Lambda \cdot v$.

If $p > 0$, we recall that $\mathrm{W}_g^{(p)}$ is the $\mathbf{SL}_g$-module given on $\mathbb{K}$-valued points by the rule: $(\Lambda, v) \mapsto \Lambda^{[p]} \cdot v$ where $\Lambda^{[p]} \in \mathbf{SL}_g(\mathbb{K})$ is the matrix obtained from $\Lambda \in \mathbf{SL}_g(\mathbb{K})$ by replacing each entry by its p-th power.

Let $\{e_1^*, \ldots, e_g^*\}$ be the standard basis of W_g^*: it is the dual of the standard basis $\{e_1, \ldots, e_g\}$ of W_g. Using the Kronecker delta, for $i, j \in \{1, \ldots, g\}$ we have $e_j^*(e_i) = \delta_{ij}$. The standard basis of the $\mathbb{K}$-vector space

$$\mathrm{Mat}_g = \mathrm{Hom}_{\mathbb{K}}(\mathrm{W}_g, \mathrm{W}_g) = \mathrm{W}_g \otimes_{\mathbb{K}} \mathrm{W}_g^*$$

is $\{e_i \otimes e_j^* | 1 \leq i, j \leq g\}$. For $m \in \{1, \ldots, g\}$ we have $(e_i \otimes e_j^*)(e_m) = \delta_{jm} e_i$. The standard basis of V_g is

$$\{e_i \otimes e_i^* | 1 \leq i \leq g\} \cup \{e_i \otimes e_j^* + e_j \otimes e_i^* | 1 \leq i < j \leq g\}.$$

For $g \geq 2$ and $i, j \in \{1, \ldots, g\}$ distinct let

$$\Lambda_{i,j} \in \mathbf{SO}_g(\mathbb{K}) \tag{4.7}$$

be such that it maps (e_i, e_j) to $(e_j, -e_i)$ and fixes each $e_{i'}$ with $i' \in \{1, \ldots, g\} \backslash \{i, j\}$. If $p = 2$, we have $\Lambda_{i,j} = \Lambda_{j,i}$.

The classification of finite dimensional simple modules over a simply connected semisimple group G over $\mathbb{K}$ in terms of weights associated to a maximal torus T_G of G (see [Ja], Part II, Chs. 2 and 5) implies the following property.

($\sharp$) If V and V' are two finite dimensional G-modules which are isomorphic as T_G-modules, then they are isomorphic as G-modules provided either both are semisimple G-modules or one is a simple G-module.

If $p = 2$, let $\mathrm{V}_g^{\mathrm{alt}}$ be the $\mathbb{K}$-vector subspace of V_g of alternating symmetric matrices and let $\mathrm{V}_g^{\mathrm{Diag}} := \mathrm{V}_g/\mathrm{V}_g^{\mathrm{alt}}$; we have

$$\dim_{\mathbb{K}}(\mathrm{V}_g^{\mathrm{alt}}) = \frac{g(g-1)}{2} \quad \text{and} \quad \dim_{\mathbb{K}}(\mathrm{V}_g^{\mathrm{Diag}}) = g.$$

The $\mathbb{K}$-vector subspace $\Delta_g \subset \mathrm{Mat}_g$ of upper triangular matrices is a direct supplement of $\mathrm{V}_g^{\mathrm{alt}}$ in Mat_g and we have a canonical $\mathbb{K}$-linear isomorphism

$$\Delta_g \to \mathrm{Mat}_g / V_g^{\mathrm{alt}}.$$

The $\mathbb{K}$-linear map $\mathrm{Mat}_g \to \mathrm{V}_g$ defined by the rule $M \mapsto M + M^{\mathrm{t}}$ is surjective if and only if $p \neq 2$: if $p = 2$, the image is $\mathrm{V}_g^{\mathrm{alt}}$.

We now study the short exact sequence $0 \to \mathrm{V}_g \to \mathrm{Mat}_g \to \mathrm{Mat}_g/\mathrm{V}_g \to 0$ of $\mathbf{SL}_g$-modules under the congruence action.

Lemma 4.2.1 *If $p = 2$, then the following properties hold.*

(a) *We have a short exact sequence of $\mathbf{SL}_g$-modules*

$$0 \to V_g^{alt} \to V_g \to V_g^{Diag} \to 0.$$

(b) *The $\mathbf{SL}_g$-module V_g^{Diag} is simple and isomorphic to $W_g^{(2)}$.*

(c) *The $\mathbf{SL}_g$-module V_g^{alt} is simple (if $g = 2$, then $V_2^{alt} = \mathbb{K}\mathcal{J} = \mathbb{K}\begin{pmatrix} 0 & 1 \\ 1 & 0 \end{pmatrix}$ is a trivial simple $\mathbf{SL}_2$-module).*

(d) *The short exact sequence of part (a) does not split.*

(e) *The* $\mathbf{SL}_g$*-modules* V_g^{alt}, $\bigwedge^2(W_g)$ *and* Mat_g/V_g *are isomorphic.*

(f) *The* $\mathbf{SL}_g$*-module* V_g *is not isomorphic to either* V_g^* *or* $\tau_g^*(V_g^*)$.

(g) *The* $\mathbf{SL}_g$*-modules* V_g *and* $Sym^2(W_g)$ *are not isomorphic.*

(h) *The only* $\mathbf{SL}_g$*-submodules of* Mat_g *are* 0, V_g^{alt}, V_g *and* Mat_g.

(i) *The* $\mathbf{SL}_g$*-modules* Mat_g^* *and* $\tau_g^*(Mat_g)$ *are isomorphic.*

(j) *The* $\mathbf{SL}_g$*-modules* Mat_g/V_g^{alt} *and* $\tau_g^*(V_g^*)$ *are isomorphic.*

Proof Part (a) holds as from [A], Thm. 1 we get that $\mathrm{V}_g^{\mathrm{alt}}$ is an $\mathbf{SL}_g$-submodule of V_g.

Part (b) follows from (♮) applied to $G = \mathbf{SL}_g$ as we have a natural identification $\mathrm{V}_g^{\mathrm{Diag}} = \mathrm{W}_g^{(2)}$ of $\mathbf{T}_g$-modules.

To prove parts (c) and (d), we first remark that the $\mathbf{SL}_g$-submodule V_g^M of Mat_g generated by a nonzero $M \in \mathrm{Mat}_g$ is the $\mathbb{K}$-span of the congruence class of M. As the congruence relation on $\mathrm{V}_g^{\mathrm{alt}}\setminus\{0\}$ has precisely $\lfloor \frac{g}{2} \rfloor$ congruence classes given by the (even) rank (see [A], Thm. 4), it is easy to see that for $Q \in V_g^{\mathrm{alt}}$ we have $\mathrm{V}_g^Q = \mathrm{V}_g^{\mathrm{alt}}$. From this part (c) follows. If $Q \in \mathrm{V}_g$ is nonalternate, then $\mathrm{V}_g^Q \cap \mathrm{Z}_g \neq 0$ (see [A], Thm. 6) and using this we easily get that $\mathrm{Z}_g \subset \mathrm{V}_g^Q$ and thus (again based on loc. cit.) $\mathrm{V}_g\setminus\mathrm{V}_g^{\mathrm{alt}} \subset \mathrm{V}_g^Q$. This implies that $\mathrm{V}_g^Q = \mathrm{V}_g$ and that (d) holds.

Part (e) follows from (♮) as the $\mathbf{T}_g$-modules $\mathrm{V}_g^{\mathrm{alt}}$, $\bigwedge^2(\mathrm{W}_g)$ and $\mathrm{Mat}_g/\mathrm{V}_g$ are isomorphic. Part (f) follows from parts (a) to (d) and their duals.

As we have an $\mathbf{SL}_g$-invariant epimorphism $\mathrm{Sym}^2(\mathrm{W}_g) \to \bigwedge^2(\mathrm{W}_g)$, part (g) follows from parts (d) and (e).

Based on parts (a) to (d) (or on the proof of parts (c) and (d)), to prove part (h) it suffices to show that if $M \in \mathrm{Mat}_g\setminus\mathrm{V}_g$, then $\mathrm{V}_g^M = \mathrm{Mat}_g$. Based on parts (c) and (e), we have an $\mathbf{SL}_g$-invariant epimorphism $\mathrm{V}_g^M \to \mathrm{Mat}_g/\mathrm{V}_g$. Thus we can assume that $M \in \mathrm{V}_g + \mathbb{K}e_1 \otimes e_2^*$. Writing $\mathrm{V}_g + \mathbb{K}e_1 \otimes e_2^*$ as a direct sum of $\mathbf{T}_g$-modules on which $\mathbf{T}_g$ acts via one character, we can assume that in fact $M = \nu e_2 \otimes e_1^* + (e_1 \otimes e_2^*)$ with $\nu \in \mathbb{K}\setminus\{1\}$. As

$$(1_g + e_2 \otimes e_1^*)M(1_g + e_2 \otimes e_1^*)^{\mathrm{t}} = (M + e_2 \otimes e_2^*)(1_g + e_1 \otimes e_2^*) = M + (1+\nu)(e_2 \otimes e_2^*)$$

is in V_g^M, we get that $(1+\nu)(e_2 \otimes e_2^*) \in V_g^M \cap (\mathrm{Z}_g\setminus\{0\})$ and from the proof of parts (c) and (d) we get that V_g^M contains V_g. From this, as V_g^M surjects onto $\mathrm{Mat}_g/\mathrm{V}_g$, we conclude that $\mathrm{V}_g^M = \mathrm{Mat}_g$. Thus part (h) holds.

Part (i) follows from the fact that the nondegenerate trace bilinear map

$$\mathrm{Mat}_g \times \tau_g^*(\mathrm{Mat}_g) \to \mathbb{K}$$

is $\mathbf{SL}_g$-invariant, i.e., for $\Lambda \in \mathbf{SL}_g(\mathbb{K})$ and $M_1, M_2 \in \mathrm{Mat}_g$, we have

$$\mathrm{Tr}(\Lambda M_1 \Lambda^{\mathrm{t}}(\Lambda^{\mathrm{t}})^{-1} M_2 \Lambda^{-1}) = \mathrm{Tr}(\Lambda M_1 M_2 \Lambda^{-1}) = \mathrm{Tr}(M_1 M_2).$$

The dual $\mathrm{Mat}_g^* \to V_g^*$ of the $\mathbf{SL}_g$-invariant monomorphism $\mathrm{V}_g \to \mathrm{Mat}_g$ induces an $\mathbf{SL}_g$-invariant epimorphism $\tau_g^*(\mathrm{Mat}_g^*) \to \tau_g^*(V_g^*)$ which, based on part (i), can be viewed as an $\mathbf{SL}_g$-invariant epimorphism $\mathrm{Mat}_g \to \tau_g^*(V_g^*)$. Thus, based on part (h) and reasons of dimensions, we get that $\tau_g^*(V_g^*)$ is isomorphic to the quotient $\mathrm{Mat}_g/\mathrm{V}_g^{\mathrm{alt}}$ of Mat_g. Thus part (j) holds. □

Lemma 4.2.2 *If $p \neq 2$, then the following properties hold.*

(a) *The $\mathbf{SL}_g$-modules V_g and $Sym^2(W_g)$ (resp. V_g^* and $Sym^2(W_g^*)$) are simple and isomorphic.*

(b) *The $\mathbf{SL}_g$-modules V_g and $\tau_g^*(V_g^*)$ (or $\tau_g^*(V_g)$ and V_g^*) are isomorphic.*

Proof We only prove part (a) for V_g and $\mathrm{Sym}^2(\mathrm{W}_g)$ as the proof for V_g^* and $\mathrm{Sym}^2(\mathrm{W}_g^*)$ is the same. The $\mathbf{T}_g$-modules V_g and $\mathrm{Sym}^2(\mathrm{W}_g)$ are isomorphic under the natural $\mathbb{K}$-linear map that maps $e_i \otimes e_j^* + e_j \otimes e_i^*$ to the image of the element $e_i \otimes e_j \in \mathrm{W}_g \otimes_{\mathbb{K}} \mathrm{W}_g$ into $\mathrm{Sym}^2(\mathrm{W}_g)$ for all $1 \leq i \leq j \leq g$. The proof of Lemma 4.2.1 (c) and (d) can be easily adapted to give that V_g is a simple $\mathbf{SL}_g$-module. From the last two sentences and (♯) we get that the $\mathbf{SL}_g$-modules V_g and $\mathrm{Sym}^2(\mathrm{W}_g)$ are isomorphic. Thus part (a) holds.

Part (b) follows from part (a) as the two $\mathbf{SL}_g$-modules W_g and $\tau_g^*(\mathrm{W}_g^*)$ are isomorphic. □

4.2.2 *Rings and Vector Spaces of Invariants: The $\mathbb{K}$ Case*

With arbitrary characteristic $p \geq 0$, we begin by giving an explicit description of the $\mathbf{SL}_g(\mathbb{K})$-action on the $\mathbb{K}$-algebra $\mathrm{Sym}((\mathrm{V}_g^{r+1})^*)$. Let $T_{ij} \in \mathrm{V}_g^*$ be defined by letting $T_{ij}(Q)$ be the ij entry of $Q \in \mathrm{V}_g$. Hence the set $\{T_{ij} | 1 \leq i \leq j \leq g\}$ is a basis of V_g^*. We write

$$\mathrm{Sym}((\mathrm{V}_g^{r+1})^*) = \mathbb{K}[T, \ldots, T^{(r)}] := \mathbb{K}[T_{ij}, \ldots, T_{ij}^{(r)} | 1 \leq i \leq j \leq g],$$

where $T = T^{(0)} = (T_{ij})_{1 \leq i,j \leq g} = (T_{ij}^{(0)})_{1 \leq i,j \leq g}, \ldots, T^{(r)} = (T_{ij}^{(r)})_{1 \leq i,j \leq g}$ are symmetric matrices of indeterminates of size $g \times g$ and for every $l \in \{0, \ldots, r\}$, the set $\{T_{ij}^{(l)} | 1 \leq i \leq j \leq g\}$ is $\{T_{ij} | 1 \leq i \leq j \leq g\}$ but viewed as a $\mathbb{K}$-basis of the l-th factor V_g^* of $(\mathrm{V}_g^{r+1})^*$. As the identifications $T_{ij}^{(l)} = T_{ij}^{(l)}(T, \ldots, T^{(r)})$ make sense for all $1 \leq i \leq j \leq g$ and all $l \in \{0, \ldots, r\}$, for every polynomial $F \in \mathbb{K}[T_{ij}, \ldots, T_{ij}^{(r)} | 1 \leq i \leq j \leq g]$ we can write

$$F = F(T, \ldots, T^{(r)}) = F(T_{11}, T_{12}, \ldots, T_{gg}^{(r)})$$

as well as we can similarly evaluate F at elements of $\mathrm{V}_g \otimes_{\mathbb{K}} \mathrm{Sym}((\mathrm{V}_g^{r+1})^*)$.

Recall the $\mathbf{SL}_g$-action on V_g given by the rule $(\Lambda, Q) \mapsto \Lambda Q \Lambda^{\mathrm{t}}$, with $\Lambda \in \mathbf{SL}_g(\mathbb{K})$ and $Q \in \mathrm{V}_g$. We claim that we have an identity

$$(\Lambda \cdot F)(T, \ldots, T^{(r)}) = F(\Lambda^{-1}T(\Lambda^{-1})^{\mathrm{t}}, \ldots, (\Lambda^{-1}T^{(r)}(\Lambda^{-1})^{\mathrm{t}}).$$

To check this we can assume $r = 0$ and $F = T_{ij}$ and we need to prove that $\Lambda \cdot T_{ij}$ is the ij entry of the symmetric matrix $\Lambda^{-1}T(\Lambda^{-1})^{\mathrm{t}}$. But this follows from the fact that for all $Q \in \mathrm{V}_g$, $(\Lambda \cdot T_{ij})(Q)$ is the ij entry of the matrix $\Lambda^{-1}Q(\Lambda^{-1})^{\mathrm{t}}$, by the very definition of the duals of $\mathbf{SL}_g$-modules.

For $l \in \{0, \ldots, r\}$ we have the matrix identity

$$(\Lambda \cdot T_{ij}^{(l)})_{1\le i\le j\le g} = \Lambda^{-1}T^{(l)}(\Lambda^{-1})^{\mathrm{t}}. \tag{4.8}$$

Recalling the notation in Display (4.7), for $g \geq 2$ and $i, j \in \{1, \ldots, g\}$ distinct we have

$$\Lambda_{ij} \cdot T_{ij}^{(l)} = -T_{ij}^{(l)}, \quad \Lambda_{ij} \cdot T_{ii}^{(l)} = T_{jj}^{(l)} \quad \text{and} \quad \Lambda_{ij} \cdot T_{jj}^{(l)} = T_{ii}^{(l)}. \tag{4.9}$$

For $g \geq 3$, $i, j \in \{1, \ldots, g\}$ distinct, and $i', j' \in \{1, \ldots, g\}\backslash\{i, j\}$ we have

$$\Lambda_{ij} \cdot T_{i'j'}^{(l)} = T_{i'j'}^{(l)}. \tag{4.10}$$

For $g \geq 3$ and $i, j, i' \in \{1, \ldots, g\}$ distinct we have

$$\Lambda_{ij} \cdot T_{ji'}^{(l)} = -T_{ii'}^{(l)} \quad \text{and} \quad \Lambda_{ij} \cdot T_{ii'}^{(l)} = T_{ji'}^{(l)}. \tag{4.11}$$

Equations (4.9)–(4.11) can be all read out from the following matrix identity

$$\begin{pmatrix} 0 & -1 & 0 \\ 1 & 0 & 0 \\ 0 & 0 & 1 \end{pmatrix} \begin{pmatrix} T_{11} & T_{12} & T_{13} \\ T_{12} & T_{22} & T_{23} \\ T_{13} & T_{23} & T_{33} \end{pmatrix} \begin{pmatrix} 0 & 1 & 0 \\ -1 & 0 & 0 \\ 0 & 0 & 1 \end{pmatrix} = \begin{pmatrix} T_{22} & -T_{12} & -T_{23} \\ -T_{12} & T_{11} & T_{13} \\ -T_{23} & T_{13} & T_{33} \end{pmatrix}.$$

We view

$$\mathrm{Sym}((\mathrm{V}_g^{r+1})^*) = \oplus_{s\in\mathbb{Z}_{\ge 0}}\mathrm{Sym}((\mathrm{V}_g^{r+1})^*)_s$$

as a $\mathbb{Z}_{\geq 0}$-graded $\mathbb{K}$-algebra by declaring each $T_{ij}^{(l)}$ with $0 \leq l \leq r$ and $1 \leq i \leq j \leq g$ to be homogeneous of degree 1. For a nonzero polynomial $h \in \mathrm{Sym}((\mathrm{V}_g^{r+1})^*)_s$ we speak about its T-degree s. If for each $l \in \{0, \ldots, r\}$, h has partial homogeneous degree s_l in the upper index $^{(l)}$, then we also say that h has partial T-degree $(s_0, \ldots, s_l)$; we have $s = \sum_{l=0}^{r} s_l$.

We consider the $\mathbb{K}$-monomorphisms (inclusions)

$$\mathrm{Sym}((\mathrm{V}_g^1)^*) \subset \mathrm{Sym}((\mathrm{V}_g^2)^*) \subset \cdots \subset \mathrm{Sym}((\mathrm{V}_g^r)^*) \subset \mathrm{Sym}((\mathrm{V}_g^{r+1})^*) \subset \cdots;$$

they correspond to natural projections $\cdots \to \mathrm{V}_g^{r+1} \to \mathrm{V}_g^r \to \cdots \to \mathrm{V}_g^2 \to \mathrm{V}_g^1$ on first coordinates and induce $\mathbb{Z}_{\geq 0}$-graded $\mathbb{K}$-monomorphisms (inclusions)

$$\mathrm{Sym}((\mathrm{V}_g^1)^*)^{\mathbf{SL}_g} \subset \mathrm{Sym}((\mathrm{V}_g^2)^*)^{\mathbf{SL}_g} \subset \cdots \subset \mathrm{Sym}((\mathrm{V}_g^r)^*)^{\mathbf{SL}_g} \subset \cdots. \tag{4.12}$$

As $\mathbf{SL}_g$ is semisimple, the rings in Display (4.12) are unique factorization domains with $\mathbb{K}^\times$ as their groups of units.

For $s \in \mathbb{W}_{\varepsilon_g}$ let

$$\mathbb{H}_g^r(s)_{\mathbb{K}} := \mathrm{Sym}((\mathrm{V}_g^{r+1})^*)_{gs}^{\mathbf{SL}_g}.$$

Thus we get a $\mathbb{W}_{\varepsilon_g}$-graded $\mathbb{K}$-algebra

$$\mathbb{H}_{g,\mathrm{tot},\mathbb{K}}^r := \bigoplus_{s\in\mathbb{W}_{\varepsilon_g}} \mathbb{H}_g^r(s)_{\mathbb{K}}$$

contained in $\mathrm{Sym}((\mathrm{V}_g^{r+1})^*)$ and a $\mathbb{Z}_{\geq 0}$-graded $\mathbb{K}$-algebra

$$\mathbb{H}_{g,\mathbb{K}}^r := \bigoplus_{s\in\mathbb{Z}_{\geq 0}} \mathbb{H}_g^r(s)_{\mathbb{K}}$$

contained in either $\mathbb{H}_{g,\mathrm{tot},\mathbb{K}}^r$ or $\mathrm{Sym}((\mathrm{V}_g^{r+1})^*)$.

If g is even, we consider the $\mathbb{H}_{g,K}^r$-module

$$\mathbb{H}_{g,\mathbb{K}}^{r,\perp} := \bigoplus_{s\in\mathbb{W}_0\setminus\mathbb{Z}_{\geq 0}} \mathbb{H}_g^r(s)_{\mathbb{K}} = \bigoplus_{s\in\frac{1}{2}+\mathbb{Z}_{\geq 0}} \mathbb{H}_g^r(s)_{\mathbb{K}}.$$

Lemma 4.2.3 *The following properties hold.*

(a) *If g is odd, then we have $\mathbb{H}_{g,\mathbb{K}}^r = \mathbb{H}_{g,tot,\mathbb{K}}^r = Sym((V_g^{r+1})^*)^{SL_g}$.*

(b) *We assume that g is even. Then we have $\mathbb{H}_{g,tot,\mathbb{K}}^r = Sym((V_g^{r+1})^*)^{SL_g}$ and the square of each homogeneous polynomial of $\mathbb{H}_{g,tot,\mathbb{K}}^r$ belongs to $\mathbb{H}_{g,\mathbb{K}}^r$ and hence the $\mathbb{K}$-monomorphism (inclusion) $\mathbb{H}_{g,\mathbb{K}}^r \to \mathbb{H}_{g,tot,\mathbb{K}}^r$ is integral. Moreover, $\mathbb{H}_{g,\mathbb{K}}^r$ is normal and exactly one of the following possibilities holds:*

(b.i) *either $\mathbb{H}_{g,\mathbb{K}}^{r,\perp} = 0$, i.e., we have an identity $\mathbb{H}_{g,\mathbb{K}}^r = \mathbb{H}_{g,tot,\mathbb{K}}^r$;*

(b.ii) *or $\mathbb{H}_{g,\mathbb{K}}^{r,\perp} \neq 0$ and then $[Frac(\mathbb{H}_{g,tot,\mathbb{K}}^r) : Frac(\mathbb{H}_{g,\mathbb{K}}^r)] = 2$ and there exists a homogeneous prime element*

$$\mathfrak{h}_{g,r,odd} \in \mathbb{H}_{g,\mathbb{K}}^{r,\perp}$$

of the smallest T-degree such that we have three identities

$$Frac(\mathbb{H}_{g,tot,\mathbb{K}}^r) = Frac(\mathbb{H}_{g,\mathbb{K}}^r) \oplus Frac(\mathbb{H}_{g,\mathbb{K}}^r)\mathfrak{h}_{g,r,odd}$$

$$= Frac(\mathbb{H}_{g,\mathbb{K}}^r)[x]/[x^2 - \mathfrak{h}_{g,r,odd}^2],$$

$$Frac(\mathbb{H}_{g,tot,\mathbb{K}}^r) \otimes_{\mathbb{H}_{g,\mathbb{K}}^r} \mathbb{H}_{g,\mathbb{K}}^{r,\perp} = Frac(\mathbb{H}_{g,tot,\mathbb{K}}^r)\mathfrak{h}_{g,r,odd},$$

hence the finite field extension $Frac(\mathbb{H}^r_{g,\mathbb{K}}) \to Frac(\mathbb{H}^r_{g,tot,\mathbb{K}})$ *is separable if and only if* $p \neq 2$.

(c) *If* $\mathbb{H}^r_{g,tot,\mathbb{K}}$ *is Cohen–Macaulay, then* $\mathbb{H}^r_{g,\mathbb{K}}$ *is also Cohen–Macaulay and* $\mathbb{H}^r_{g,tot,\mathbb{K}}$ *is in fact Gorenstein.*

(d) *If* $p = 0$*, then* $\mathbb{H}^r_{g,tot,\mathbb{K}}$ *is Cohen–Macaulay.*

Proof If there exists a nonzero polynomial $h \in \mathrm{Sym}((V_g^{r+1})^*)_s^{\mathbf{SL}_g}$, then for a $\mathbb{K}$-algebra $\mathbb{B}$, an element $\lambda \in \boldsymbol{\mu}_g(\mathbb{B}) = Z(\mathbf{SL}_g)(\mathbb{B})$ acts on h as the multiplication by $\lambda^{2s} = 1$ and thus g divides $2s$. From this the part (a) and the first two parts of (b) follow.

As $\mathbf{SL}_g$ is semisimple, from [HR], Cor. 1.9 (see [Mur], Thm. for the Gorenstein conclusion) we get that part (d) holds and part (c) holds for $\mathbb{H}^r_{g,\mathrm{tot},\mathbb{K}}$. Thus parts (b) and (c) hold if $\mathbb{H}^r_{g,\mathbb{K}} = \mathbb{H}^r_{g,\mathrm{tot},\mathbb{K}}$.

To end the proof it suffices to show that if $\mathbb{H}^r_{g,\mathbb{K}} \neq \mathbb{H}^r_{g,\mathrm{tot},\mathbb{K}}$ (so g is even), then $[\mathrm{Frac}(\mathbb{H}^r_{g,\mathrm{tot},\mathbb{K}}) : \mathrm{Frac}(\mathbb{H}^r_{g,\mathbb{K}})] = 2$, there exists $\mathfrak{h}_{g,r,\mathrm{odd}} \in \mathbb{H}^{r,\perp}_{g,\mathbb{K}}$ such that (b.ii) holds, $\mathbb{H}^r_{g,\mathbb{K}}$ is normal, and, if $\mathbb{H}^r_{g,\mathrm{tot},\mathbb{K}}$ is Cohen–Macaulay, so is $\mathbb{H}^r_{g,\mathbb{K}}$.

Let $\mathfrak{h}_{g,r,\mathrm{odd}} \in \mathbb{H}^{r,\perp}_{g,\mathbb{K}}$ be a nonzero homogeneous element of the smallest (positive) T-degree. As all prime divisors of $\mathfrak{h}_{g,r,\mathrm{odd}}$ in the unique factorization domain $\mathbb{H}^r_{g,\mathrm{tot},\mathbb{K}}$ are homogeneous of positive T-degree, from the smallest T-degree property of $\mathfrak{h}_{g,r,\mathrm{odd}}$, we get that $\mathfrak{h}_{g,r,\mathrm{odd}}$ is a prime element of $\mathbb{H}^r_{g,\mathrm{tot},\mathbb{K}}$.

As $\mathbb{H}^r_{g,\mathrm{tot},\mathbb{K}}$ is an integral domain, 1 and $\mathfrak{h}_{g,r,\mathrm{odd}}$ generate an $\mathbb{H}^r_{g,\mathbb{K}}$-submodule of $\mathbb{H}^r_{g,\mathrm{tot},\mathbb{K}}$ which is free of rank 2. If the $\mathbb{K}$-algebra $\mathbb{H}^r_{g,\mathrm{tot},\mathbb{K}}$ is generated by homogeneous elements $h_1, \ldots, h_m \in \mathbb{H}^r_{g,\mathbb{K}}$ and $\mathfrak{h}_{g,r,\mathrm{odd}}, h_1^\perp, \ldots, h_{m^\perp}^\perp \in \mathbb{H}^{r,\perp}_{g,\mathbb{K}}$, then the $\mathbb{K}$-algebra $\mathbb{H}^r_{g,\mathbb{K}}$ is generated by $h_1, \ldots, h_m$ and by products of two elements in the set $\{\mathfrak{h}_{g,r,\mathrm{odd}}, h_1^\perp, \ldots, h_{m^\perp}^\perp\}$, and moreover $\mathfrak{h}_{g,r,\mathrm{odd}}, h_1^\perp, \ldots, h_{m^\perp}^\perp \in \mathbb{H}^{r,\perp}_{g,\mathbb{K}}$ are homogeneous generators of the $\mathbb{H}^r_{g,\mathbb{K}}$-module $\mathbb{H}^{r,\perp}_{g,\mathbb{K}}$. Thus, as

$$\mathfrak{h}^2_{g,r,\mathrm{odd}}, \mathfrak{h}_{g,r,\mathrm{odd}}h_1^\perp, \ldots, \mathfrak{h}_{g,r,\mathrm{odd}}h_{m^\perp}^\perp \in \mathbb{H}^r_{g,\mathbb{K}},$$

the three identities of part (b) follow from the identities

$$\begin{aligned}\mathrm{Frac}(\mathbb{H}^r_{g,\mathrm{tot},\mathbb{K}}) &= \mathrm{Frac}(\mathbb{H}^r_{g,\mathbb{K}})(\mathfrak{h}_{g,r,\mathrm{odd}}, h_1^\perp, \ldots, h_m^\perp)\\ &= \mathrm{Frac}(\mathbb{H}^r_{g,\mathbb{K}})(\mathfrak{h}_{g,r,\mathrm{odd}})\\ &= \mathrm{Frac}(\mathbb{H}^r_{g,\mathbb{K}})[x]/(x^2 - \mathfrak{h}^2_{g,r,\mathrm{odd}})\\ &= \mathrm{Frac}(\mathbb{H}^r_{g,\mathbb{K}}) \oplus \mathrm{Frac}(\mathbb{H}^r_{g,\mathbb{K}})\mathfrak{h}_{g,r,\mathrm{odd}}.\end{aligned} \tag{4.13}$$

As $\mathbb{H}^r_{g,\mathrm{tot},\mathbb{K}} = \mathbb{H}^r_{g,\mathbb{K}} \oplus \mathbb{H}^{r,\perp}_{g,\mathbb{K}}$ is normal (being a unique factorization domain) and as each nonzero element of $\mathbb{H}^{r,\perp}_{g,\mathbb{K}}$ belongs to $\mathrm{Frac}(\mathbb{H}^r_{g,\mathbb{K}})\mathfrak{h}_{g,r,\mathrm{odd}}$, $\mathbb{H}^r_{g,\mathbb{K}}$ equals to its normalization in $\mathbb{H}^r_{g,\mathrm{tot},\mathbb{K}}$ and hence is normal.

We use standard equivalent definitions of the notion Cohen–Macaulay ring for finitely generated $\mathbb{Z}_{\geq 0}$-graded algebras over fields (see [Cho], Thm. 1). As $\mathbb{H}^r_{g,\text{tot},\mathbb{K}}$ is Cohen–Macaulay of dimension $(r+1)\frac{g(g+1)}{2} - g^2 + 1$ (see (a) and Theorem 4.1.1), there exists a homogeneous system of parameters $h_1, \ldots, h_{(r+1)\frac{g(g+1)}{2} - g^2 + 1}$ which is a regular sequence of $\mathbb{H}^r_{g,\text{tot},\mathbb{K}}$.

Part (b) implies that $h_1^2, \ldots, h^2_{(r+1)\frac{g(g+1)}{2} - g^2 + 1}$ is a homogeneous system of parameters which is a regular sequence of $\mathbb{H}^r_{g,\mathbb{K}}$, hence $\mathbb{H}^r_{g,\mathbb{K}}$ is Cohen–Macaulay. □

4.2.3 Rings and Vector Spaces of Invariants: The General Case

To list more applications needed in Chaps. 2 and 3, let

$$\mathbb{H}^{r,\mathbb{Z}}_{g,\text{tot}} := \mathbb{H}^r_{g,\text{tot},\mathbb{C}} \cap (\mathbb{Z}[T, \ldots, T^{(r)}]) \quad \text{and} \quad \mathbb{H}^{r,\mathbb{Z}}_{g} := \mathbb{H}^r_{g,\mathbb{C}} \cap (\mathbb{Z}[T, \ldots, T^{(r)}]).$$

For $s \in \mathbb{W}_{\varepsilon_g}$ let $\mathbb{H}^{r,\mathbb{Z}}_{g,\text{tot}}(s) := \mathbb{H}^r_{g,\text{tot}}(s)_{\mathbb{C}} \cap \mathbb{H}^{r,\mathbb{Z}}_{g,\text{tot}}$. Also, for $s \in \mathbb{Z}_{\geq 0}$ we define $\mathbb{H}^{r,\mathbb{Z}}_{g}(s) := \mathbb{H}^r_{g}(s)_{\mathbb{C}} \cap \mathbb{H}^{r,\mathbb{Z}}_{g} = \mathbb{H}^{r,\mathbb{Z}}_{g,\text{tot}}(s)$. We have direct sum decompositions

$$\mathbb{H}^{r,\mathbb{Z}}_{g} = \oplus_{s \in \mathbb{Z}_{\geq 0}} \mathbb{H}^{r,\mathbb{Z}}_{g}(s) \subset \mathbb{H}^{r,\mathbb{Z}}_{g,\text{tot}} = \oplus_{s \in \mathbb{W}_{\varepsilon_g}} \mathbb{H}^{r,\mathbb{Z}}_{g,\text{tot}}(s).$$

For an arbitrary $\mathbb{Z}$-algebra B we set

$$\mathbb{H}^{r,\mathbb{Z}}_{g,\text{tot},B} := \mathbb{H}^{r,\mathbb{Z}}_{g,\text{tot}} \otimes_{\mathbb{Z}} B, \quad \mathbb{H}^{r,\mathbb{Z}}_{g,\text{tot}}(s)_B := \mathbb{H}^{r,\mathbb{Z}}_{g,\text{tot}}(s) \otimes_{\mathbb{Z}} B, \quad \mathbb{H}^{r,\mathbb{Z}}_{g,B} := \mathbb{H}^{r,\mathbb{Z}}_{g} \otimes_{\mathbb{Z}} B \quad \text{and}$$

$$\mathbb{H}^{r,\mathbb{Z}}_{g}(s)_B := \mathbb{H}^{r,\mathbb{Z}}_{g}(s) \otimes_{\mathbb{Z}} B.$$

We get direct sum decompositions

$$\mathbb{H}^{r,\mathbb{Z}}_{g,B} = \oplus_{s \in \mathbb{Z}_{\geq 0}} \mathbb{H}^{r,\mathbb{Z}}_{g}(s)_B \subset \mathbb{H}^{r,\mathbb{Z}}_{g,\text{tot},B} = \oplus_{s \in \mathbb{W}_{\varepsilon_g}} \mathbb{H}^{r,\mathbb{Z}}_{g,\text{tot}}(s)_B.$$

From [Se], Thm. 2 we get that the $\mathbb{Z}$-algebra $\mathbb{H}^{r,\mathbb{Z}}_{g,\text{tot}}$ is finitely generated and arguments similar to the ones of the proof of Lemma 4.2.3(a) and (b) show that the same holds for $\mathbb{H}^{r,\mathbb{Z}}_{g}$. Thus the B-algebras $\mathbb{H}^{r,\mathbb{Z}}_{g,\text{tot},B}$ and $\mathbb{H}^{r,\mathbb{Z}}_{g,B}$ are finitely presented.

Moreover, if B is a field of characteristic 0, then the B-algebras $\mathbb{H}^{r,\mathbb{Z}}_{g,\text{tot},B}$ and $\mathbb{H}^{r,\mathbb{Z}}_{g,B}$ are Cohen–Macaulay unique factorization domains and hence Gorenstein (see [HR], Cor. 1.9) and their groups of units are $B^{\times}$.

If $p = 0$, then we have canonical identifications

$$\mathbb{H}^r_{g,\text{tot},\mathbb{K}} = \mathbb{H}^{r,\mathbb{Z}}_{g,\text{tot},\mathbb{K}} \quad \text{and} \quad \mathbb{H}^r_{g,\mathbb{K}} = \mathbb{H}^{r,\mathbb{Z}}_{g,\mathbb{K}} \tag{4.14}$$

compatible with the gradings but for $p > 0$ we only have canonical inclusions ($\mathbb{K}$-algebra monomorphisms compatible with the gradings)

$$\mathbb{H}^{r,\mathbb{Z}}_{g,\text{tot},\mathbb{K}} \subset \mathbb{H}^{r}_{g,\text{tot},\mathbb{K}} \quad \text{and} \quad \mathbb{H}^{r,\mathbb{Z}}_{g,\mathbb{K}} \subset \mathbb{H}^{r}_{g,\mathbb{K}}. \tag{4.15}$$

Based on the last two paragraphs, we can set

$$D(g,r) := (r+1)\frac{g(g+1)}{2} - g^2 + 1,$$

$$D_p(g,r,s) := \dim_{\mathbb{K}}(\mathbb{H}^r_g(s)_{\mathbb{K}}) \text{ for } s \in \mathbb{W}_{\varepsilon_g},$$

$$D(g,r,s) := D_0(g,r,s) \qquad \text{for } s \in \mathbb{W}_{\varepsilon_g}.$$

For instance, $D(1,r) = r+1$, $\mathbb{H}^r_{1,\mathbb{K}}$ is a polynomial $\mathbb{K}$-algebra in $r+1$ indeterminates of degree 1 and therefore $D_p(1,r,s) = \binom{r+s}{s}$ for all $s \in \mathbb{Z}_{\geq 0}$. Similarly, for $g \geq 2$ we have $\mathbb{H}^0_{g,\mathbb{K}} = \mathbb{K}$. From the inclusions of Display (4.15) we get that

$$D_p(g,r,s) \geq D(g,r,s).$$

Once an identity $D_p(g,r,s) = D(g,r,s)$ is proved, afterwards we use the simpler notation $D(g,r,s)$ instead of $D_p(g,r,s)$.

From Theorem 4.1.1 and Lemma 4.2.3(a) and (b) we get directly the following consequence.

Corollary 4.2.4 *For all integers $r \geq 1$ we have*

$$\dim(\mathbb{H}^r_{g,tot,\mathbb{K}}) = \dim(\mathbb{H}^r_{g,\mathbb{K}}) = D(g,r).$$

4.2.4 The Case $s = \frac{1}{2}$

To compute the numbers $D_p(g,r,s)$ in the case $s = \frac{1}{2}$ we recall the computation of all $\mathbf{T}_g$-invariants of $\mathrm{Sym}((\mathrm{V}^{r+1}_g)^*) = \mathbb{K}[T,\ldots,T^{(r)}]$.

Remark 4.2.5 The torus $\mathbf{T}_g$ leaves invariant the $\mathbb{K}$-span of each monomial

$$\mu := \prod_{1\leq i\leq j\leq g} \prod_{l=0}^{r} (T^{(l)}_{ij})^{n_{ij,l}}$$

in the $T_{ij}^{(l)}$s, where each $n_{ij,l} \in \mathbb{N} \cup \{0\}$. For $m \in \{1, \ldots, g\}$, let

$$\mathrm{Index}_m(\mu) := \sum_{1 \leq i \leq j \leq g} \sum_{l=0}^{r} n_{ij,l}(\delta_{mi} + \delta_{mj}).$$

We have

$$\sum_{m=1}^{g} \mathrm{Index}_m(\mu) = 2\deg(\mu). \tag{4.16}$$

For $\lambda \in \mathbb{K}^\times \setminus \{1\}$, the matrix $\Lambda := \lambda^{-1}\mathrm{Diag}(1, \ldots, 1, \lambda^g, 1, \ldots, 1) \in \mathbf{T}_g(\mathbb{K})$ with λ^{g-1} as the mm entry, maps μ to $\lambda^{g\mathrm{Index}_m(\mu)-2\deg(\mu)}\mu$. Thus, as $\mathbf{T}_g(\mathbb{K})$ is generated by such Λs, the monomial μ is $\mathbf{T}_g$-invariant if and only if $\mathrm{Index}_m(\mu)$ is independent on $m \in \{1, \ldots, g\}$, in which case it is denoted simply as $\mathrm{Index}(\mu)$ and from Eq. (4.16) we get that

$$2\deg(\mu) = g\mathrm{Index}(\mu);$$

thus, if g is even, we have $\deg(\mu) = \frac{g}{2}$ if and only if $\mathrm{Index}(\mu) = 1$.

Proposition 4.2.6 *If g is even, then we have $D_p(g, r, \frac{1}{2}) = 0$.*

Proof If $g = 2$, we have $\mathbb{H}_2^r(\frac{1}{2})_{\mathbb{K}} \subset \oplus_{l=0}^{r} \mathbb{K}T_{12}^{(l)}$ (see Remark 4.2.5) and $T_{12}^{(l)}$ is not an $\mathbf{SL}_2$-invariant (for $p = 2$, see Lemma 4.2.1(a)–(d) in the dual context or see Lemma 4.2.1(j)). Thus $\mathbb{H}_2^r(\frac{1}{2})_{\mathbb{K}} = 0$ and $D_p(2, r, \frac{1}{2}) = 0$.

For $g \geq 4$ we show that the assumption that there exists $h \in \mathbb{H}_g^r(\frac{1}{2})_{\mathbb{K}} \setminus \{0\}$ leads to a contradiction. We know h is in the $\mathbb{K}$-span of monomials $\prod_{m=1}^{\frac{g}{2}} T_{i_m, j_m}^{(l_m)}$ with $i_m < j_m$, $\{i_1, j_1, \ldots, i_{\frac{g}{2}}, j_{\frac{g}{2}}\} = \{1, \ldots, g\}$ and $l_1, \ldots, l_{\frac{g}{2}} \in \{0, \ldots, r\}$, see Remark 4.2.5. We fix such a monomial $\mu := \prod_{m=1}^{\frac{g}{2}} T_{i_m, j_m}^{(l_m)}$ which appears with a nonzero coefficient c in h.

For $p \neq 2$, the coefficient of μ in $\Lambda_{i_1, j_2} \cdot h$ is $-c$ (see Eqs. (4.9)–(4.11)) which contradicts the fact that h is an $\mathbf{SL}_g$-invariant.

If $p = 2$, let $\Lambda \in \mathbf{SL}_g(\mathbb{K})$ be such that it normalizes $\mathrm{Span}(\{e_{i_1}, e_{j_1}\})$, fixes each $e_{i'}$ with $i' \in \{1, \ldots, g\} \setminus \{i_1, j_1\}$ and (by the case $g = 2$) does not fix $T_{i_1, j_1}^{(l_1)}$. Based on Eq. (4.8), similar to Eqs. (4.9)–(4.11) we argue that for each $l \in \{0, \ldots, r\}$, Λ leaves invariant $\mathrm{Span}(\{T_{i_1,i_1}^{(l)}, T_{i_1,j_1}^{(l)}, T_{j_1,j_1}^{(l)}\})$ and fixes each indeterminate $T_{ij}^{(l)}$ with $i, j \in \{1, \ldots, g\} \setminus \{i_1, j_1\}$. From the last two sentences we get that $\Lambda \cdot h \neq h$ which contradicts the fact that h is an $\mathbf{SL}_g$-invariant. □

4.2.5 *The Case $r = 1$*

To compute the numbers $D_p(g, r, s)$ in the case $r = 1$ and $p \neq 2$ we first recall the following well-known result.

Lemma 4.2.7 *Let $y_0, \ldots, y_g$ be indeterminates. Let $\sigma_0(y_1, \ldots, y_g) := 1$ and*

$$\sigma_1(y_1, \ldots, y_g) := \sum_{m=1}^{g} y_m, \cdots, \sigma_g(y_1, \ldots, y_g) := \prod_{m=1}^{g} y_m$$

be the elementary symmetric polynomials in the indeterminates $y_1, \ldots, y_g$. Then the $\mathbb{K}$-algebra endomorphism

$$\theta_g : \mathbb{K}[y_0, \ldots, y_g] \to \mathbb{K}[y_0, \ldots, y_g]$$

given for $m \in \{0, \ldots, g\}$ by the rule

$$\theta_g(y_m) := y_0^{g-m} \sigma_m(y_1, \ldots, y_g)$$

becomes integral after inverting y_0 (i.e., the induced $\mathbb{K}$-algebra endomorphism $(\theta_g)_{y_0}$ of $\mathbb{K}[y_0, \ldots, y_g]_{y_0}$ is integral).

Proof Let $z_1 := y_0^{-1} y_1, \ldots, z_g := y_0^{-1} y_g \in \mathbb{K}[y_0, \ldots, y_g][y_0^{-1}]$. The $\mathbb{K}$-algebra monomorphism $\mathbb{K}[\sigma_1(z_1, \ldots, z_g), \ldots, \sigma_g(z_1, \ldots, z_g)] \to \mathbb{K}[z_1, \ldots, z_g]$ is integral as $z_1, \ldots, z_g$ are roots of the monic polynomial

$$\sum_{m=0}^{g} (-1)^m z^{g-m} \sigma_m(z_1, \ldots, z_m)$$

in the indeterminate z, and thus by tensoring it with $\mathbb{K}[y_0, y_0^{-1}]$ over $\mathbb{K}$ we get that $(\theta_g)_{y_0}$ is integral. □

The $p = 0$ case of the next theorem is classical (for instance, it is proved in [T], Ch. 20, Sect. 6, p. 304).

Theorem 4.2.8 *Let $\mathbb{H}^{1,\Theta}_{g,\mathbb{K}}$ be the $\mathbb{K}$-subalgebra of $\mathbb{H}^{1}_{g,tot,\mathbb{K}}$ generated by the polynomials $\Theta_0, \ldots, \Theta_g$ of Eq. (4.3). Then the following properties hold.*

(a) *The polynomials $\Theta_0, \ldots, \Theta_g$ are algebraically independent over $\mathbb{K}$, i.e., $\mathbb{H}^{1,\Theta}_{g,\mathbb{K}}$ is a polynomial $\mathbb{K}$-algebra in $g + 1$ indeterminates.*

(b) *If $p \neq 2$, then we have*

$$\mathbb{H}^{1,\Theta}_{g,\mathbb{K}} = \mathbb{H}^{1}_{g,\mathbb{K}} = \mathbb{H}^{1}_{g,tot,\mathbb{K}} = Sym((V_g^2)^*)^{\boldsymbol{SL}_g} = \mathcal{O}(\boldsymbol{V}_g^2)^{\boldsymbol{SL}_g}$$

(thus, if g is even, we have $\mathbb{H}^{1,\perp}_{g,\mathbb{K}} = 0$ and $D_p(g, 1, s) = 0$ if $s \in \mathbb{W}_0 \setminus \mathbb{Z}_{\geq 0}$) and $D_p(g, 1, s) = \binom{g+s}{s}$ for all $s \in \mathbb{Z}_{\geq 0}$.

(c) *If $p = 2$, then the $\mathbb{K}$-monomorphisms $\mathbb{H}^{1,\Theta}_{g,\mathbb{K}} \to \mathbb{H}^{1}_{g,\mathbb{K}} \to \mathbb{H}^{1}_{g,tot,\mathbb{K}}$ are integral and purely inseparable (but $\mathbb{H}^{1,\Theta}_{g,\mathbb{K}} \to \mathbb{H}^{1}_{g,\mathbb{K}}$ is not an isomorphism for $g > 1$).*

Proof With $F_0 \in \mathbb{H}^{1,\Theta}_{g,\mathbb{K}}$ as in the proof of Theorem 4.1.1, we claim that by inverting F_0 the $\mathbb{K}$-algebra monomorphism $\mathbb{H}^{1,\Theta}_{g,\mathbb{K},F_0} \to \mathbb{H}^{1}_{g,\mathrm{tot},\mathbb{K},F_0}$ is integral and purely inseparable for $p = 2$ and is an isomorphism if $p \neq 2$.

The fibers of the morphism $\varpi_{F_0} : \mathrm{Spec}(\mathbb{K}[T, T^{(1)}]_{F_0}) \to \mathrm{Spec}(\mathbb{H}^{1,\Theta}_{g,\mathbb{K},F_0})$ were in essence computed in the proof of Theorem 4.1.1. To explain this we consider a point $P \in \mathrm{Spec}(\mathbb{H}^{1,\Theta}_{g,\mathbb{K},F_0})(\mathbb{K})$ given by a surjective $\mathbb{K}$-homomorphism $\mathbb{H}^{1,\Theta}_{g,\mathbb{K},F_0} \to \mathbb{K}$ which for $i \in \{0, \ldots, g\}$ maps Θ_i to some $\nu_i \in \mathbb{K}$. Let $\lambda_0 \in \mathbb{K}$ be such that $\lambda_0^g = \nu_0$ and let $\lambda_1, \ldots, \lambda_g \in \mathbb{K}$ be such that (see Eq. (4.4))

$$\nu_0 \prod_{i=1}^{g} (x - \lambda_i) = \sum_{i=0}^{g} x^{g-i} \nu_i .$$

As $F_0(\nu_0, \ldots, \nu_g) \neq 0$, $\lambda_0 \neq 0$, $\lambda_1, \ldots, \lambda_g$ are distinct, and, if $p = 2$, we have $\sum_{i=1}^{g} \lambda_i \neq 0$. Let $D_0 := \lambda_0 1_g$, $D_1 := \mathrm{Diag}(\lambda_0\lambda_1, \ldots, \lambda_0\lambda_g) \in \mathrm{Z}_g$. The pair (D_0, D_1) defines a $\mathbb{K}$-valued point of $\varpi_{F_0}^{-1}(P)$ and let $\mathbf{o}$ be its orbit in $\mathbf{V}_g^2$; it is a smooth, closed subscheme of $\mathbf{V}_g^2$ of dimension $g^2 - 1$. We have an $\mathbf{SL}_g$-invariant closed embedding $\mathbf{o} \subset \varpi_{F_0}^{-1}(P)$.

In this paragraph we check that $\mathbf{o} = (\varpi_{F_0}^{-1}(P))_{\mathrm{red}}$, i.e., we have an identity $\mathbf{o}(\mathbb{K}) = (\varpi_{F_0}^{-1}(P))(\mathbb{K})$. It suffices to show that if $\lambda'_0, \lambda'_1, \ldots, \lambda'_g \in \mathbb{K}$ are such that $(\lambda'_0)^g = \nu_0 = \lambda_0^g$ and $\{\lambda_0\lambda_1, \ldots, \lambda_0\lambda_g\} = \{\lambda'_0\lambda'_1, \ldots, \lambda'_0\lambda'_g\}$, then (D_0, D_1) is $\mathbf{SL}_g$-equivalent to $(\lambda'_0 1_g, \mathrm{Diag}(\lambda'_0\lambda'_1, \ldots, \lambda'_0\lambda'_g))$. Based on the proof of Theorem 4.1.1, we can assume that $\lambda'_0 = \lambda_0$ and it suffices to show that D_1 and $\mathrm{Diag}(\lambda_0\lambda'_1, \ldots, \lambda_0\lambda'_g)$ are $\mathbf{SO}_g$-equivalent. But this follows by repeatedly applying the fact that for $g \geq 2$ and integers $1 \leq i < j \leq g$, similarly to Eq. (4.9) we get that $\Lambda_{ij} \cdot D_1$ is obtained from D_1 by interchanging its ii and jj entries.

To prove that $\varpi_{F_0}^{-1}(P) = \mathbf{o}$ is reduced if $p \neq 2$, it suffices to show that if $\mathbb{B} := \mathbb{K}[x]/(x^2) = \mathbb{K} \oplus \mathbb{K}\epsilon$, with $\epsilon := x + (x^2)$, then the injection

$$\mathbf{o}(\mathbb{B}) \to \varpi_{F_0}^{-1}(P)(\mathbb{B})$$

is in fact a bijection. To check this, it suffices to show that if $(D'_0, D'_1) \in \mathbb{B} \otimes_{\mathbb{K}} \mathrm{V}_g$ is such that modulo ϵ is (D_0, D_1), $\det(D'_0) = \det(D_0)$ and $(D'_0)^{-1} D'_1$ has eigenvalues $\lambda_1, \ldots, \lambda_g$, then there exists $\Lambda' \in \mathrm{Ker}(\mathbf{SL}_g(\mathbb{B}) \to \mathbf{SL}_g(\mathbb{K}))$ such that we have $(\Lambda' D_0 (\Lambda')^{\mathrm{t}}, \Lambda' D_1 (\Lambda')^{\mathrm{t}}) = (D'_0, D'_1)$. To check this we note that, as $p \neq 2$, each $Q \in \mathrm{V}_g$ of trace zero is of the form $\lambda_0(M_0 + M_0^{\mathrm{t}})$ with $M_0 \in \mathrm{Mat}_g$ of trace zero (for instance, we can take $M_0 = \frac{1}{2\lambda_0} Q$) and thus we can assume that $D'_0 = D_0$.[6]

We show that there exists $\Lambda' \in \mathrm{Ker}(\mathbf{SO}_g(\mathbb{B}) \to \mathbf{SO}_g(\mathbb{K}))$ such that we have $\Lambda' D_1 (\Lambda')^{\mathrm{t}} = \Lambda' D_1 (\Lambda')^{-1} = D'_1$. As D_1 and D'_1 have the same distinct eigenvalues, there exists a matrix $M \in \mathrm{Mat}_g$ such that

$$D'_1 = (1_g + \epsilon M) D_1 (1_g - \epsilon M) = D_1 + \epsilon(M D_1 - D_1 M)$$

[6] If $g \geq 2$, for $p = 2$ we cannot assume that $D'_0 = D_0$.

As $MD_1 - D_1M$ and D_1 are symmetric, D_1 commutes with $M + M^t$. As $D_1 \in \mathrm{Z}_g$ has distinct eigenvalues, we get that $M + M^t \in \mathrm{Z}_g$. This implies that there exists $M_1 \in \mathrm{Mat}_g$ with $M_1^t + M_1 = 0$ (so $M_1 \in \mathrm{Lie}(\mathbf{SO}_g)$) such that $M - M_1 \in \mathrm{Z}_g$. We have $MD_1 - D_1M = M_1D_1 - D_1M_1$ and hence we can take $\Lambda' = 1_g + \epsilon M_1$.

The morphism $\mathrm{Spec}(\mathbb{H}^1_{g,\mathrm{tot},\mathbb{K},F_0}) \to \mathrm{Spec}(\mathbb{H}^{1,\Theta,F_0}_{g,\mathbb{K}})$ induces a bijection at the level of $\mathbb{K}$-valued points and has generically reduced fibers if $p \neq 2$ by the previous paragraph. This implies that this morphism is finite, purely inseparable if $p = 2$ and birational if $p \neq 2$, and from Zariski's Main Theorem we get that it is an isomorphism if $p \neq 2$, i.e., the claim holds.

From previous paragraph and Corollary 4.2.4 we get that we have an equality $\dim(\mathbb{H}^{1,\Theta}_{g,\mathbb{K}}) = g + 1$. As $\mathbb{H}^{1,\Theta}_{g,\mathbb{K}}$ is a quotient of a polynomial $\mathbb{K}$-algebra in $\dim(\mathbb{H}^{1,\Theta}_{g,\mathbb{K}}) = g + 1$ indeterminates, it is a polynomial $\mathbb{K}$-algebra in $g + 1$ indeterminates and $\Theta_0, \ldots, \Theta_g$ are algebraically independent over $\mathbb{K}$. Thus $\mathbb{H}^{1,\Theta}_{g,\mathbb{K}} \subset \mathbb{H}^1_{g,\mathrm{tot},\mathbb{K}}$ is an inclusion between two noetherian unique factorization (hence normal) domains such that for $p \neq 2$ we have $\mathrm{Frac}(\mathbb{H}^{1,\Theta}_{g,\mathbb{K}}) = \mathrm{Frac}(\mathbb{H}^1_{g,\mathrm{tot},\mathbb{K}})$.

Recall that $\mathrm{L}_g = \mathbb{K}1_g \times \mathrm{Z}_g \subset \mathrm{V}_g^2$ and we consider the composite morphism

$$\theta_g^* : \mathbf{L}_g \to \mathbf{V}_g^2 \to (\mathbf{V}_g^2)^{\mathbf{SL}_g} = \mathrm{Spec}(\mathbb{H}^r_{g,\mathrm{tot},\mathbb{K}}) \to \mathrm{Spec}(\mathbb{K}[\Theta_0, \ldots, \Theta_g]).$$

As for $(D_0, D_1) := (\lambda_0 1_g, \mathrm{Diag}(\lambda_1, \ldots, \lambda_g)) \in \mathrm{L}_g$ and $m \in \{1, \ldots, g\}$ we have $\Theta_m(D_0, D_1) = \lambda_0^{g-m}\sigma_m(\lambda_1, \ldots, \lambda_g)$ (see Eq. (4.3)), the notation matches with the one of Lemma 4.2.7, i.e., θ_g^* can be identified with the morphism between spectra defined by θ_g. From this and Lemma 4.2.7, as we have inclusions $\mathbb{H}^{1,\Theta}_{g,\mathbb{K},\Theta_0} \subset \mathbb{H}^1_{g,\mathrm{tot},\mathbb{K},\Theta_0} \subset \mathrm{Frac}(\mathbb{H}^{1,\Theta}_{g,\mathbb{K},\Theta_0})$, we get an integral $\mathbb{K}$-algebra monomorphism $\mathbb{H}^{1,\Theta}_{g,\mathbb{K},\Theta_0} \to \mathbb{H}^1_{g,\mathrm{tot},\mathbb{K},\Theta_0}$ which for $p = 2$ is purely inseparable and for $p \neq 2$ is an isomorphism as we know it is birational (see previous paragraph) and as $\mathbb{H}^{1,\Theta}_{g,\mathbb{K},\Theta_0}$ is normal. Switching the roles of $T^{(0)}$ and $T^{(1)}$, we similarly get an integral $\mathbb{K}$-algebra monomorphism $\mathbb{H}^{1,\Theta}_{g,\mathbb{K},\Theta_g} \to \mathbb{H}^1_{g,\mathrm{tot},\mathbb{K},\Theta_g}$ which for $p = 2$ is purely inseparable and for $p \neq 2$ is an isomorphism. Thus $\mathbb{H}^1_{g,\mathrm{tot},\mathbb{K}} \subset \mathbb{H}^{1,\Theta}_{g,\mathbb{K},\Theta_0} \cap \mathbb{H}^{1,\Theta}_{g,\mathbb{K},\Theta_g} = \mathbb{H}^{1,\Theta}_{g,\mathbb{K}}$ if $p \neq 2$. Hence $\mathbb{H}^{1,\Theta}_{g,\mathbb{K}} = \mathbb{H}^1_{g,\mathrm{tot},\mathbb{K}}$ if $p \neq 2$ from which part (a) follows.

If $p = 2$, then for the normalization $\mathbb{H}^{1,\Theta,\mathrm{n}}_{g,\mathbb{K}}$ of $\mathbb{H}^{1,\Theta}_{g,\mathbb{K}}$ in $\mathbb{H}^1_{g,\mathrm{tot},\mathbb{K}}$ we have the identity $\mathbb{H}^{1,\Theta,\mathrm{n}}_{g,\mathbb{K}} = \mathbb{H}^{1,\Theta,\mathrm{n}}_{g,\mathbb{K},\Theta_0} \cap \mathbb{H}^{1,\Theta,\mathrm{n}}_{g,\mathbb{K},\Theta_g}$ and similarly we argue that $\mathbb{H}^{1,\Theta,\mathrm{n}}_{g,\mathbb{K}} = \mathbb{H}^1_{g,\mathrm{tot},\mathbb{K}}$ from which part (b) follows. □

4.2.6 Basic $\mathbf{SL}_g$-Invariants: Θs and Υs

In order to present a generalization of Theorem 4.2.8(a) we first introduce the index set

$$\Delta(g, r) := \{(m_0, \ldots, m_r) \in \mathbb{Z}_{\geq 0}^{r+1} | \sum_{i=0}^{r} m_i = g\}$$

that indexes polynomials generalizing $\Theta_0, \ldots, \Theta_g$. If

$$\Delta^{+1}(g,r) := \{(n_0, \ldots, n_r) \in \mathbb{N}^{r+1} | \sum_{i=0}^{r} n_i = g + r + 1\},$$

then the rule $(m_0, \ldots, m_r) \to (m_0 + 1, \ldots, m_r + 1)$ defines a bijection

$$\Delta(g,r) \to \Delta^{+1}(g,r).$$

It is well-known that $\Delta^{+1}(g,r)$ has as many elements as the number of ways of inserting r commas among the $g + r$ spaces (places) between two consecutive 1s of the sequence 1 1 $\cdots$ 1 1 formed by $g + r + 1$ numbers 1, and thus it is equal to $\binom{g+r}{r}$. For instance, the 4-tuple $(1, 2, 1, 2) \in \Delta^{+1}(2, 3)$ corresponds to the following insertion of commas 1, 1 1, 1, 1 1. We conclude that $\Delta(g,r)$ has $\binom{g+r}{r}$ elements.

For $r \in \{1, 2\}$ or $g = 1$ we have $\binom{g+r}{r} = D(g,r)$ and for $r \geq 3$ and $g \geq 2$ we have $\binom{g+r}{r} > D(g,r)$. Moreover, for $r \geq 3$ we have

$$\lim_{g\to\infty} \frac{\binom{g+r}{r}}{D(g,r)} = \infty.$$

We define polynomials

$$\Theta_{m_0,\ldots,m_r} \in \mathbb{H}_g^{r,\mathbb{Z}}(1) \subset \mathbb{H}_g^{r,\mathbb{Z}} \subset \mathbb{H}_{g,\mathrm{tot}}^{r,\mathbb{Z}}$$

indexed by $(m_0, \ldots, m_r) \in \Delta(g,r)$ via the identity

$$\det(\sum_{i=0}^{r} y_i Q_i) = \sum_{(m_0,\ldots,m_r)\in\Delta(g,r)} (\prod_{i=0}^{r} y_i^{m_i})\Theta_{m_0,\ldots,m_r}(Q_0, Q_2, \ldots, Q_r),$$

where $y_0, \ldots, y_r$ are $r + 1$ indeterminates; so the $\Theta_{m_0,\ldots,m_r}$s are partial polarizations of the invariant polynomial det.

We also denote by $\Theta_{m_0,\ldots,m_r}$ its image in $\mathcal{O}(\mathbf{V}_g^{r+1})^{\mathbf{SL}_g} = \mathcal{O}(\mathbf{V}_g^{r+1}/\mathbf{SL}_g)$. Note that for all $i \in \{0, \ldots, g\}$ we have $\Theta_i = \Theta_{g-i,i}$ (see Eq. (4.3)) and for $r \geq 2$ inside $\mathcal{O}(\mathbf{V}_g^{r+1})^{\mathbf{SL}_g}$ we identify it as well with $\Theta_{g-i,i,0,0,\ldots,0}$ (the number of 0s being $r - 1$), via the inclusions of Display (4.12).

Let $\mathbb{H}_g^{r,\Theta}$ be the $\mathbb{Z}_{\geq 0}$-graded $\mathbb{Z}$-subalgebra of $\mathbb{H}_g^{r,\mathbb{Z}}$ (or $\mathbb{H}_{g,\mathrm{tot}}^{r,\mathbb{Z}}$) generated by all $\Theta_{m_0,\ldots,m_r}$s with $(m_0, \ldots, m_r) \in \Delta(g,r)$ (the case $r = 1$ was first introduced and used in Theorem 4.2.8). For a $\mathbb{Z}$-algebra B, we define its Θ-rings of invariants $\mathbb{H}_{g,B}^{r,\Theta} := \mathbb{H}_g^{r,\Theta} \otimes_{\mathbb{Z}} B$. We have B-algebra monomorphisms compatible with the gradings

$$\mathbb{H}_{g,B}^{r,\Theta} \subset \mathbb{H}_{g,B}^{r,\mathbb{Z}} \subset \mathbb{H}_{g,\mathrm{tot},B}^{r,\mathbb{Z}}. \tag{4.17}$$

For an integer $l \geq 0$, let

$$c(T^{(l)}) := (T_{11}^{(l)}\ T_{12}^{(l)}\ \cdots\ T_{1g}^{(l)}\ T_{22}^{(l)}\ T_{23}^{(l)}\ \cdots\ T_{gg}^{(l)})^{\mathrm{t}}$$

be the column vector whose entries are $T_{ij}^{(l)}$ with $1 \leq i \leq j \leq g$ listed increasingly in the lexicographic order of the indexes ij. If the inequality $r \geq \frac{g(g+1)}{2} - 1$ holds, then for each increasing sequence $q_1 < q_2 < \cdots < q_{\frac{g(g+1)}{2}}$ of integers in the set $\{0, 1, \ldots, r\}$, let

$$\Upsilon_{q_1,\ldots,q_{\frac{g(g+1)}{2}}} := \det([c(T^{(q_1)})| \cdots |c(T^{(q_{\frac{g(g+1)}{2}})})]).$$

If g is odd, then $\frac{g(g+1)}{2} \in \frac{g}{2}\mathbb{Z}$ and hence we have

$$\Upsilon_{q_1,\ldots,q_{\frac{g(g+1)}{2}}} \in \mathbb{H}_g^{q_{\frac{g(g+1)}{2}}}(\frac{g(g+1)}{2})_{\mathbb{K}} \subset \mathbb{H}_{g,\mathbb{K}}^{q_{\frac{g(g+1)}{2}}} \subset \mathbb{H}_{g,\mathbb{K}}^{r}.$$

If g is even, then $\frac{g(g+1)}{2} \notin \frac{g}{2}\mathbb{Z}$ and hence we have

$$\Upsilon_{q_1,\ldots,q_{\frac{g(g+1)}{2}}} \in \mathrm{Sym}((V_g^{1+q_{\frac{g(g+1)}{2}}})^*)_{\frac{g(g+1)}{2}}^{\mathbf{SL}_g} \subset \mathbb{H}_{g,\mathbb{K}}^{q_{\frac{g(g+1)}{2}},\perp} \subset \mathbb{H}_{g,\mathbb{K}}^{r,\perp}.$$

For instance, if $g = 2$ (so $r \geq 2$), then we can take $\mathfrak{h}_{2,r,\mathrm{odd}} = \Upsilon_{0,1,2}$.

The number of such $\Upsilon_{q_1,\ldots,q_{\frac{g(g+1)}{2}}}$ with $q_{\frac{g(g+1)}{2}} \leq r$ is $\binom{r+1}{\frac{g(g+1)}{2}}$ (which makes sense even when $r < \frac{g(g+1)}{2} - 1$, being 0 in such a case).

Remark 4.2.9 As for given integers $r > l \geq 1$ the $\mathbf{SL}_g$-invariant projections $\mathbb{V}_g^{r+1} \to \mathbb{V}_g^{l+1}$ admit $\mathbf{SL}_g$-invariant sections, for the $\mathbb{Z}_{\geq 0}$-graded $\mathbb{K}$-monomorphisms of Display (4.12) we have retractions $\mathbb{H}_{g,\mathrm{tot},\mathbb{K}}^{r} \to \mathbb{H}_{g,\mathrm{tot},\mathbb{K}}^{l}$ which map $\mathbb{H}_{g,\mathbb{K}}^{r}$ onto $\mathbb{H}_{g,\mathbb{K}}^{l}$ and $\mathbb{H}_{g,\mathbb{K}}^{r,\Theta}$ onto $\mathbb{H}_{g,\mathbb{K}}^{l,\Theta}$. Thus, if $\mathbb{H}_{g,\mathbb{K}}^{r} = \mathbb{H}_{g,\mathrm{tot},\mathbb{K}}^{r}$, then we also have $\mathbb{H}_{g,\mathbb{K}}^{l} = \mathbb{H}_{g,\mathrm{tot},\mathbb{K}}^{l}$. Similarly, if $\mathbb{H}_{g,\mathbb{K}}^{r} = \mathbb{H}_{g,\mathbb{K}}^{r,\Theta}$, then we also have $\mathbb{H}_{g,\mathbb{K}}^{l} = \mathbb{H}_{g,\mathbb{K}}^{l,\Theta}$, and if $\mathbb{H}_{g,\mathbb{K}}^{r,\Theta} = \mathbb{H}_{g,\mathrm{tot},\mathbb{K}}^{r}$, then we also have $\mathbb{H}_{g,\mathbb{K}}^{l,\Theta} = \mathbb{H}_{g,\mathrm{tot},\mathbb{K}}^{l}$.

4.2.7 The Case $s = 1$

Let $\Gamma_p(g, r)$ (resp. $\Gamma_p(g, \mathrm{tot}, r)$) be the smallest number of homogeneous generators of the $\mathbb{Z}_{\geq 0}$-graded $\mathbb{K}$-algebra $\mathbb{H}_{g,\mathbb{K}}^{r}$ (resp. $\mathbb{W}_{\varepsilon_g}$-graded $\mathbb{K}$-algebra $\mathbb{H}_{g,\mathrm{tot},\mathbb{K}}^{r}$).

Proposition 4.2.10 *For all $g, r \in \mathbb{N}$ the following properties hold.*

(a) *We have an inclusion*

$$\mathbb{H}^r_g(1)_{\mathbb{K}} \supset Span(\{\Theta_{m_0,\dots,m_r} | (m_0, \dots, m_r) \in \Delta(g, r)\})$$

as well as inequalities

$$\min\{\Gamma_p(g, r), \Gamma_p(g, tot, r)\} \geq D_p(g, r, 1) \geq \binom{g+r}{r} \geq D(g, r). \tag{4.18}$$

Thus, for $r \geq 3$ we have

$$\lim_{g\to\infty} \frac{\min\{\Gamma_p(g, r), \Gamma_p(g, tot, r)\}}{D(g, r)} = \infty.$$

(b) *If $p \nmid g!$, then we have*

$$\mathbb{H}^r_g(1)_{\mathbb{K}} = Span(\{\Theta_{m_0,\dots,m_r} | (m_0, \dots, m_r) \in \Delta(g, r)\})$$

as well as

$$D_p(g, r, 1) = \binom{g+r}{r}.$$

Proof As $\Theta_{m_0,\dots,m_r}$ has partial T-degree $(m_0, \dots, m_r)$, the set of polynomials $\{\Theta_{m_0,\dots,m_r} | (m_0, \dots, m_r) \in \Delta(g, r)\}$ has $\binom{g+r}{r}$ elements and is linearly independent over $\mathbb{K}$. Thus the inclusion of part (a) holds and we have $D_p(g, r, 1) \geq \binom{g+r}{r}$. Referring to Inequalities (4.18), its first inequality is obvious and its last inequality was checked above. So to end the proof of both parts (a) and (b) it suffices to show that $D_p(g, r, 1) \leq \binom{g+r}{r}$ if $p \nmid g!$.

To check this, it suffices to show that for all $(m_0, \dots, m_r) \in \Delta(g, r)$, the $\mathbb{K}$-vector space of polynomials in $\mathbb{H}^r_{g,\text{tot},\mathbb{K}}$ of partial T-degree $(m_0, \dots, m_r)$ is one dimensional (generated by $\Theta_{m_0,m_1,\dots,m_r}$). Equivalently, we have to show that for all $(m_0, \dots, m_r) \in \Delta(g, r)$, $(\otimes_{l=0}^r \text{Sym}^{m_l}(V_g^*))^{\mathbf{SL}_g}$ has dimension at most 1. As we have natural $\mathbf{SL}_g$-epimorphisms

$$(V_g^*)^{\otimes m_l} \to \text{Sym}^{m_l}(V_g^*)$$

which admit sections (as m_l, if nonzero, is invertible in $\mathbb{K}$) it suffices to show that $\dim_{\mathbb{K}}(((V_g^*)^{\otimes g})^{\mathbf{SL}_g}) \leq 1$ (recall that $\sum_{l=0}^r m_l = g$).

The case $g = 1$ is trivial and hence we can assume $g \geq 2$. This implies that $p \neq 2$. Therefore we can identify canonically $\mathrm{V}_g = \text{Sym}^2(\mathrm{W}_g)$ (see Lemma 4.2.2(a)) with the direct summand of $\mathrm{W}_g \otimes \mathrm{W}_g$ formed by vectors fixed by the natural 'symmetrization' action of the group $(\mathbb{Z}/2\mathbb{Z})$ on $\mathrm{W}_g \otimes \mathrm{W}_g$ (which interchanges the two factors). Taking duals we can also identify $\mathrm{V}_g^* = \text{Sym}^2(\mathrm{W}_g^*)$ with the direct summand of $\mathrm{W}_g^* \otimes \mathrm{W}_g^*$ fixed by the natural 'symmetrization' action of the group

$(\mathbb{Z}/2\mathbb{Z})$ on $W_g^* \otimes W_g^*$. Implicitly, we identify $(V_g^*)^{\otimes g}$ with the direct summand of $(W_g^* \otimes_{\mathbb{K}} W_g^*)^{\otimes g}$ formed by elements fixed by the natural diagonal action of the group $(\mathbb{Z}/2\mathbb{Z})^g$ on $(W_g^* \otimes_{\mathbb{K}} W_g^*)^{\otimes g}$.

It is well-known that we can identify (see [HR], Sect. 1, Applic. (a), Case (ii) and Ex. after Cor. 1.9)

$$\mathrm{Sym}((W_g^*)^{2g}) = \mathbb{K}[\underline{x}_1, \ldots, \underline{x}_{2g}] := \mathbb{K}[x_{j,i} | 1 \le i \le g, 1 \le j \le 2g]$$

with $\underline{x}_j := (x_{j,1}\ x_{j,2} \cdots x_{j,g})^{\mathrm{t}}$, in such a way that $\mathrm{Sym}((W_g^*)^{2g})^{\mathbf{SL}_g}$ is the $\mathbb{K}$-subalgebra generated by the determinants $d_{\underline{j}}$ of the $g \times g$ matrices $\underline{x}_{\underline{j}}$ whose columns are $\underline{x}_{j_1}, \ldots, \underline{x}_{j_g}$, where $\underline{j} = (j_1, \ldots, j_g)$ is an arbitrary g-tuple of integers satisfying $1 \le j_1 < j_2 < \cdots < j_g \le 2g$.

Let $\nabla(g)$ be the set with $\binom{2g}{g}$ elements formed by such g-tuples $\underline{j} = (j_1, j_2, \ldots, j_g)$. Let $\nabla_1(g)$ be the subset of $\nabla(g)$ formed by all g-tuples $\underline{j} = (j_1, j_2, \ldots, j_g)$ with $j_1 = 1$; it has $\binom{2g-1}{g}$ elements. For $\underline{j} \in \nabla_1(g)$, let $\underline{j}' = (j_1', j_2', \ldots, j_g') \in \nabla(g)$ be such that $\sqcup_{i=1}^{g}\{j_i, j_i'\} = \{1, \ldots, 2g\}$. Let $\underline{j}_0 := (1, 3, \ldots, 2g-1) \in \nabla_1(g)$.

We know that

$$((W_g^* \otimes_{\mathbb{K}} W_g^*)^{\otimes g})^{\mathbf{SL}_g} = \{ \sum_{\underline{j} \in \nabla_1(g)} c_{\underline{j}} d_{\underline{j}} d_{\underline{j}'} | c_{\underline{j}} \in \mathbb{K} \text{ for all } \underline{j} \in \nabla_1(g)\}.$$

An element $\underline{\varepsilon} = (\varepsilon_1, \ldots, \varepsilon_g) \in (\mathbb{Z}/2\mathbb{Z})^g$ acts naturally on $((W_g^* \otimes_{\mathbb{K}} W_g^*)^{\otimes g})^{\mathbf{SL}_g}$ via the $\mathbb{K}$-automorphism $\iota_{\underline{\varepsilon}}$ of $\mathbb{K}[\underline{x}_1, \ldots, \underline{x}_{2g}]$ which interchanges $\underline{x}_{2l-1} \longleftrightarrow \underline{x}_{2l}$ for every l with $\varepsilon_l = 1 + 2\mathbb{Z}$ and which fixes both $\underline{x}_{2l-1}$ and $\underline{x}_{2l}$ if $\varepsilon_l = 2\mathbb{Z}$. To end the proof it suffices to show that the $\mathbb{K}$-vector space

$$(((W_g^* \otimes_{\mathbb{K}} W_g^*)^{\otimes g})^{\mathbf{SL}_g})^{(\mathbb{Z}/2\mathbb{Z})^g}$$

has dimension at most 1.

If $\theta = \sum_{\underline{j} \in \nabla_1(g)} c_{\underline{j}} d_{\underline{j}} d_{\underline{j}'} \in (((W_g^* \otimes_{\mathbb{K}} W_g^*)^{\otimes g})^{\mathbf{SL}_g})^{(\mathbb{Z}/2\mathbb{Z})^g}$, then clearly we have $c_{\underline{j}} = 0$ if there exists $l \in \{1, 2, \ldots, g\}$ such that both $2l-1$ and $2l$ are entries of either $\underline{j}$ or $\underline{j}'$; we call such a $\underline{j} \in \nabla_1(g)$ *unmixed*. As the set of products $d_{\underline{j}} d_{\underline{j}'}$ indexed by all mixed (i.e., not unmixed) $\underline{j} \in \nabla_1(g)$ is permuted transitively by $(\mathbb{Z}/2\mathbb{Z})^g$, we get that θ is uniquely determined by $c_{\underline{j}_0}$, hence $(((W_g^* \otimes_{\mathbb{K}} W_g^*)^{\otimes g})^{\mathbf{SL}_g})^{(\mathbb{Z}/2\mathbb{Z})^g}$ has dimension at most 1. □

4.2.8 On Singular Loci

We call a subspace W of V_g *singular* if for all $C \in W$ we have $\det(C) = 0$.

Lemma 4.2.11 *Let $Q_1, Q_2 \in V_g$ be such that Span($\{Q_1, Q_2\}$) is singular. Then the intersection Ker(Q_1) $\cap$ Ker(Q_2) is nontrivial (i.e., it is not 0) and there exists $Q \in$ Span($\{Q_1, Q_2\}$) such that Ker(Q) = Ker(Q_1) $\cap$ Ker(Q_2).*

Proof We define $d_0 := \dim_{\mathbb{K}}(\mathrm{Ker}(Q_1) \cap \mathrm{Ker}(Q_2))$, $d_1 := \dim_{\mathbb{K}}(\mathrm{Ker}(Q_1))$, and $d_2 := \dim_{\mathbb{K}}(\mathrm{Ker}(Q_2))$. We have the following relations $d_1, d_2 \in \{1, \ldots, g\}$ and $d_0 \in \{0, 1, \ldots, \min\{d_1, d_2\}\}$. We can assume that $d_1 < g$ as the case $d_1 = g$, i.e., $Q_1 = 0$, is trivial.

Up to $\mathbf{SL}_g$-equivalence, we can assume that $Q_1 = \mathrm{Diag}(0, \ldots, 0, 1, \ldots, 1)$ (the number of 0s being d_1), that

$$\mathrm{Ker}(Q_2) = \mathrm{Span}(\{e_1, \ldots, e_{d_0}, e_{d_1+1}, \ldots, e_{d_1+d_2-d_0}\}),$$

and that Q_2 has four (possible empty) diagonal blocks which are $0_{d_0}, 1_{d_1-d_0}, 0_{d_2-d_0}, 1_{g-d_1-d_2+d_0}$, where 0_m denotes the zero matrix of size $m \times m$.

If $d_0 = 0$, then $\det(y_1 Q_1 + Q_2)$ is a monic polynomial in y_1 of degree $g - d_1 > 0$ and thus there exists $y_1 \in \mathbb{K}$ such that $y_1 Q_1 + Q_2$ is nonsingular. This contradicts the fact that $\mathrm{Span}(\{Q_1, Q_2\})$ is singular. Therefore $d_0 > 0$, i.e., the intersection $\mathrm{Ker}(Q_1) \cap \mathrm{Ker}(Q_2)$ is nontrivial.

If $M^{\wr}$ is the matrix obtained from a square $g \times g$ matrix M by removing its first d_0 rows and columns, then a similar argument shows that there exist $y_1 \in \mathbb{K}$ such that $y_1 Q_1^{\wr} + Q_2^{\wr}$ is a nonsingular matrix of size $(g - d_0) \times (g - d_0)$ and therefore the rank of $Q := y_1 Q_1 + Q_2$ is $g - d_0$, i.e., $\mathrm{Ker}(Q)$ has dimension d_0 and is equal to $\mathrm{Ker}(Q_1) \cap \mathrm{Ker}(Q_2) = \mathrm{Span}(\{e_1, \ldots, e_{d_0}\})$. □

Proposition 4.2.12 *Let $(Q_0, \ldots, Q_r) \in V_g^{r+1} = \mathbf{V}_g^{r+1}(\mathbb{K})$ be such that the $\mathbb{K}$-vector subspace Span($\{Q_0, \ldots, Q_r\}$) of V_g is singular. Then the following properties hold.*

(a) *The intersection $\cap_{i=0}^{r}$Ker(Q_i) is nontrivial.*

(b) *The closure in $\mathbf{V}_g^{r+1}$ of the $\mathbf{SL}_g$-orbit of $(Q_0, \ldots, Q_r)$ contains the zero vector of $V_g^{r+1} = \mathbf{V}_g^{r+1}(\mathbb{K})$.*

Proof We first prove (a). As the case $r = 0$ is trivial we can assume that $r > 0$. We prove by induction on $j \in \{1, \ldots, r\}$ that there exists $\widetilde{Q}_j \in \mathrm{Span}(\{Q_0, \ldots, Q_j\})$ such that we have $0 \neq \mathrm{Ker}(\widetilde{Q}_j) = \cap_{i=0}^{j}\mathrm{Ker}(Q_i)$. The base of the induction for $j = 1$ follows from Lemma 4.2.11 applied to $Q_0, Q_1 \in V_g$. For $j \in \{1, \ldots, r-1\}$, assuming that $\widetilde{Q}_j$ exists, from Lemma 4.2.11 applied to $\widetilde{Q}_j, Q_{j+1} \in V_g$, it follows that there exists a suitable symmetric matrix $\widetilde{Q}_{j+1} \in \mathrm{Span}(\{\widetilde{Q}_j, Q_{j+1}\}) \subset \mathrm{Span}(\{Q_0, \ldots, Q_{j+1}\})$ such that we have an identity $0 \neq \mathrm{Ker}(\widetilde{Q}_{j+1}) = \mathrm{Ker}(\widetilde{Q}_j) \cap \mathrm{Ker}(Q_{j+1}) = \cap_{i=0}^{j+1}\mathrm{Ker}(Q_i)$. This ends the induction. Part (a) follows from the existence of $\widetilde{Q}_r$.

To prove part (b), as $\cap_{i=0}^{r}\mathrm{Ker}(Q_i)$ is nontrivial, up to an $\mathbf{SL}_g$-equivalence, we can assume that $e_g \in \cap_{i=0}^{r}\mathrm{Ker}(Q_i)$. This implies that the last rows (and columns) of each Q_i are 0. For $\lambda \in \mathbb{K}^{\times}$, for the diagonal matrix

$$\Lambda_\lambda := \mathrm{Diag}(\lambda, \lambda, \ldots, \lambda, \lambda^{-g+1}) \in \mathbf{SL}_g(\mathbb{K})$$

we compute

$$\Lambda_\lambda \cdot (Q_0, \ldots, Q_g) = (\Lambda_\lambda Q_0 \Lambda_\lambda^{\mathrm{t}}, \ldots, \Lambda_\lambda Q_g \Lambda_\lambda^{\mathrm{t}}) = (\lambda^2 Q_0, \ldots, \lambda^2 Q_g),$$

and by taking $\lambda \to 0$ we conclude that part (b) holds. □

4.2.9 On Θ-Rings of SL_g-Invariants

The following theorem complements and extends Theorem 4.2.8.

Theorem 4.2.13 *For all integers $r \geq 1$ the following properties hold.*

(a) *The $\mathbb{W}_{\varepsilon_g}$-graded $\mathbb{K}$-algebra*

$$\mathbb{J}^r_{g,\mathbb{K}} := \mathbb{H}^r_{g,tot,\mathbb{K}}/(\Theta_{m_0,\ldots,m_r} | (m_0, \ldots, m_r) \in \Delta(g, r))$$

is local artinian and therefore we have integral $\mathbb{K}$-monomorphisms

$$\mathbb{H}^{r,\Theta}_{g,\mathbb{K}} \to \mathbb{H}^r_{g,tot,\mathbb{K}} \quad \text{and} \quad \mathbb{H}^{r,\Theta}_{g,\mathbb{K}} \to \mathbb{H}^r_{g,\mathbb{K}}.$$

(b) *The $\mathbb{K}$-algebra monomorphisms of Display (4.15) are integral.*

(c) *There exists a subset with $D(g, r)$ elements*

$$\mathcal{T}_0 \subset \mathbb{H}^r_g(1)_{\mathbb{K}} = Span(\{\Theta_{m_0,\ldots,m_r} | (m_0, \ldots, m_r) \in \Delta(g, r)\})$$

which is algebraically independent, i.e., the $\mathbb{K}$-subalgebra

$$\mathbb{K}[\mathcal{T}_0] := \mathbb{K}[\Theta | \Theta \in \mathcal{T}_0]$$

of $\mathbb{H}^r_{g,\mathbb{K}}$ is a polynomial $\mathbb{K}$-algebra in $D(g, r)$ indeterminates, and for which the $\mathbb{K}$-monomorphisms $\mathbb{K}[\mathcal{T}_0] \to \mathbb{H}^r_{g,tot,\mathbb{K}}$ and $\mathbb{K}[\mathcal{T}_0] \to \mathbb{H}^r_{g,\mathbb{K}}$ are finite.

(d) *Referring to part (b), if $\mathbb{H}^r_{g,tot,\mathbb{K}}$ is Cohen–Macaulay, then the $\mathbb{K}[\mathcal{T}_0]$-modules $\mathbb{H}^r_{g,tot,\mathbb{K}}$ and $\mathbb{H}^r_{g,\mathbb{K}}$ are free of finite rank.*

(e) *If $r \in \{1, 2\}$, then we have $\mathbb{H}^{r,\Theta}_{g,\mathbb{K}} = \mathbb{K}[\mathcal{T}_0]$ and thus $\mathbb{H}^{r,\Theta}_{g,\mathbb{K}}$ is a polynomial $\mathbb{K}$-algebra (equal for $r = 1$ and $p \neq 2$ with $\mathbb{H}^1_{g,\mathbb{K}} = \mathbb{H}^1_{g,tot,\mathbb{K}}$, see Theorem 4.2.8(b)).*

Proof We recall that $\mathbf{V}_g^{r+1} = \mathrm{Spec}(\mathbb{K}[T, T', \ldots, T^{(r)}])$ and we define

$$\mathbb{B}^r_{g,\mathbb{K}} := \mathbb{K}[T, T', \ldots, T^{(r)}]/(\Theta_{m_0,\ldots,m_r} | (m_0, \ldots, m_r) \in \Delta(g, r)),$$

$$\mathbb{N}^r_{g,\mathbb{K}} := \mathrm{Ker}(\mathbb{J}^r_{g,\mathbb{K}} \to \mathbb{B}^r_{g,\mathbb{K}}).$$

We consider the $\mathbf{SL}_g$-invariant closed subscheme

$$\mathbf{V}_g^{r+1,\Theta} := \mathrm{Spec}(\mathbb{B}^r_{g,\mathbb{K}}) \subset \mathbf{V}_g^{r+1}.$$

If $p = 0$, then, as $\mathbf{V}_g^{r+1} \to \mathbf{V}_g^{r+1}/\mathbf{SL}_g$ is a universal categorical quotient, we have $(\mathbf{V}_g^{r+1,\Theta})^{\mathbf{SL}_g} = \mathrm{Spec}(\mathbb{J}^r_{g,\mathbb{K}})$ and thus $\mathbb{N}^r_{g,\mathbb{K}} = 0$.

If $p > 0$, then, as $\mathbf{V}_g^{r+1} \to \mathbf{V}_g^{r+1}/\mathbf{SL}_g$ is surjective (being submersive), the natural morphism $\mathbf{V}_g^{r+1,\Theta} \to \mathrm{Spec}(\mathbb{J}^r_{g,\mathbb{K}})$ is also surjective and therefore $\mathbb{N}^r_{g,\mathbb{K}}$ is a nilpotent ideal of $\mathbb{J}^r_{g,\mathbb{K}}$.

Note that if $(Q_0, \ldots, Q_r) \in \mathbf{V}_g^{r+1,\Theta}(\mathbb{K}) \subset \mathrm{V}_g^{r+1}$, then, with $y_0, \ldots, y_r$ as indeterminates, we have $\det(\sum_{i=0}^r y_i Q_i) = 0$, and hence the $\mathbb{K}$-vector subspace $\mathrm{Span}(\{Q_0, \ldots, Q_r\})$ of V_g is singular. Therefore, if $h \in \mathbb{J}^r_{g,\mathbb{K}}$ and if $\nu \in \mathbb{K}$ is its value at the zero vector of $\mathbf{V}_g^{r+1,\Theta}(\mathbb{K}) \subset \mathrm{V}_g^{r+1}$, from Proposition 4.2.12(b) we get that $h - \nu$ has value 0 at each $(Q_0, \ldots, Q_r) \in \mathbf{V}_g^{r+1,\Theta}(\mathbb{K})$ and thus its image in $\mathbb{J}^r_{g,\mathbb{K}}/\mathbb{N}^r_{g,\mathbb{K}} \subset \mathbb{B}^r_{g,\mathbb{K}}$ is a nilpotent element, hence $h - \nu$ itself is a nilpotent element. This implies that part (a) holds.

Part (b) follows from part (a) and inclusions of Display (4.17) with $B = \mathbb{K}$.

As the field $\mathbb{K}$ is infinite and all $\Theta_{m_0,\ldots,m_r}$s are homogeneous of the same T-degree g, from part (a) and the homogeneous form of Noether's normalization theorem (see [E], Part II, Thm. 13.3), we get that part (c) holds.

If $\mathbb{H}^r_{g,\mathrm{tot},\mathbb{K}}$ is Cohen–Macaulay, then so is $\mathbb{H}^r_{g,\mathbb{K}}$ (see Lemma 4.2.3(c)) and this implies that the $\mathbb{K}[\mathcal{T}_0]$-modules $\mathbb{H}^r_{g,\mathrm{tot},\mathbb{K}}$ and $\mathbb{H}^r_{g,\mathbb{K}}$ are free (see either [Ke], Cor. 4.2, or [Cho], Thm.). Thus part (d) holds.

For $r \in \{1, 2\}$ we have $D(g, r) = D(g, r, 1) = \binom{g+r}{r}$ and therefore we have $\mathbb{H}^{r,\Theta}_{g,\mathbb{K}} = \mathbb{K}[\mathcal{T}_0]$; hence part (e) follows from part (c). □

4.2.10 Applications of the Cohen–Macaulay Property

We recall that the partially-ordered semiring of Hilbert (Poincaré) series is the semiring $\mathbb{Z}_{\geq 0}[[x]]$ endowed with the following partial order $\preceq$: for two series $\wp_1(x), \wp_2(x) \in \mathbb{Z}_{\geq 0}[[x]]$ we say that $\wp_1(x) \preceq \wp_2(x)$ if there exists $\wp_0(x) \in \mathbb{Z}_{\geq 0}[[x]]$ such that $\wp_2(x) = \wp_1(x) + \wp_0(x)$.

Corollary 4.2.14 *Let $g, r \in \mathbb{N}$. We assume that $\mathbb{H}^r_{g,tot,\mathbb{K}}$ is Cohen–Macaulay. Then the following properties hold.*

(a) *There exists a uniquely determined nondecreasing sequence*

$$(c_p(g,r,l))_{l=1}^{\varrho_p^-(g,r)}$$

of integers ≥ 2 *with* $\varrho_p^-(g,r) \in \mathbb{N} \cup \{0\}$ *such that the Hilbert series*

$$\wp_p(g,r)(x) := \sum_{s=0}^{\infty} D_p(g,r,s)x^s = \sum_{s=0}^{\infty} \dim_{\mathbb{K}}(\mathbb{H}_g^r(s)_{\mathbb{K}})x^s$$

of $\mathbb{H}_{g,\mathbb{K}}^r$ *is the rational polynomial function*

$$\frac{1 + [D_p(g,r,1) - D(g,r)]x + \sum_{l=1}^{\varrho_p^-(g,r)} x^{c_p(g,r,l)}}{(1-x)^{D(g,r)}}.$$

Thus

$$\wp_p(g,r)(x) \succeq \frac{1 + [D_p(g,r,1) - D(g,r)]x}{(1-x)^{D(g,r)}} =: \wp_p(g,r;1)(x)$$

$$\succeq \frac{1 + \left[\binom{g+r}{r} - D(g,r)\right]x}{(1-x)^{D(g,r)}} =: \wp_0(g,r;1)(x),$$

and the equality $\wp_p(g,r;1)(x) = \wp_0(g,r;1)(x)$ *holds if* $p \nmid g!$*. Moreover, for all integers* $s \geq \max\{0, c_p(g,r,\varrho_p^-(g,r)) - D(g,r) + 1\}$ *we have an inequality*

$$D_p(g,r,s) - \binom{D(g,r)+s-1}{D(g,r)-1} - [D_p(g,r,1) - D(g,r)]\binom{D(g,r)+s-2}{D(g,r)-1} \geq 0$$

whose left-hand side is equal to the sum

$$\sum_{l=1}^{\varrho_p^-(g,r)} \binom{D(g,r)+s-1-c_p(g,r,l)}{D(g,r)-1}.$$

(b) *Let*

$$\varrho_p(g,r) := 1 + \varrho_p^-(g,r) + D_p(g,r,1) - D(g,r) \in \mathbb{N}.$$

If $\mathbb{H}_{g,\mathbb{K}}^{r,\perp} \neq 0$*, then there exists a uniquely determined nondecreasing sequence*

$$(c_p(g,odd,r,l))_{l=1}^{\varrho_p(g,r)}$$

of integers ≥ 1 *such that the Hilbert series (in* $\mathbb{Z}_{\geq 0}[[x^{\frac{1}{2}}]]$*)*

$$\wp_p(g, tot, r)(x^{\frac{1}{2}}) := \sum_{s \in \mathbb{W}_0} D_p(g, r, s)x^s = \sum_{s \in \mathbb{W}_0} \dim_{\mathbb{K}}(\mathbb{H}^r_g(s)_{\mathbb{K}})x^s$$

of $\mathbb{H}^r_{g,tot,\mathbb{K}}$ *is the rational polynomial function in* $x^{\frac{1}{2}}$

$$\wp_p(g, r)(x) + \frac{x^{\frac{1}{2}} \sum_{l=1}^{\varrho_p(g,r)} x^{c_p(g,odd,r,l)}}{(1-x)^{D(g,r)}}.$$

Proof Let the subset $\mathcal{T}_0 = \{h_1, \ldots, h_{D(g,r)}\} \subset \mathbb{H}^r_g(1)_{\mathbb{K}}$ be such that the $\mathbb{K}$-homomorphism $\mathbb{K}[\mathcal{T}_0] = \mathbb{K}[h_1, \ldots, h_{D(g,r)}] \to \mathbb{H}^r_{g,\mathrm{tot},\mathbb{K}}$ is integral by Theorem 4.2.13(c). Let $h_{D(g,r)+1}, \ldots, h_{D_p(g,r,1)} \in \mathbb{H}^r_g(1)_{\mathbb{K}}$ be such that $\{h_l | 1 \leq l \leq D_p(g, r, 1)\}$ is a $\mathbb{K}$-basis of $\mathbb{H}^r_g(1)_{\mathbb{K}}$. We take the nondecreasing sequence $(c_p(g, r, l))_{l=1}^{\varrho_p^-(g,r)}$ in $\mathbb{N}\backslash\{1\}$ such that there exist nonzero polynomials

$$h_{D_p(g,r,1)+l} \in \mathbb{H}^r_g(c_p(g, r, l))_{\mathbb{K}}$$

indexed by $l \in \{1, \ldots, \varrho_p^-(g, r)\}$, with the property that we have a direct sum decomposition of $\mathbb{Z}_{\geq 0}$-graded $\mathbb{K}[\mathcal{T}_0]$-modules

$$\mathbb{H}^r_{g,\mathbb{K}} = \mathbb{K}[\mathcal{T}_0] \oplus \left(\bigoplus_{l=D(g,r)+1}^{D_p(g,r,1)+\varrho_p^-(g,r)} \mathbb{K}[\mathcal{T}_0]h_l \right), \tag{4.19}$$

by the proof of [Ke], Cor. 4.2 or [Cho], Thm. 1 and the fact that $\mathbb{H}^r_{g,\mathbb{K}}$ is Cohen–Macaulay (see Lemma 4.2.3(c)).

If $\mathbb{H}^{r,\perp}_{g,\mathbb{K}} \neq 0$, i.e., g is even and $[\mathrm{Frac}(\mathbb{H}^r_{g,\mathrm{tot},\mathbb{K}}) : \mathrm{Frac}(\mathbb{H}^r_{g,\mathbb{K}})] = 2$, then the $\mathbb{K}[\mathcal{T}_0]$-module $\mathbb{H}^{r,\perp}_{g,\mathbb{K}}$ is also free (by the Quillen–Suslin's theorem, see [La], Ch. XXI, Sect. 3, Thm. 3.5) and it has the same rank $\varrho_p(g, r)$ as $\mathbb{H}^r_{g,\mathbb{K}}$ (see Eqs. (4.13)) and thus there exist nonzero homogeneous polynomials

$$h_{D_p(g,r,1)+\varrho_p^-(g,r)+1} \in \mathbb{H}^r_g(\frac{1}{2} + c_p(g, \mathrm{odd}, r, 1))_{\mathbb{K}} \subset \mathbb{H}^{r,\perp}_{g,\mathbb{K}}, \ldots,$$

$$h_{D_p(g,r,1)+\varrho_p^-(g,r)+\varrho_p(g,r)} \in \mathbb{H}^r_g(\frac{1}{2} + c_p(g, \mathrm{odd}, r, \varrho_p(g, r)))_{\mathbb{K}} \subset \mathbb{H}^{r,\perp}_{g,\mathbb{K}}$$

such that we have a direct sum decomposition

$$\mathbb{H}^r_{g,\mathrm{tot},\mathbb{K}} = \mathbb{K}[\mathcal{T}_0] \ominus \left(\bigoplus_{l=D(g,r)+1}^{D_p(g,r,1)+\varrho_p^-(g,r)+\varrho_p(g,r)} \mathbb{K}[\mathcal{T}_0]h_l \right) \tag{4.20}$$

of $\mathbb{W}_0$-graded $\mathbb{K}[\mathcal{T}_0]$-modules.

The corollary follows from [Ke], Cor. 4.2 and the last two direct sum decompositions. □

From Eqs. (4.19) and (4.20) and from Theorem 4.2.13(d) and (e) we get directly the following consequence (for $r = 1$, cf. also Theorem 4.2.8(b), and for $r = 2$, cf. also Proposition 4.2.10(b)).

Corollary 4.2.15 *For all $g, r \in \mathbb{N}$, if $\mathbb{H}^r_{g,tot,\mathbb{K}}$ is Cohen–Macaulay, then the following properties hold.*

(a) *The integers $\varrho_p^-(g, r) \geq 0$ and $\varrho_p(g, r) \geq 1$ of Corollary 4.2.14 have the property that for each $\mathbb{K}$-algebra $\mathbb{K}[\mathcal{T}_0]$ as in Theorem 4.2.13(c) we have*

$$\varrho_p(g, r) = rank_{\mathbb{K}[\mathcal{T}_0]}(\mathbb{H}^r_{g,\mathbb{K}}) = 1 + D_p(g, r, 1) - D(g, r) + \varrho_p^-(g, r).$$

If $p \neq 2$ or $g = 1$, then for $r = 1$ we have

$$\varrho_p(g, 1) = rank_{\mathbb{H}^{1,\ominus}_{g,\mathbb{K}}}(\mathbb{H}^1_{g,\mathbb{K}}) = 1.$$

If moreover $p \nmid g!$, then for $r = 2$ we have

$$\varrho_p(g, 2) = rank_{\mathbb{H}^{2,\ominus}_{g,\mathbb{K}}}(\mathbb{H}^2_{g,\mathbb{K}}) = 1 + \varrho_p^-(g, 2).$$

(b) *If g is even and $\mathbb{H}^r_{g,tot,\mathbb{K}} \neq \mathbb{H}^r_{g,\mathbb{K}}$, then we have*

$$rank_{\mathbb{K}[\mathcal{T}_0]}(\mathbb{H}^r_{g,tot,\mathbb{K}}) = 2rank_{\mathbb{K}[\mathcal{T}_0]}(\mathbb{H}^r_{g,\mathbb{K}}) = 2rank_{\mathbb{K}[\mathcal{T}_0]}(\mathbb{H}^{r,\perp}_{g,\mathbb{K}}) = 2\varrho_p(g, r).$$

If moreover $p \nmid g!$, then for $r = 2$ we have

$$rank_{\mathbb{H}^{2,\ominus}_{g,\mathbb{K}}}(\mathbb{H}^2_{g,tot,\mathbb{K}}) = 2rank_{\mathbb{H}^{2,\ominus}_{g,\mathbb{K}}}(\mathbb{H}^2_{g,\mathbb{K}}) = 2\varrho_p(g, 2) = 2 + 2\varrho_p^-(g, 2).$$

4.2.11 Transcendence Bases

It is convenient to work this subsection with

$$\mathbf{SL}'_g := \mathbf{SL}_g/\mathrm{Ker}(\mathbf{SL}_g \to \mathbf{GL}_{V_g})$$

instead of with $\mathbf{SL}_g$. This is so as the induced homomorphism $\mathbf{SL}'_g \to \mathbf{GL}_{V_g}$ is a faithful representation (closed embedding). If g is odd we have $\mathbf{SL}'_g = \mathbf{SL}_g$ and if g is even we have a short exact sequence $1 \to \mu_{2,\mathbb{K}} \to \mathbf{SL}_g \to \mathbf{SL}'_g \to 1$. So the order of $\mathrm{Ker}(\mathbf{SL}_g \to \mathbf{SL}'_g)$ is $2^{1-\varepsilon_g}$.

Let $(\mathbf{V}^{r+1}_g/\mathbf{SL}'_g)_{\mathrm{sm}}$ be the largest open subscheme of $\mathbf{V}^{r+1}_g/\mathbf{SL}'_g$ with the property that the quotient morphism $\mathbf{V}^{r+1}_g \to \mathbf{V}^{r+1}_g/\mathbf{SL}'_g$ is smooth above it, i.e., if we denote by $\mathbf{V}^{r+1}_{g,\mathrm{sm}}$ the inverse image of $(\mathbf{V}^{r+1}_g/\mathbf{SL}'_g)_{\mathrm{sm}}$ in $\mathbf{V}^{r+1}_g$, then the induced morphism $\mathbf{V}^{r+1}_{g,\mathrm{sm}} \to (\mathbf{V}^{r+1}_g/\mathbf{SL}'_g)_{\mathrm{sm}}$ is smooth. As $\mathrm{Frac}(\mathbb{H}^r_{g,\mathrm{tot},\mathbb{K}})$ is equal to its purely inseparable closure in $\mathrm{Frac}(\mathbb{K}[T, \ldots, T^{(r)}])$ (see [DK], Thm. 2.3.12), the field extension

$\mathrm{Frac}(\mathbb{H}'^{r}_{g,\mathrm{tot},\mathbb{K}}) \to \mathrm{Frac}(\mathbb{K}[T, \ldots, T^{(r)}])$ is separably generated (see [Mat], Thm. 26.2 and 26.4). From this and the fact that $(\mathrm{V}_g^{r+1})^{\mathrm{s}}$ is nonempty if $r \geq 1$ (see Claim 4.1.13), we get that $(\mathbf{V}_g^{r+1}/\mathbf{SL}'_g)_{\mathrm{sm}}$ is nonempty if $r \geq 1$.

We recall that $(\mathbf{V}_{g,\mathrm{sm}}^{r+1})^{\mathrm{s}}$ is the stable locus of $\mathbf{V}_{g,\mathrm{sm}}^{r+1}$ and we have an identity $\mathbf{V}_{g,\mathrm{sm}}^{r+1}/\mathbf{SL}'_g = (\mathbf{V}_g^{r+1}/\mathbf{SL}'_g)_{\mathrm{sm}}$. Let $F_0 = \Theta_0 \cdot F_1 \cdot J \in \mathcal{O}(\mathbf{V}_g^2)^{\mathbf{SL}'_g} = \mathcal{O}(\mathbf{V}_g^2/\mathbf{SL}'_g)$ be as in the proof of Theorem 4.1.1.

Lemma 4.2.16 *For integers $r \geq l \geq 2$ we consider the natural $\boldsymbol{SL}'_g$-morphism*

$$\varpi_g^{r+1,l} : \boldsymbol{V}_g^{r+1} \to \boldsymbol{V}_g^l \times_{(\boldsymbol{V}_g^l/\boldsymbol{SL}'_g)} (\boldsymbol{V}_g^{r+1}/\boldsymbol{SL}'_g).$$

The following properties hold.

(a) *The $\boldsymbol{SL}'_g$-morphism $\varpi_g^{3,2}$ is generically finite of degree $2^{g-2+\varepsilon_g}$ and it is generically étale if and only if $p \neq 2$.*

(b) *If $l \geq 3$, then the $\boldsymbol{SL}'_g$-morphism $\varpi_g^{r+1,l}$ is generically finite of degree 1. More precisely, if $\boldsymbol{X}_g^l$ is the intersection of $(\boldsymbol{V}_{g,sm}^l)^s$ with the nonempty $\boldsymbol{SL}'_g$-invariant open subscheme of $(\boldsymbol{V}_g^l)^s$ whose $\mathbb{K}$-valued points are those l-tuples*

$$(Q_0, \ldots, Q_{l-1}) \in V_g^l = \boldsymbol{V}_g^l(\mathbb{K})$$

for which $F_0(Q_0, Q_1) \neq 0$ and the group $Stab_{\boldsymbol{SL}'_g}((Q_0, \ldots, Q_{l-1}))$ is trivial, and if $\boldsymbol{X}_g^l \times \boldsymbol{V}_g^{r+1-l}$ is the open subscheme of $\boldsymbol{V}_g^{r+1}$ which is the pullback of $\boldsymbol{X}_g^l$ via the projection morphism $\boldsymbol{V}_g^{r+1} \to \boldsymbol{V}_g^l$, then the $\boldsymbol{SL}'_g$-morphism

$$\varpi_g^{r+1,l,-} : \boldsymbol{X}_g^l \times \boldsymbol{V}_g^{r+1-l} \to \boldsymbol{X}_g^l \times_{(\boldsymbol{V}_g^l/\boldsymbol{SL}'_g)} (\boldsymbol{V}_g^{r+1}/\boldsymbol{SL}'_g)$$

induced naturally by $\varpi_g^{r+1,l}$ is an isomorphism.[7]

(c) *If $l \geq 3$, the field extension $Frac(\boldsymbol{V}_g^l/\boldsymbol{SL}'_g) \to Frac(\boldsymbol{V}_g^{r+1}/\boldsymbol{SL}'_g)$ (induced naturally by the projection morphism $\boldsymbol{V}_g^{r+1} \to \boldsymbol{V}_g^l$) is purely transcendental of transcendence degree equal to $(r+1-l)\frac{g(g+1)}{2}$. Moreover, a subset of $Frac(\boldsymbol{V}_g^{r+1}/\boldsymbol{SL}'_g)$ with $(r+1-l)\frac{g(g+1)}{2}$ elements is a generating transcendence basis (resp. a transcendence basis) of $Frac(\boldsymbol{V}_g^{r+1}/\boldsymbol{SL}'_g)$ over $Frac(\boldsymbol{V}_g^l/\boldsymbol{SL}'_g)$ if and only if it is a generating transcendence basis (resp. a transcendence basis) of $Frac(\mathbb{K}[T, T', \ldots, T^{(r)}])$ over $Frac(\mathbb{K}[T, T', \ldots, T^{(l)}])$.*

Proof If $P_2 := (Q_0, Q_1, Q_2) \in (\mathbf{V}_{g,\mathrm{sm}}^3)^{\mathrm{s}}(\mathbb{K})$ maps to a $P_1 := (Q_0, Q_1)$ in $(\mathbf{V}_{g,\mathrm{sm}}^2)^{\mathrm{s}}(\mathbb{K})$ with $F_0(Q_0, Q_1) \neq 0$, then the fiber $(\varpi_g^{3,2})^{-1}(\varpi_g^{3,2}(P_2))$ is naturally identified with the quotient group scheme

---

[7] The $\mathbf{SL}'_g$-invariant morphism $\mathbf{X}_g^l \to \mathbf{X}_g^l/\mathbf{SL}'_g$ is a geometric quotient (see Lemma 4.1.6(a)) and the $\mathbf{SL}'_g$-action on $\mathbf{X}_g^l$ is free. Thus [MFK], Ch. 0, Prop. 0.9 implies that $\mathbf{X}_g^l \to \mathbf{X}_g^l/\mathbf{SL}'_g$ is a fiber bundle in the sense of [MFK], Ch. 0, Def. 0.10.

$$\mathrm{Stab}_{\mathbf{SL}'_g}(P_1)/\mathrm{Stab}_{\mathbf{SL}'_g}(P_2) \simeq \mathrm{Stab}_{\mathbf{SL}_g}(P_1)/\mathrm{Stab}_{\mathbf{SL}_g}(P_2).$$

Based on the proof of Theorem 4.1.1 we can assume that P_1 is in $\mathbb{K}1_g \times Z_g$ with $\mathrm{Stab}_{\mathbf{SL}_g}(P_1) \simeq \boldsymbol{\mu}_{2,\mathbb{K}}^{g-1}$. But choosing Q_2 to be generic (such as all its ij entries with $1 \leq i \leq j \leq g$ are nonzero), we have $\mathrm{Stab}_{\mathbf{SL}'_g}(P_2) = \mathrm{Spec}(\mathbb{K})$, equivalently we have $\mathrm{Stab}_{\mathbf{SL}_g}(P_2) = \mathrm{Ker}(\mathbf{SL}_g \to \mathbf{GL}_{\mathrm{V}_g}) \simeq \boldsymbol{\mu}_{2,\mathbb{K}}^{1-\varepsilon_g}$. We conclude that for such generic Q_2, the fiber $(\varpi_g^{3,2})^{-1}(\varpi_g^{3,2}(P_2))$ is naturally identified with $\boldsymbol{\mu}_{2,\mathbb{K}}^{g-2+\varepsilon_g}$ from which part (a) follows.

Let $P_r = (Q_0, \ldots, Q_r) \in (\mathbf{X}_g^l \times \mathbf{V}_g^{r+1-l})(\mathbb{K})$; so $P_{l-1} := (Q_0, Q_1, \ldots, Q_{l-1})$ is in $(\mathbf{X}_g^l)(\mathbb{K})$. If $P_r \in (\mathbf{V}_{g,\mathrm{sm}}^{r+1})^s(\mathbb{K})$, then the fiber $(\varpi_g^{r+1,l})^{-1}(\varpi_g^{r+1,l}(P_r))$ is naturally identified with the quotient group scheme $\mathrm{Stab}_{\mathbf{SL}'_g}(P_l)/\mathrm{Stab}_{\mathbf{SL}'_g}(P_r)$ and thus it is $\mathrm{Spec}(\mathbb{K})$.

Similarly, if $P_r \notin (\mathbf{V}_{g,\mathrm{sm}}^{r+1})^s(\mathbb{K})$, then $[(\varpi_g^{r+1,l})^{-1}(\varpi_g^{r+1,l}(P_r))]_{\mathrm{red}}$ is $\mathrm{Spec}(\mathbb{K})$.

From the last two paragraphs we get that the morphism $\varpi_g^{r+1,l,-}$ is birational and induces a bijection at the level of $\mathbb{K}$-valued points. As the target of $\varpi_g^{r+1,l,-}$ is normal, being smooth over the normal scheme $(\mathbf{V}_g^{r+1})/\mathbf{SL}'_g$, from Zariski's Main Theorem we get that $\varpi_g^{r+1,l,-}$ is an isomorphism. Thus part (b) holds.

Part (c) is a standard 'descent and vector bundle' consequence of part (b). We only recall the 'descent and vector bundle' argument for the first part of part (c). Let $H_l \in \mathbb{H}_{g,\mathrm{tot},\mathbb{K}}^{l-1}$ be such that the pullback to $\mathbf{V}_g^l$ of the principal open $D(H_l)$ of $\mathrm{Spec}(\mathbb{H}_{g,\mathrm{tot},\mathbb{K}}^{l-1}) = \mathbf{V}_g^l/\mathbf{SL}'_g$ is contained in $\mathbf{X}_g^l$ and thus it is $(\mathbf{X}_g^l)_{H_l}$. Inverting H_l, the $\mathbb{H}_{g,\mathrm{tot},\mathbb{K},H_l}^{l-1}$-algebra $\mathbb{H}_{g,\mathrm{tot},\mathbb{K},H_l}^{r}$ is $\mathbb{Z}_{\geq 0}$-graded by the total degree in just the upper indexes ${}^{(l)}, \ldots, {}^{(r)}$. The fact that this $\mathbb{Z}_{\geq 0}$-graded structure makes the morphism

$$\mathrm{Spec}(\mathbb{H}_{g,\mathrm{tot},\mathbb{K},H_l}^{r}) \to \mathrm{Spec}(\mathbb{H}_{g,\mathrm{tot},\mathbb{K},H_l}^{l-1})$$

get the structure of a locally free vector bundle over $\mathrm{Spec}(\mathbb{H}_{g,\mathrm{tot},\mathbb{K},H_l}^{l-1})$ follows after pullback via the (smooth) faithfully flat morphism $(\mathbf{X}_g^l)_{H_l} \to \mathrm{Spec}(\mathbb{H}_{g,\mathrm{tot},\mathbb{K},H_l}^{l-1})$ from part (b). This implies that the first part of (c) holds. □

For an integer $r \geq 2$, we define the sets

$$\Delta_3(g,r) := \{(m_0, \ldots, m_r) \in \Delta(g,r) | \exists i \in \{2, \ldots, r\} \text{ such that } m_i = g - m_0 - m_1\},$$
$$\Delta_3^*(g,r) := \{(m_0, \ldots, m_r) \in \Delta(g,r) | m_r = g - m_0 - m_1 > 0\}.$$

Clearly, $\Delta_3^*(g,r) \subset \Delta_3(g,r) \subset \Delta(g,r)$ and $\Delta_3(g,2) = \Delta(g,2)$.

To ease the notation, we denote an arbitrary element of $\Delta_3(g,r)$ by an $r+1$-tuple $(a, b, 0, \ldots, 0, \ldots, 0, c, 0, \ldots, 0)$, where $(a,b,c) \in \Delta(g,2)$ and c is on the i-th position with $i \in \{2, \ldots, r\}$ arbitrary. Similarly, we denote an arbitrary element of $\Delta_3^*(g,r)$ by an $r+1$-tuple $(a, b, 0, \ldots, 0, \ldots, 0, c)$, where $(a,b,c) \in \Delta(g,2)$ with $c > 0$ (equivalently, where $(a,b,c) \in \Delta_3^*(g,2)$).

As we have a bijection $\Delta_3^*(g,r) \to \sqcup_{i=0}^{g-1} \Delta(i,1)$, the set $\Delta_3^*(g,r)$ has $\frac{g(g+1)}{2}$ elements. Similarly, as we have a bijection $\Delta_3(g,r) \to \Delta_3(g,2) \sqcup_{l=3}^{r} \Delta_3^*(g,2) \times \{l\}$, the set $\Delta_3(g,r)$ has $\frac{(g+1)(g+2)}{2} + (r-2)\frac{g(g+1)}{2} = D(g,r)$ elements.

Theorem 4.2.17 *Let $r \geq 2$ be an integer. Then the set*

$$\mathcal{T}_{g,r} := \{\Theta_{a,b,0,\ldots,0,\ldots,0,c,0,\ldots,0} | (a,b,0,\ldots,0,\ldots,0,c,0,\ldots,0) \in \Delta_3(g,r)\}$$

is a transcendence basis of $Frac(\mathbb{H}^r_{g,tot,\mathbb{K}})$ over $\mathbb{K}$.

Proof As $D(g,2) = D(g,2,1) = \binom{g+2}{2}$ is the number of elements of the set $\Delta_3(g,2) = \Delta(g,2)$, for $r=2$ we have $\mathrm{Span}(\mathcal{T}_{g,r}) = \mathrm{Span}(\mathcal{T}_0)$ and the theorem follows from Theorem 4.2.13(c).

We assume now that $r > 2$. The finite field extension

$$\eta_1 := \mathrm{Frac}(\mathbb{H}^1_{g,\mathrm{tot},\mathbb{K}}) \supset \mathbb{K}(\Theta_0, \Theta_1, \ldots, \Theta_g) = \mathbb{K}(\Theta_{g,0}, \Theta_{g-1,1}, \ldots, \Theta_{0,g}) \quad (4.21)$$

is an isomorphism if $p \neq 2$ (see Theorem 4.2.8(b)). Let $\eta_2 := \mathrm{Frac}(\mathbb{H}^2_{g,\mathrm{tot},\mathbb{K}})$. As the theorem holds for $r=2$, the field extension

$$\eta_1(\Theta_{a,b,c} | (a,b,c) \in \Delta_3^*(g,2)) \to \eta_2$$

is finite. From this and Lemma 4.2.16(a), by denoting for $2 \leq l \leq r$

$$\kappa_l := \mathbb{K}(T_{ij}, T_{i,j}^{(1)} | 1 \leq i \leq j \leq g)(\Theta_{a,b,0,\ldots,0,c} | (a,b,0,\ldots,0,c) \in \Delta_3^*(g,l)),$$

we get a finite field extension

$$\kappa_2 \to \mathbb{K}(T_{ij}, T_{ij}^{(1)}, T_{ij}^{(2)} | 1 \leq i \leq j \leq g) = \mathrm{Frac}(\mathcal{O}(\mathbf{V}_g^3)).$$

Replacing the index $^{(2)}$ by the index $^{(l)}$ with $2 < l \leq r$, we get a finite field extension

$$\kappa_l \to \mathbb{K}(T_{ij}, T_{ij}^{(1)}, T_{ij}^{(l)} | 1 \leq i \leq j \leq g).$$

From the last two sentences, if κ denotes the composite field of $\kappa_2, \ldots, \kappa_r$, we get a finite field extension

$$\kappa \to \mathbb{K}(T_{ij}, T_{ij}^{(1)}, \ldots, T_{ij}^{(r)} | 1 \leq i \leq j \leq g).$$

As κ is $\mathbb{K}(T_{ij}, T_{i,j}^{(1)} | 1 \leq i \leq j \leq g)$-generated by the family of $(r-1)\frac{g(g+1)}{2}$ polynomials $(\Theta_\square)_{\square \in \sqcup_{l=2}^{r} \Delta_3^*(g,l)}$, by reasons of transcendence degrees we get that this family is algebraically independent over $\mathbb{K}(T_{ij}, T_{i,j}^{(1)} | 1 \leq i \leq j \leq g)$ and thus also over η_1. So the family of $(r-2)\frac{g(g+1)}{2}$ polynomials $(\Theta_\square)_{\square \in \sqcup_{l=3}^{r} \Delta_3^*(g,l)}$ is algebraically over κ_2. From this and Lemma 4.2.16(c) applied to $l=3$ we get that the theorem holds for $r > 2$. □

We define

$$\mathbb{K}_{g,r} := \mathrm{Frac}(\mathbb{H}^r_{g,\mathrm{tot},\mathbb{K}}) \supset \mathrm{Frac}(\mathbb{H}^r_{g,\mathbb{K}}) =: \mathbb{K}_{g,\mathrm{even},r},$$

$$\mathbb{K}_{g,\infty} := \bigcup_{r\geq 1} \mathbb{K}_{g,r} \supset \bigcup_{r\geq 1} \mathbb{K}_{g,\mathrm{even},r} =: \mathbb{K}_{g,\mathrm{even},\infty},$$

$$\mathbb{H}_{g,\mathrm{tot},\mathbb{K}} := \bigcup_{r\geq 1} \mathbb{H}^r_{g,\mathrm{tot},\mathbb{K}} \quad \text{and} \quad \mathbb{H}_{g,\mathbb{K}} := \bigcup_{r\geq 1} \mathbb{H}^r_{g,\mathbb{K}},$$

the unions being taken inside the field of fractions of

$$\mathbb{T}_g := \bigcup_{r\geq 0} \mathbb{K}[T^{(0)}, T^{(1)}, \ldots, T^{(r)}].$$

We consider the shift $\mathbb{K}$-monomorphism

$$\sigma : \mathbb{T}_g \to \mathbb{T}_g$$

that sends $\sigma(T^{(r)}) = T^{(r+1)}$ for all integers $r \geq 0$ (i.e., $\sigma(T^{(r)}_{ij}) = T^{(r+1)}_{ij}$ for all integers $1 \leq i \leq j \leq g$ and $r \geq 0$).

Definition 4.2.18 Let κ be a subfield of $\mathbb{K}_{g,\infty}$ that contains $\mathbb{K}$. A subset $\bowtie \subset \kappa$ is said to σ-generate κ or is called a *σ-generating set* of κ if the subset

$$\kappa \cap \{\sigma^i(h) | i \geq 0, \ h \in \bowtie\}$$

generates κ over $\mathbb{K}$.[8] In such a case, we also say that κ is σ-generated by $\bowtie$.

Corollary 4.2.19 *The following properties hold.*

(a) *The set of distinct polynomials*

$$\{\sigma^i(\Theta_{a,b,c}) | (a,b,c) \in \Delta_3^*(g,2), i \geq 0\} \subset \mathbb{T}_g$$

is algebraically independent over $\mathbb{K}$, i.e., its subset $\{\Theta_{a,b,c})_{(a,b,c)\in\Delta_3^*(g,2)}\}$ is σ-algebraically independent in the sense of Sect. 2.2.

(b) *The 'big field' $\mathbb{K}_{g,\infty}$ is σ-generated by a finite subset of $\mathbb{K}_{g,3}$, i.e., there exists $N \in \mathbb{N}$ and $h_1, \ldots, h_N \in \mathbb{K}_{g,3}$ such that we have*

$$\mathbb{K}_{g,\infty} = \mathbb{K}(\sigma^i(h_j) | 1 \leq j \leq N, i \geq 0).$$

(c) *If g is even, then we have $\mathbb{H}^{3,\perp}_{g,\mathbb{K}} \neq 0$.*

---

[8] This is a more general notion than the 'ϕ-generate' notion of Sect. 2.2 as we do not requite that $\sigma(\kappa) \subset \kappa$.

(d) *The 'big even field' $\mathbb{K}_{g,even,\infty}$ is σ-generated by a finite subset of $\mathbb{K}_{g,even,4}$, i.e., there exists $\tilde{N} \in \mathbb{N}$ and $\tilde{h}_1, \ldots, \tilde{h}_{\tilde{N}}$ in $\mathbb{K}_{g,even,4}$ such that we have*

$$\mathbb{K}_{g,even,\infty} = \mathbb{K}(\sigma^i(\tilde{h}_j)|1 \leq j \leq \tilde{N}, i \geq 0).$$

Proof For part (a) it suffices to show that for all integers $i \geq 0$

$$\{\sigma^i(\Theta_{a,b,c})|(a,b,c) \in \Delta_3^*(g,2)\} = \{\Theta_{0,0,\ldots,0,a,b,c}|(a,b,c) \in \Delta_3^*(g,2)\}$$

(where the number of consecutive 0s is i) is a transcendence basis of $\mathbb{K}_{g,i+2}$ over $\mathbb{K}_{g,i+1}$. For $i = 0$ this follows from Theorem 4.2.17 applied to $r = 2$ and Display (4.21). For $i \geq 2$, by considering a $\mathbb{K}$-automorphism $\iota_{0,1;i+1,i+2}$ of

$$\mathbb{K}[T^{(0)}, T^{(1)}, \ldots, T^{(i+2)}]$$

which interchanges $T_{ij}^{(0)} \longleftrightarrow T_{ij}^{(i)}$ and $T_{ij}^{(1)} \longleftrightarrow T_{ij}^{(i+1)}$ and which fixes each $T_{ij}^{(l)}$ with $l \in \{2, 3, \ldots, i-1, i+2\}$, it suffices to show that, when inserting i consecutive 0s, the family of polynomials

$$(\Theta_{a,b,0,\ldots,0,c})_{(a,b,c)\in\Delta_3^*(g,2)} = (\iota_{0,1;i+1,i+2}(\Theta_{0,\ldots,0,a,b,c}))_{(a,b,c)\in\Delta_3^*(g,2)}$$

is a transcendence basis of $\mathbb{K}_{g,i+2}$ over $\mathbb{K}_{g,i+1}$, but this follows from Theorem 4.2.17 applied to $r \in \{i+1, i+2\}$. For $i = 1$, the same argument applies but by considering the $\mathbb{K}$-automorphism $\iota_{0,1,2}$ of $\mathbb{K}[T^{(0)}, T^{(1)}, T^{(2)}, T^{(3)}]$ that fixes $T_{ij}^{(3)}$ and circularly maps $T_{ij}^{(0)} \mapsto T_{ij}^{(1)} \mapsto T_{ij}^{(2)} \mapsto T_{ij}^{(0)}$ for each index ij.

To prove part (b), let $\{h_1, \ldots, h_{N'}\} \subset \mathbb{K}_{g,2}$ be a σ-generating set of $\mathbb{K}_{g,2}$. We take $N \in \{N', \ldots, N' + \frac{g(g+1)}{2}\}$ to be the smallest number such that there exists a generating transcendence basis $\mathcal{T}_3$ of $\mathbb{K}_{g,3}$ over $\mathbb{K}_{g,2}$ which is the union of $\{h_{N'+1}, \ldots, h_N\}$ and

$$\bigcup_{q=N+1}^{N'+\frac{g(g+1)}{2}} \{\sigma^{i_q}(h_{j_q})|(i_q, j_q) \in \{0,1,2,3\} \times \{1, \ldots, N'\}\} \cap \mathbb{K}_{g,3}.$$

The fact that $\{h_1, \ldots, h_N\} \subset \mathbb{K}_{g,3}$ is a σ-generating set of $\mathbb{K}_{g,\infty}$ follows from the fact (see Lemma 4.2.16(c)) that for each integer $i \geq 0$, $\sigma^i(\mathcal{T}_3)$ is a generating transcendence basis of $\mathbb{K}_{g,3+i}$ over $\mathbb{K}_{g,2+i}$.

Let now g be even. Part (c) follows from the inclusion $\{h_1, \ldots, h_N\} \subset \mathbb{K}_{g,3}$ and the fact that $\mathbb{H}_{g,\mathbb{K}}^{-1+\frac{g(g+1)}{2},\perp} \neq 0$ (by the existence of $\Upsilon_{1,2,\ldots,\frac{g(g+1)}{2}}$).

To prove part (d), based on part (c), let $r \in \{3,4\}$ be the smallest integer such that $\mathbb{H}_{g,\mathbb{K}}^{r-1,\perp} \neq 0$. Let $\{\tilde{h}_1, \ldots, \tilde{h}_{\tilde{N}'}\} \subset \mathbb{H}_{g,\mathbb{K}}^{r-1}$ be a σ-generating set of $\mathrm{Frac}(\mathbb{H}_{g,\mathbb{K}}^{r-1})$. A 'descent and vector bundle' argument similar to the one of the proof of Lemma 4.2.16(c) shows that for each integer $i \geq 0$, as $\mathbb{K}_{g,r+i-1} \to \mathbb{K}_{g,r+i}$ is a purely transcendental field extension, the field extension $\mathrm{Frac}(\mathbb{H}_{g,\mathbb{K}}^{r-1+i}) \to \mathrm{Frac}(\mathbb{H}_{g,\mathbb{K}}^{r+i})$ is also purely transcendental, and a subset of $\mathrm{Frac}(\mathbb{H}_{g,\mathbb{K}}^{r+i})$ with $\frac{g(g+1)}{2}$ elements is a generating transcendence basis (or a transcendence basis) over $\mathrm{Frac}(\mathbb{H}^{r-1+i}_{g,\mathbb{K}})$ if and only if it is so over $\mathbb{K}_{g,r+i-1}$. The rest of the argument is as in the proof of part (b) and gives that we can choose $\tilde{N} \in \{\tilde{N}', \ldots, \tilde{N}' + \frac{g(g+1)}{2}\}$. □

Remark 4.2.20 The fact that the 'big field' and the 'big even field' of Corollary 4.2.19(b) and (d) are σ-finitely generated follows from the following general fact of difference algebra applied to the subfields $\mathbb{K}_{g,\infty}$ and $\mathbb{K}_{g,\mathrm{even},\infty}$ of $\mathrm{Frac}(\mathbb{T}_g)$ (see [Le], Ch. 4, Thm. 4.4.1): each σ-subfield of a σ-finitely generated field is σ-finitely generated. The novel part of Corollary 4.2.19(b) and (d) are the bounds 3 and 4 for the 'order' of the σ-generating set of $\mathbb{K}_{g,\infty}$ and $\mathbb{K}_{g,\mathrm{even},\infty}$ (respectively).

4.2.12 *Additional Invariants*

Similar to Lemma 4.2.16, for $p \neq 2$ we consider the natural $\mathbf{SL}_g$-morphism

$$\theta^{3,2}_g : \mathbf{V}^3_g \to \mathbf{V}^2_g \times_{(\mathbf{V}^2_g/\mathbf{SL}_g)} \mathrm{Spec}(\mathbb{H}^{2,\Theta}_{g,\mathbb{K}}).$$

This makes sense as (see Theorem 4.2.8(b)) $\mathbf{V}^2_g/\mathbf{SL}_g = \mathrm{Spec}(\mathbb{H}^1_{g,\mathbb{K}}) = \mathrm{Spec}(\mathbb{H}^{1,\Theta}_{g,\mathbb{K}})$. From Theorem 4.2.13(c) and Lemma 4.2.16(a) we get that $\theta^{3,2}_g$ is generically finite of some degree

$$b_p(g, 2) \in \mathbb{N},$$

which we can compute in many cases as follows.

Lemma 4.2.21 *We assume that $p \nmid g!$ and $p \neq 2$. If $\mathbb{H}^2_{g,tot,\mathbb{K}}$ is Cohen–Macaulay, then the following properties hold.*

(a) *If g is odd or if g is even but $\mathbb{H}^{2,\perp}_{g,\mathbb{K}} \neq 0$, then we have $b_p(g, 2) = 2^{g-1}\varrho_p(g, 2)$.*

(b) *If g is even and $\mathbb{H}^{2,\perp}_{g,\mathbb{K}} = 0$ (so $g \geq 4$ as $\Upsilon_{1,2,3} \in \mathbb{H}^{2,\perp}_{2,\mathbb{K}}$ is nonzero), then we have $b_p(g, 2) = 2^{g-2}\varrho_p(g, 2)$.*

Proof We have $b_p(g, 2) = 2^{g-2+\varepsilon_g}\mathrm{rank}_{\mathbb{H}^{2,\Theta}_{g,\mathbb{K}}}(\mathbb{H}^2_{g,\mathrm{tot},\mathbb{K}})$, see Lemma 4.2.16(a). From this and Corollary 4.2.15(a) and (b) applied to $r = 2$ we get that lemma holds. □

For $p \neq 2$, if the finite field extension $\mathrm{Frac}(\mathbb{H}^{2,\Theta}_{g,\mathbb{K}}) \to \mathrm{Frac}(\mathbb{H}^2_{g,\mathbb{K}})$ is separable, then we can interpret $b_p(g, 2)$ as follows.[9]

[9] One can introduce the separable version $b^{\mathrm{sep}}_p(g, r)$ of $b_p(g, r)$ which is a divisor of $b_p(g, r)$ and then with no assumption on the finite field extension $\mathrm{Frac}(\mathbb{H}^{2,\Theta}_{g,\mathbb{K}}) \to \mathrm{Frac}(\mathbb{H}^2_{g,\mathbb{K}})$ we get the same interpretation for $b^{\mathrm{sep}}_p(g, r)$. Moreover, by considering the smooth locus of the morphism $\mathrm{Spec}(\mathbb{Z}[T, \ldots, T^{(r)}]) \to \mathrm{Spec}(\mathbb{H}^{r,\mathbb{Z}}_{g,\mathrm{tot}})$, it is a standard piece of algebraic geometry to get that for $p >> 1$ we have $b_p(g, r) = b^{\mathrm{sep}}_p(g, r)$.

For each triple $(D_0, D_1, D_2) \in \mathbb{K}^\times 1_g \times Z_g \times V_g$ such that $F_0(D_0, D_1) \neq 0$ (see proof of Theorem 4.1.1 for F_0), $D_1 = \mathrm{Diag}(\nu_1, \ldots, \nu_g)$ is generic in a sense which can be made precise (mainly just that $\prod_{i=1}^{g} \nu_i \prod_{1\leq i<j\leq g}(\nu_i + \nu_j) \neq 0$), and D_2 is also generic in a sense which can be made precise, then $b_p(g, 2)$ is the number of $\mathbb{K}$-valued points of

$$(\theta_g^{3,2})^{-1}(\theta_g^{3,2}(D_0, D_1, D_2))$$

and thus it is the number of solutions $Q_2 \in V_g$ of the system of equations

$$\Theta_{a,b,c}(D_0, D_1, Q_2) = \Theta_{a,b,c}(D_0, D_1, D_2) \text{ for all } (a, b, c) \in \Delta(g, 2).$$

This system of equations is equivalent to the fact that the identity

$$\det(y_0 D_0 + y_1 D_1 + y_2 Q_2) = \det(y_0 D_0 + y_1 D_1 + y_2 D_2) \tag{4.22}$$

holds for all $y_0, y_1, y_2 \in \mathbb{K}$. This equation clearly holds if $y_2 = 0$ and thus we can assume that $y_2 \neq 0$ and we denote $\lambda_0 := y_0 y_2^{-1}$ and $\lambda_1 := y_0 y_2^{-1}$.

Equation (4.22) is equivalent to the characteristic polynomials equation

$$\chi_{\lambda_1 D_1 + Q_2}(x) = \chi_{\lambda_1 D_1 + D_2}(x) \text{ for all } \lambda_1 \in \mathbb{K}, \tag{4.23}$$

which implies the equation

$$\mathrm{Tr}((\lambda_1 D_1 + Q_2)^i) = \mathrm{Tr}((\lambda_1 D_1 + D_2)^i) \text{ for all } i \in \{1, \ldots, g\}. \tag{4.24}$$

If moreover $p \nmid g!$, then in fact Systems of Eqs. (4.23) and (4.24) are equivalent.

Writing

$$\mathrm{Tr}((\lambda_1 D_1 + Q_2)^v) = \sum_{m=0}^{v} \lambda_1^m P_{g,v,m}(\nu_1, \ldots, \nu_g, Q_2),$$

each $P_{g,v,m}(w_1, \ldots, w_g, T^{(2)}) = P_{g,v,m}(w_1, \ldots, w_g, T_{11}^{(2)}, \ldots, T_{gg}^{(2)})$ is a homogeneous polynomial in $g + \frac{g(g+1)}{2}$ indeterminates of total degree v and of partial degree $v - m$ in the index $^{(2)}$. For $m = v$ we have

$$P_{g,v,v}(w_1, \ldots, w_g, T_{11}^{(2)}, \ldots, T_{gg}^{(2)}) = \sum_{i=1}^{g} w_i^v.$$

System of Eqs. (4.24) is equivalent to the system

$$P_{g,v,m}(\nu_1, \ldots, \nu_g, T^{(2)}) = P_{g,v,m}(\nu_1, \ldots, \nu_g, D_2) \text{ for all } 0 \leq m < v \leq g \tag{4.25}$$

of $\frac{g(g+1)}{2}$ equations in the $\frac{g(g+1)}{2}$ indeterminates $T_{ij}^{(2)}$ with $1 \leq i \leq j \leq g$.

Theorem 4.2.22 *If $p = 0$, then the following properties hold.*

(a) *For $g \leq 3$ we have $b_0(g, 2) = \prod_{i=1}^{g} i^{1+g-i}$.*

(b) *We have $\varrho_0(3, 2) = 3$ and $\varrho_0(2, 2) = 1$.*

Proof As $g \leq 3$, if D_2 is the zero matrix 0_g of size $g \times g$, it is easy to see that System of Eqs. (4.25) has the unique solution $Q_2 = 0_g$. This implies that System of Eqs. (4.25) has the same number of $\mathbb{K}$-valued points as the projective scheme in the $1 + \frac{g(g+1)}{2}$ indeterminates $T_{00}^{(2)}$ and $T_{i,j}^{(2)}$ with $1 \leq i \leq j \leq g$ which is given by the system of homogeneous equations

$$P_{g,v,m}(\nu_1, \dots, \nu_g, T^{(2)}) = (T_{00}^{(2)})^{v-m} P_{g,v,m}(\nu_1, \dots, \nu_g, D_2) \ \text{for all} \ 0 \leq m < v \leq g. \tag{4.26}$$

Choosing D_2 to be generic so that the scheme $(\theta_g^{3,2})^{-1}(\theta_g^{3,2}(D_0, D_1, D_2))$ is étale isomorphic to $\mathrm{Spec}(\mathbb{K}^{b_0(g,2)})$, from Bézout's Theorem (for instance, see [F], Ch. 12, Thm. 12.3 or see [H], Ch. 1, Thm. 7.7 applied repeatedly), we get that $b_0(g, 2)$ is the product of the degrees in the index $^{(2)}$ of the homogeneous forms of System of Equations (4.26) and thus part (a) holds.

Part (b) follows from part (a) and Lemma 4.2.21. □

Corollary 4.2.23 *Let $r \geq 2$ be a natural number. Then the finite morphism $Spec(\mathbb{H}_{3,\mathbb{K}}^{r}) \to Spec(\mathbb{H}_{3,\mathbb{K}}^{r,\Theta})$ is not an isomorphism and thus we have an inequality $\Gamma_p(3, r) > D(3, r, 1)$.*

Proof We first assume that $p = 0$. We have $\varrho_0(3, 2) = 3$ (see Theorem 4.2.22(b)). For $r = 2$ we have $\mathbb{H}_{3,\mathbb{K}}^{r,\Theta} = \mathbb{K}[\mathcal{T}_0]$ (see Theorem 4.2.13(e)) and thus the case $r = 2$ follows from Eq. (4.19). The passage from $r = 2$ to $r > 2$ follows from Remark 4.2.9.

The case $p > 0$ follows from the previous paragraph, Eq. (4.14) and the existence of the second $\mathbb{K}$-algebra monomorphism of Display (4.15). □

Example 4.2.24 We assume that $g = 3$. Our six polynomials in the nine indeterminates $w_1, w_2, w_3, T_{11}^{(2)}, T_{12}^{(2)}, T_{13}^{(2)}, T_{22}^{(2)}, T_{23}^{(2)}$, and $T_{33}^{(2)}$ are:

$$P_{3,1,0} = T_{11}^{(2)} + T_{22}^{(2)} + T_{33}^{(2)} = \mathrm{Tr}(T^{(2)}),$$

$$P_{3,2,0} = \mathrm{Tr}((T^{(2)})^2),$$

$$P_{3,2,1} = 2(w_1 T_{11}^{(2)} + w_2 T_{22}^{(2)} + w_3 T_{33}^{(2)}),$$

$$P_{3,3,0} = \mathrm{Tr}((T^{(2)})^3),$$

$$\begin{aligned} P_{3,3,1} = {} & 3w_1[(T_{11}^{(2)})^2 + (T_{12}^{(2)})^2 + (T_{13}^{(2)})^2] + 3w_2[(T_{12}^{(2)})^2 + (T_{22}^{(2)})^2 + (T_{23}^{(2)})^2] \\ & + 3w_3[(T_{13}^{(2)})^2 + (T_{23}^{(2)})^2 + (T_{33}^{(2)})^2], \end{aligned}$$

$$P_{3,3,2} = 3(w_1^2 T_{11}^{(2)} + w_2^2 T_{22}^{(2)} + w_3^2 T_{33}^{(2)}).$$

We check that if $D_2 = (d_{ij})_{1 \leq i,j \leq 3}$ is generic, then for $p \notin \{2, 3\}$, System of Eqs. (4.25) has 12 solutions. As ν_1, ν_2, and ν_3 are distinct, the Vandermonde 3×3 matrix V_{ν_1,ν_2,ν_3} whose ij entry for $1 \leq i, j \leq 3$ is ν_i^{j-1} is invertible. Thus, as the 3 equations in $Q_2 = (q_{ij})_{1 \leq i,j \leq 3} \in \mathrm{V}_g$

$$P_{3,1+i,i}(\nu_1, \nu_2, \nu_3, Q_2) = P_{3,1+i,i}(\nu_1, \nu_2, \nu_3, D_2)$$

for $i \in \{0, 1, 2\}$ can be substituted (as $p \notin \{2, 3\}$) by one matrix identity

$$(q_{11}\ q_{22}\ q_{33})V_{\nu_1,\nu_2,\nu_3} = (d_{11}\ d_{22}\ d_{33})V_{\nu_1,\nu_2,\nu_3},$$

we conclude that $q_{11} = d_{11}$, $q_{22} = d_{22}$ and $q_{33} = d_{33}$. We are left to show that by denoting $\alpha := d_{12} = d_{21}$, $\beta := d_{13} = d_{31}$, $\gamma := d_{23} = d_{32}$, $x := q_{12} = q_{21}$, $y := q_{13} = q_{31}$, and $z := q_{23} = q_{32}$, the system of 3 equations

$$x^2 + y^2 + z^2 = \alpha^2 + \beta^2 + \gamma^2, \tag{4.27}$$

$$\nu_1(y^2 + z^2) + \nu_2(x^2 + z^2) + \nu_3(x^2 + y^2) = \nu_1(\beta^2 + \gamma^2) + \nu_2(\alpha^2 + \gamma^2) + \nu_3(\alpha^2 + \beta_2), \tag{4.28}$$

$$2xyz - d_{33}x^2 - d_{22}y^2 - d_{11}z^2 = 2\alpha\beta\gamma - d_{33}\alpha^2 - d_{22}\beta^2 - d_{11}\gamma^2 \tag{4.29}$$

in the 3 indeterminates x, y and z has 12 solutions. Note that here we have reinterpreted the equations $P_{3,i,0}(\nu_1, \nu_2, \nu_3, Q_2) = P_{3,i,0}(\nu_1, \nu_2, \nu_3, D_2)$ for $i \in \{0, 1, 2\}$ as a single polynomial identity $\chi_{Q_2}(x) = \chi_{D_2}(x)$ and accordingly we have replaced the equation $P_{3,3,0}(\nu_1, \nu_2, \nu_3, Q_2) = P_{3,3,0}(\nu_1, \nu_2, \nu_3, D_2)$ by the equation $\det(Q_2) - q_{11}q_{22}q_{33} = \det(D_2) - d_{11}d_{22}d_{33}$ (in view of the identities $q_{jj} = d_{jj}$ for $j \in \{1, 2, 3\}$). Based on Eqs. (4.27) and (4.28) can be replaced by the equation

$$y^2 + \nu z^2 = \beta^2 + \nu\gamma^2, \tag{4.30}$$

where $\nu := (\nu_3 - \nu_1)(\nu_2 - \nu_1)^{-1} \in \mathbb{K}\backslash\{0, 1\}$. Based on Eqs. (4.27) and (4.30), by taking D_2 generic so that the d_{11}, d_{22}, d_{33} are distinct, the Eq. (4.29) can be replaced, after a scalar multiplication of $(x, y, z, \alpha, \beta, \gamma)$ by an element of $\mathbb{K}^\times$, by the equation

$$xyz - z^2 = \alpha\beta\gamma - \gamma^2. \tag{4.31}$$

By denoting

$$\mathfrak{P} := xyz,$$

the system of 3 Eqs. (4.27), (4.30), and (4.31) gives identities

$$z^2 = \mathfrak{P} - \alpha\beta\gamma + \gamma^2,$$

$$y^2 = -\nu\mathfrak{P} + \nu\alpha\beta\gamma - \nu\gamma^2 + \beta^2, \tag{4.32}$$

$$x^2 = (\nu - 1)\mathfrak{P} + \alpha^2 + \alpha\beta\gamma(1 - \nu) + \nu\gamma^2 \tag{4.33}$$

and thus $\mathfrak{P}$ is a root of the polynomial

$$w^2 - (w - \alpha\beta\gamma + \gamma^2)[-\nu w + \nu\alpha\beta\gamma - \nu\gamma^2 + \beta^2][(\nu - 1)w + \alpha^2 + \alpha\beta\gamma(1 - \nu) + \nu\gamma^2], \tag{4.34}$$

which is a polynomial of degree 3 in w as $\nu \in \mathbb{K}^\times \backslash \{0, 1\}$. It is easy to see that for generic $\alpha, \beta, \gamma, \nu \in \mathbb{K}$, the polynomial of Diagram (4.34) has 3 distinct and nonzero solutions and that Eqs. (4.33) and (4.32) and $z = (xy)^{-1}\mathfrak{P}$ produce 4 solutions for each value of $\mathfrak{P}$. For $p = 0$, this implies that $b_0(3, 2) = 12$. From this and Lemma 4.2.16(a) we get that for $p = 0$ the integral $\mathbb{K}$-monomorphism

$$\mathbb{H}^{2,\Theta}_{3,\mathbb{K}} \to \mathbb{H}^2_{3,\mathbb{K}} = \mathbb{H}^2_{3,\mathrm{tot},\mathbb{K}} = \mathrm{Sym}((\mathrm{V}^3_3)^*)^{\mathbf{SL}_3}$$

induces at the level of field of fractions a field extension of degree $3 = 12/4$ (which also follows from Theorem 4.2.22(2)).

4.3 Finer Results on Quadratics for the Case $g = 2$

In this section we assume that $g = 2$. After we exemplify all prior notions, we include generalizations of the polynomials $\Theta_{2,0}, \Theta_{1,1,}, \Theta_{0,2}, \Upsilon_{0,1,2}$ in order to present a more detailed description of the $\mathbb{K}$-algebra monomorphism $\mathbb{H}^r_{2,\mathbb{K}} \to \mathbb{H}^r_{2,\mathrm{tot},\mathbb{K}}$ for each integer $r \geq 1$.

Example 4.3.1 In this paragraph we assume $p \neq 2$. The three polynomials in the five indeterminates $\nu_1, \nu_2, T^{(2)}_{11}, T^{(2)}_{12}, T^{(2)}_{22}$ of the proof of Theorem 4.2.22 (a) are:

$$P_{2,1,0} = T^{(2)}_{11} + T^{(2)}_{22} = \mathrm{Tr}(T^{(2)}),$$

$$P_{2,2,0} = (T^{(2)}_{11})^2 + 2(T^{(2)}_{12})^2 + (T^{(2)}_{22})^2 = \mathrm{Tr}((T^{(2)})^2) = \mathrm{Tr}(T^{(2)})^2 - 2\det(T^{(2)}),$$

$$P_{2,2,1} = 2(\nu_1 T^{(2)}_{11} + \nu_2 T^{(2)}_{22}).$$

Thus, if $D_2 = (d_{ij})_{1\leq i,j\leq 2}$ is generic in the sense that $d_{12} = d_{21} \neq 0$, then System of Eqs. (4.25) has 2 solutions $Q_2 = (q_{ij})_{1\leq i,j\leq 2} \in \mathrm{V}_2$ given by $q_{11} = d_{11}, q_{22} = d_{22}$, and $q_{12} = q_{21} \in \{\sqrt{d_{12}}, -\sqrt{d_{12}}\}$. We get that $b_0(2, 2) = 2$.[10] As $0 \neq \Upsilon_{0,1,2} \in \mathbb{H}^{2,\perp}_{2,\mathbb{K}}$, the $\mathbb{K}$-monomorphism $\mathbb{H}^2_{2,\mathbb{K}} \to \mathbb{H}^2_{2,\mathrm{tot},\mathbb{K}}$ is not birational (see Lemma 4.2.3(b)).

In this paragraph we assume $p = 0$. From the last property and the equality $b_0(2, 2) = 2$ we get that the integral $\mathbb{K}$-monomorphism $\mathbb{H}^{2,\Theta}_{2,\mathbb{K}} \to \mathbb{H}^2_{2,\mathbb{K}}$ is birational and thus (as $\mathbb{H}^{2,\Theta}_{2,\mathbb{K}}$ is normal being a polynomial $\mathbb{K}$-algebra, see Theorem 4.2.13(e)) it is an isomorphism, i.e., we have

[10] More generally, we get that $b^{\mathrm{sep}}_p(2, 2) = 2$ for $p > 2$.

$$\mathbb{H}^2_{2,\mathbb{K}} = \mathbb{H}^{2,\Theta}_{2,\mathbb{K}} = \mathbb{K}[\Theta_{2,0,0}, \Theta_{1,1,0}, \Theta_{1,0,1}, \Theta_{0,2,0}, \Theta_{0,1,1}, \Theta_{0,0,2}]. \tag{4.35}$$

As $\mathbb{H}^{2,\perp}_{2,\mathbb{K}} \neq 0$, its rank as a free $\mathbb{H}^{2,\Theta}_{2,\mathbb{K}}$-module is also 1 (see Corollary 4.2.15(b)) and thus is generated by each nonzero homogeneous polynomial of the smallest degree, i.e., we have

$$\mathbb{H}^{2,\perp}_{2,\mathbb{K}} = \mathbb{H}^2_{2,\mathbb{K}}\Upsilon_{0,1,2} = \mathbb{H}^{2,\Theta}_{2,\mathbb{K}}\Upsilon_{0,1,2}.$$

This implies that we have

$$\mathbb{H}^2_{2,\mathrm{tot},\mathbb{K}} = \mathbb{H}^2_{2,\mathbb{K}} \oplus \mathbb{H}^2_{2,\mathbb{K}}\Upsilon_{0,1,2} = \mathbb{H}^{2,\Theta}_{2,\mathbb{K}} \oplus \mathbb{H}^{2,\Theta}_{2,\mathbb{K}}\Upsilon_{0,1,2}.$$

As we have $\sigma(\Theta_{2,0,0}) = \Theta_{0,2,0}$, $\sigma^2(\Theta_{2,0,0}) = \Theta_{0,0,2}$ and $\sigma(\Theta_{1,1,0}) = \Theta_{0,1,1}$, the set $\{\Theta_{2,0,0}, \Theta_{1,1,0}, \Theta_{1,0,1}\}$ is a σ-generating set of $\mathbb{K}_{2,\mathrm{even},2}$. The proof of Corollary 4.2.19(d) for $g = 2$ gives that we can actually take $\tilde{N} = 4$ as it is easy to check that $\{\Theta_{1,0,0,1}, \sigma(\Theta_{1,0,1}), \sigma^2(\Theta_{1,1})\}$ is a generating transcendence basis of $\mathrm{Frac}(\mathbb{K}[T, T', T'', T'''])$ over $\mathrm{Frac}(\mathbb{K}[T, T', T''])$.

4.3.1 Singular Loci and Homogeneous Rings of Grassmannians

We consider the singular locus $S_2^{r+1} := \prod_{l=0}^{r} S_2^{(l)}$ of $\mathbf{V}_2^{r+1}$. We have

$$S_2^{(l)} = \mathrm{Spec}(\mathbb{K}[T_{11}^{(l)}, T_{22}^{(l)}, T_{12}^{(l)}]/(T_{11}^{(l)}T_{22}^{(l)} - (T_{12}^{(l)})^2)) = \mathrm{Spec}(\mathbb{K}[u_l^2, v_l^2, u_l v_l]),$$

where $u_0, v_0, u_1, v_1, \ldots$ are indeterminates and where the identification

$$\mathbb{K}[T_{11}^{(l)}, T_{22}^{(l)}, T_{12}^{(l)}]/(T_{11}^{(l)}T_{22}^{(l)} - (T_{12}^{(l)})^2) = \mathbb{K}[u_l^2, v_l^2, u_l v_l] \tag{4.36}$$

is given by First Isomorphism Theorem applied to the $\mathbb{K}$-homomorphism

$$j_l : \mathbb{K}[T_{11}^{(l)}, T_{22}^{(l)}, T_{12}^{(l)}] \to \mathbb{K}[u_l, v_l]$$

that maps $T_{11}^{(l)}, T_{22}^{(l)}, T_{12}^{(l)}$ respectively to $u_l^2, v_l^2, u_l v_l$. We recall that we have an identity $T_{11}^{(l)}T_{22}^{(l)} - (T_{12}^{(l)})^2 = \Theta_{(0,\ldots,0,2,0,\ldots,0)}$, where the 2 in the $r+1$-tuple lower index $(0, \ldots, 0, 2, 0, \ldots, 0)$ is on the $l+1$-th position. The group $\mathbf{SL}_2$ acts on $\mathbb{K}[u_l, v_l]$ via its action on $\mathrm{Span}(u_l, v_l)$ given by the rule: $\Lambda \in \mathbf{SL}_2(\mathbb{K})$ maps $(u_l\ v_l)^{\mathrm{t}}$ to $\Lambda(u_l\ v_l)^{\mathrm{t}}$.

We consider the $\mathbb{Z}_{\geq 0}$-graded $\mathbb{K}$-algebra

$$\mathbb{Y}_r := \mathbb{K}[u_0, v_0, \ldots, u_r, v_r],$$

where the degree of each u_l and v_l is 1. The total degree is called the uv-degree and we speak about the partial uv-degree $(m_0, m_1, \ldots, m_r)$ of nonzero polynomials in the $\mathbb{K}$-span of monomials $\prod_{l=0}^{r} u_l^{a_l} v_l^{m_l - a_l}$ with each $a_l \in \{0, \ldots, m_l\}$.

Equation (4.36) is $\mathbf{SL}_2$-invariant and when l varies they combine together to induce a finite $\mathbf{SL}_2$-morphism of degree 2^{r+1}

$$\mathrm{Spec}(\mathbb{Y}_r) \to S_2^{r+1} = \mathrm{Spec}(\mathbb{Y}_r^{\text{all-even}})$$

that identifies S_2^{r+1} with the spectrum of the $\mathbb{Z}_{\geq 0}$-graded $\mathbb{K}$-subalgebra $\mathbb{Y}_r^{\text{all-even}}$ of $\mathbb{K}[u_0, v_0, \ldots, u_r, v_r]$ generated by homogeneous polynomials whose partial uv-degrees are $r+1$-tuples of $(2\mathbb{Z}_{\geq 0})^{r+1}$ and that is associated to the image of the $\mathbb{K}$-homomorphism

$$\jmath : \mathbb{K}[T, \ldots, T^{(r)}] \to \mathbb{Y}_r$$

which extends each $\jmath_l$ with $l \in \{0, \ldots, r\}$ and which maps homogeneous polynomials of partial T-degree $(m_0, \ldots, m_r)$ to homogeneous polynomials of partial uv-degree $(2m_0, \ldots, 2m_r)$. We have $\mathrm{Ker}(\jmath) = (\det(T), \ldots, \det(T^{(r)}))$.

For distinct integers $i, j \in \{0, \ldots, r\}$, let

$$y_{i,j} := u_i v_j - v_i u_j \in \mathbb{Y}_r.$$

Thus $y_{i,j} = -y_{j,i}$. If $0 \leq i < j \leq r$, the $\mathbb{K}$-homomorphism $\jmath$ maps

$$\Theta_{(0,\ldots,1,0,\ldots,0,1,0,\ldots,0)} = T_{11}^{(i)} T_{22}^{(j)} + T_{22}^{(i)} T_{11}^{(j)} - 2T_{12}^{(i)} T_{12}^{(j)},$$

where the 1s in $(0, \ldots, 1, 0, \ldots, 0, 1, 0, \ldots, 0)$ are on the $i+1$ and $j+1$ positions, to

$$u_i^2 v_j^2 + v_i^2 u_j^2 - 2u_i v_i u_j v_j = y_{ij}^2.$$

We define

$$\mathcal{I}_r := \{(i, j) \in \mathbb{Z}^2 | 0 \leq i < j \leq r\},$$

$$\mathcal{J}_r := \{(i, j, n, s) \in \mathbb{Z}^4 | 0 \leq i < j < n < s \leq r\},$$

$$\mathrm{Det}_{2,r} := \{\det(T^{(i)}) | i \in \{0, \ldots, r\}\} \text{ and}$$

$$I_{\Delta(2,r)} := \{\Theta_{m_0,\ldots,m_r} | (m_0, \ldots, m_r) \in \Delta(2, r)\}.$$

We get an identity of $\mathbb{Z}_{\geq 0}$-graded $\mathbb{K}$-algebras induced by $\jmath$:

$$\mathbb{K}[T, \ldots, T^{(r)}]/(I_{\Delta(2,r)}) = \mathbb{Y}_r^{\text{all-even}}/(y_{i,j}^2 | (i, j) \in \mathcal{I}_r). \tag{4.37}$$

It is well-known that the $\mathbb{K}$-subalgebra $\mathbb{Y}_r^{\mathbf{SL}_2}$ of $\mathbb{Y}_r$ is generated by the $y_{i,j}$s and is the homogeneous ring of the Grassmannian $\mathrm{Gr}(2, r+1)$ of $\mathbb{P}^1_{\mathbb{K}}$ lines in $\mathbb{P}^r_{\mathbb{K}}$ (see

[HR], Sect 1., Applic. (a), Case (ii) and Ex. after Cor. 1.9). From this and the Plücker relations we get that

$$\mathbb{Y}_r^{\mathbf{SL}_2} = \mathbb{K}[y_{i,j}|(i,j) \in \mathcal{I}_r] = \mathbb{K}[Y_{i,j}|(i,j) \in \mathcal{I}_r]/(Y_{i,j,n,s}|(i,j,n,s) \in \mathcal{J}_r),$$

where $Y_{i,j}$s are indeterminates, where for all $(i,j,n,s) \in \mathcal{J}_r$ we define

$$Y_{i,j,n,s} := Y_{i,j}Y_{n,s} - Y_{i,n}Y_{j,s} + Y_{i,s}Y_{j,n},$$

and where $y_{i,j} = Y_{i,j} + (Y_{i',j',n,s}|(i',j',n,s) \in \mathcal{J}_r)$ for all $(i,j) \in \mathcal{I}_r$. It is convenient to add extra indeterminates $Y_{i,i}$ for all $i \in \{0,\ldots,r\}$. We speak about (total) Y-degrees: each $Y_{i,j}$ has Y-degree 1. Hence each $Y_{i,j,n,s}$ has Y-degree 2.

4.3.2 Basic SL$_2$-Invariants: Ξs and ξs

For $g = 2$ to generalize the homogeneous polynomials $\Theta_{2,0}, \Theta_{1,1,}, \Theta_{0,2}, \Upsilon_{0,1,2}$ we first introduce the appropriate indexing sets.

For an integer m such that $2 \leq m \leq r+1$ let $\mathcal{C}_{r,m}$ be the set of all m-cycles of the permutation group Perm($\{0,\ldots,r\}$). Also, let $\mathcal{C}_r := \cup_{m=2}^{r+1}\mathcal{C}_{r,m}$. We call an m-cycle $\omega = (q_1,\ldots,q_m) \in \mathcal{C}_{r,m}$ increasing or unimodular if when viewed as a permutation of the set $\{q_1,\ldots,q_m\}$ it has exactly $m-1$ inversions, i.e, if up to circular rearrangement we have $q_1 < q_2 < \cdots < q_m$. Let $\mathcal{C}_{r,m}^{\mathrm{uni}}$ be the subset of $\mathcal{C}_{r,m}$ formed by all increasing m-cycles. Let $\mathcal{C}_r^{\mathrm{uni}} := \cup_{m=2}^{r+1}\mathcal{C}_{r,m}^{\mathrm{uni}}$. Let $\mathcal{C}_r^{\mathrm{uni,even}}$ be the subset of $\mathcal{C}_r^{\mathrm{uni}}$ formed by all cycles of even orders (lengths) and let $\mathcal{C}_r^{\mathrm{uni,odd}}$ be the subset of $\mathcal{C}_r^{\mathrm{uni}}$ formed by all cycles of odd orders (lengths). The inclusion $\mathcal{C}_{r,m}^{\mathrm{uni}} \subset \mathcal{C}_{r,m}$ is an equality precisely when $m = 2$.

Let $\mathcal{P}_r$ be the power set of $\{0,\ldots,r\}$ and let $\mathcal{P}_{r,m}$ be its subset formed by all subsets with m elements. We have a natural bijection $\mathcal{P}_{r,m} \to \mathcal{C}_{r,m}^{\mathrm{uni}}$ which maps a subset $\{q_1,q_2,\ldots,q_m\} \in \mathcal{P}_{r,m}$ with $q_1 < q_2 < \cdots < q_m$ to the increasing m-cycle $(q_1,\ldots,q_m)$. This implies that $\mathcal{C}_{r,m}^{\mathrm{uni}}$ has $\binom{r+1}{m}$ elements and that $\mathcal{C}_r^{\mathrm{uni}}$ has $\sum_{m=2}^{r+1}\binom{r+1}{m} = 2^{r+1} - r - 2$ elements.

We now introduce the special Ξ *invariants*. For $\omega = (q_1,\ldots,q_m) \in \mathcal{C}_{r,m}$ and $q_{m+1} := q_1$, we check that there exists a unique homogeneous polynomial

$$\Xi_\omega \in \mathbb{H}_{2,\mathrm{tot},\mathbb{K}}^r \subset \mathbb{K}[T,\ldots,T^{(r)}]$$

which is of partial T-degree $(m_0,\ldots,m_r)$ with each $m_l \in \{0,1\}$, m_l being 1 if and only if $l \in \{q_1,\ldots,q_m\}$ (hence it is of T-degree m), such that the element $\jmath(\Xi_\omega) \in \mathbb{Y}_r$ is the circular product

$$\xi_\omega := -\prod_{i=1}^{m}(u_{q_i}v_{q_{i+1}} - v_{q_i}u_{q_{i+1}}) = -\prod_{i=1}^{m} y_{q_i,q_{i+1}} \in (\mathbb{Y}_r^{\text{all-even}})^{\mathbf{SL}_2}.$$

The uniqueness follows from the fact that the ideal $(\mathrm{Det}_{2,r}) = \mathrm{Ker}(\jmath)$ contains no element of partial T-degree $(m_0, \ldots, m_r)$.

For $p = 0$, the existence of Ξ_ω follows from the fact that $\mathbf{V}_2^{r+1} \to \mathbf{V}_2^{r+1}/\mathbf{SL}_2$ is a universal categorical quotient (see Sect. 4.1.2) which implies that

$$(\mathbb{Y}_r^{\text{all-even}})^{\mathbf{SL}_2} = \{\mathbb{K}[T, \ldots, T^{(r)}]/(\mathrm{Det}_{2,r})\}^{\mathbf{SL}_2} = \mathbb{H}^r_{2,\mathrm{tot},\mathbb{K}}/(\mathrm{Det}_{2,r}). \tag{4.38}$$

Taking $\mathbb{K}$ to be an algebraic closure of $\mathbb{Q}$, due to its uniqueness, Ξ_ω is fixed by the Galois group of $\mathbb{Q}$ and thus, with the notation of Sect. 4.2.3, we have $\Xi_\omega \in \mathbb{H}^{r,\mathbb{Z}}_{2,\mathrm{tot},\mathbb{Q}}$. Considering the $\mathbb{Z}$ version

$$\jmath^{\mathbb{Z}} : \mathbb{Z}[T, \ldots, T^{(r)}] \to \mathbb{Y}_r^{\mathbb{Z}} := \mathbb{Z}[u_0, v_0, \ldots, u_r, v_r]$$

of $\jmath : \mathbb{K}[T, \ldots, T^{(r)}] \to \mathbb{Y}_r$, the only homogeneous polynomials of $\mathbb{Q}[T, \ldots, T^{(r)}]$ of partial T-degree $(m_0, \ldots, m_r)$ which map to $\mathbb{Y}_r^{\mathbb{Z}}$ are the ones of $\mathbb{Z}[T, \ldots, T^{(r)}]$. This implies that we have $\Xi_\omega \in \mathbb{H}^{r,\mathbb{Z}}_{2,\mathrm{tot}}$. Hence, based on Display (4.15), using scalar extensions we get that Ξ_ω exists even if $p > 0$.

For instance, as an explanation of the choice of sign in the ξ_ws, we have

$$\Xi_{(0,1)} = \Theta_{1,1} \quad \text{and} \quad \Xi_{(0,1,2)} = -\Xi_{(0,2,1)} = \Upsilon_{0,1,2}$$

as well as

$$\xi_{(0,1)} = y_{0,1}^2 \quad \text{and} \quad \xi_{(0,1,2)} = -\xi_{(0,2,1)} = y_{0,1}y_{1,2}y_{0,2}.$$

For $\omega = (q_1, \ldots, q_m) \in \mathcal{C}_{r,m}$, let

$$\xi_\omega^Y := -\prod_{i=1}^{m} Y_{q_i,q_{i+1}} \in \mathbb{K}[Y_{i,j} | (i, j) \in \mathcal{I}_r].$$

It is easy to see that a monomial in the y_{ij}s has a partial uv-degree in $(2\mathbb{Z}_{\geq 0})^{r+1}$ if and only if, up to sign, it is a monomial in the ξ_ωs with $\omega \in \mathcal{C}_r$. From this we get that

$$\begin{aligned}(\mathbb{Y}_r^{\text{all-even}})^{\mathbf{SL}_2} &= \mathbb{K}[\xi_\omega | \omega \in \mathcal{C}_r] = \mathbb{K}[y_{i,j}^2, \xi_\omega | (i, j) \in \mathcal{I}_r, \omega \in \mathcal{C}_r \backslash \mathcal{C}_{r,2}].\\ &= \mathbb{K}[\xi_\omega^Y | \omega \in \mathcal{C}_r]/\{(Y_{i,j,n,s} | (i, j, n, s) \in \mathcal{J}_r) \cap \mathbb{K}[\xi_\omega^Y | \omega \in \mathcal{C}_r]\}.\end{aligned} \tag{4.39}$$

From Eqs. (4.37) and (4.39) we get natural $\mathbb{K}$-algebra homomorphisms

$$\begin{aligned}&\mathbb{H}^r_{2,\mathrm{tot},\mathbb{K}}/(I_{\Delta(2,r)}) \xrightarrow{\zeta} \mathbb{K}[\xi_\omega | \omega \in \mathcal{C}_r]/(y_{i,j}^2 | (i, j) \in \mathcal{I}_r)\\ &= \mathbb{K}[\xi_\omega^Y | \omega \in \mathcal{C}_r]/\{[(Y_{i,j}^2 | (i, j) \in \mathcal{I}_r) + (Y_{i,j,n,s} | (i, j, n, s) \in \mathcal{J}_r)] \cap \mathbb{K}[\xi_\omega^Y | \omega \in \mathcal{C}_r]\}.\end{aligned} \tag{4.40}$$

As for $\omega = (q_1, \ldots, q_m) \in \mathcal{C}_{r,m}$ we have $\Xi_\omega \in \mathbb{H}^r_{2,\mathrm{tot},\mathbb{K}}$, ζ is an epimorphism. Moreover, if $p = 0$, then from universal categorical quotient property of the morphism $\mathbf{V}_2^{r+1} \to \mathbf{V}_2^{r+1}/\mathbf{SL}_2$, we get that ζ is an isomorphism.

Let $\bar{\xi}_\omega \in \mathbb{H}^r_{2,\mathrm{tot},\mathbb{K}}/(I_{\Delta(2,r)})$ be the image of Ξ_ω; so $\zeta(\bar{\xi}_\omega)$ is the image of ξ_ω.

Lemma 4.3.2 *Let $m \geq 4$ be an integer. The following properties hold.*

(a) *For an integer $r \geq 3$, $\{q_1, \ldots, q_m\} \in \mathcal{P}_{r,m}$, and $s \in \{3, \ldots, m-1\}$ we have an identity*

$$\xi_{(q_1,q_2,\ldots,q_m)} + \xi_{(q_1,q_{s+1},q_{s+2},\ldots,q_m)}\xi_{(q_2,q_3,\ldots,q_s)} = (-1)^s \xi_{(q_1,q_s,q_{s-1},\ldots,q_3,q_2,q_{s+1},\ldots,q_{m-1},q_m)}.$$

(b) *If r and the set $\{q_1, \ldots, q_m\}$ are as in part (a), then for each $p \neq 2$ we have an identity $\zeta(\bar{\xi}_{(q_1,q_2,q_3,q_4,\ldots,q_m)}) = 0$.*

(c) *If $r \geq 2$ is an integer and $\{q_1, q_2, q_3\}, \{q_1', q_2', q_3'\} \in \mathcal{P}_{r,3}$, then for $p \neq 2$ we have an identity $\zeta(\bar{\xi}_{(q_1,q_2,q_3)}\bar{\xi}_{(q_1',q_2',q_3')}) = 0$.*

Proof Part (a) just translates in the language of ξs the fact that

$$(y_{q_1,q_2}y_{q_{s+1},q_s} - y_{q_1,q_{s+1}}y_{q_2,q_s} + y_{q_1,q_s}y_{q_2,q_{s+1}})y_{q_2,q_3}\cdots y_{q_{s-1},q_s}y_{q_{s+1},q_{s+2}}\cdots y_{q_m,q_1} = 0.$$

For parts (b) and (c), from part (a) by taking $s = 3$ and working modulo $I_{\Delta(2,r)}$ and $(y_{i,j}^2 | (i,j) \in \mathcal{I}_r)$ and from the identity $\bar{\xi}_{(q_2,q_3)} = 0$ we get that

$$\zeta(\bar{\xi}_{(q_1,q_2,q_3,q_4,\ldots,q_m)}) = -\zeta(\bar{\xi}_{(q_1,q_3,q_2,q_4,\ldots,q_m)}). \tag{4.41}$$

Applying Eq. (4.41) repeatedly by shifting q_2 we get that:

$$\zeta(\bar{\xi}_{(q_1,q_2,q_3,q_4,\ldots,q_m)}) = -\zeta(\bar{\xi}_{(q_1,q_3,q_2,q_4,\ldots,q_m)}) = \cdots = (-1)^{m-1}\zeta(\bar{\xi}_{(q_1,q_2,q_3,q_4,\ldots,q_m)}). \tag{4.42}$$

If m is even, as $p \neq 2$, from Eq. (4.42) we get that $\zeta(\bar{\xi}_{(q_1,q_2,q_3,q_4,\ldots,q_m)}) = 0$.

If $m \geq 5$ is odd, then from part (a) by taking $s = 4$, by working modulo $I_{\Delta(2,r)}$ and $(y_{i,j}^2 | (i,j) \in \mathcal{I}_r)$, and by using that part (b) holds in the even case (so $\zeta(\bar{\xi}_{(q_1,q_5,q_6,\ldots,q_m)}) = 0$), we get that $\zeta(\bar{\xi}_{(q_1,q_2,q_3,q_4,\ldots,q_m)}) = \zeta(\bar{\xi}_{(q_1,q_4,q_3,q_2,q_5,q_6,\ldots,q_m)})$. From Eq. (4.41) applied thrice we get that

$$\zeta(\bar{\xi}_{(q_1,q_4,q_3,q_2,q_5,q_6,\ldots,q_m)}) = -\zeta(\bar{\xi}_{(q_1,q_2,q_3,q_4,\ldots,q_m)}).$$

Thus, as $p \neq 2$, we get that $\zeta(\bar{\xi}_{(q_1,q_2,q_3,q_4,\ldots,q_m)}) = 0$. Therefore part (b) holds.

To prove (c), we can assume that $r \geq 5$ and the set $\{q_1, q_2, q_3, q_1', q_2', q_3'\}$ has 6 elements. Not to carry the upper indexes $'$, to prove (c) we take $m = 6$ and it suffices to check that $\zeta(\bar{\xi}_{(q_1,q_5,q_6)}\bar{\xi}_{(q_2,q_3,q_4)}) = 0$. But this follows from part (a) applied to $(s, m) = (4, 6)$ and part (b) applied to $m = 6$. Thus part (c) holds. □

4.3.3 Generators

Proposition 4.3.3 *If* $\mathbb{H}^r_{2,tot,\mathbb{K}}/(I_{\Delta(2,r)}) \xrightarrow{\zeta} \mathbb{K}[\xi_\omega|\omega \in \mathcal{C}_r]/(y^2_{i,j}|(i,j) \in \mathcal{I}_r)$ *is an isomorphism and* $p \neq 2$ *(for instance, if* $p = 0$*), then the following properties hold.*

(a) *The smallest number of homogeneous generators of the* $\mathbb{Z}_{\geq 0}$*-graded* $\mathbb{K}$*-algebra* $\mathbb{H}^r_{2,tot,\mathbb{K}}$ *is*

$$\Gamma_p(2, tot, r) = \frac{(r+1)(r^2+2r+6)}{6}.$$

More precisely, the set

$$\{\Theta_{m_0,\ldots,m_r}|(m_0,\ldots,m_r) \in \Delta(2,r)\} \cup \{\Xi_\omega|\omega \in \mathcal{C}^{uni}_{r,3}\}$$

is a minimal homogeneous set of generators of $\mathbb{H}^r_{2,tot,\mathbb{K}}$.

(b) *The set* $\{\Theta_{m_0,\ldots,m_r}|(m_0,\ldots,m_r) \in \Delta(2,r)\}$ *is a minimal homogeneous set of generators of* $\mathbb{H}^r_{2,\mathbb{K}}$*, i.e., we have* $\mathbb{H}^{r,\Theta}_{2,\mathbb{K}} = \mathbb{H}^r_{2,\mathbb{K}}$ *and (with the notation of Proposition 4.2.10)* $\Gamma_p(2,r) = D(2,r,1) = \binom{r+2}{2}$.

Proof The case $r = 1$ follows from Theorem 4.2.8 and the fact that $\mathcal{C}^{\text{uni}}_{r,3}$ is the empty set. Thus we can assume that $r \geq 2$.

The set $\Delta(2,r)$ has $D(2,r,1) = \binom{r+2}{2}$ elements and the set $\mathcal{C}^{\text{uni}}_{r,3}$ has $\binom{r+1}{3}$ elements. Thus the union of part (a) has $\frac{(r+1)(r^2+2r+6)}{6}$ elements.

We get a minimal homogeneous set $\Gamma_{\min}$ of generators of $\mathbb{H}^r_{2,\text{tot},\mathbb{K}}$ as follows. We write $\Gamma_{\min} = \sqcup_{s=0}^{\infty}\Gamma_{\min,s}$, where $\Gamma_{\min,s}$ is the subset of $\Gamma_{\min}$ formed by all homogeneous elements of T-degree s and is obtained (constructed) inductively as follows. First, $\Gamma_{\min,1}$ is a basis of $\text{Sym}((\text{V}_2^{r+1})^*)_1^{\mathbf{SL}_2}$ and thus is the empty set (see Proposition 4.2.6). Second, once $\Gamma_{\min,1},\ldots,\Gamma_{\min,s}$ are obtained (constructed), $\Gamma_{\min,s+1}$ is a basis of a subspace of $\text{Sym}((\text{V}_2^{r+1})^*)_{s+1}^{\mathbf{SL}_2}$ which is a direct supplement in $\text{Sym}((\text{V}_2^{r+1})^*)_{s+1}^{\mathbf{SL}_2}$ of the intersection $\text{Sym}((\text{V}_2^{r+1})^*)_{s+1}^{\mathbf{SL}_2} \cap \mathbb{K}[\sqcup_{i=1}^{s}\Gamma_{\min,s}]$. Thus $\Gamma_{\min,2}$ is a basis of

$$\text{Sym}((\text{V}_2^{r+1})^*)_2^{\mathbf{SL}_2} = \mathbb{H}^r_2(1)_{\mathbb{K}} = \text{Span}(\{\Theta_{m_0,\ldots,m_r}|(m_0,\ldots,m_r) \in \Delta(2,r)\})$$

(see Proposition 4.2.10 (b) and the fact that $p \neq 2$ does not divide $2! = 2$) and therefore it has $\binom{r+2}{2}$ elements. From Diagram (4.40) we get that we can assume that $\Gamma_{\min,s} \subset \{\Xi_\omega|\omega \in \mathcal{C}_{r,s}\}$. From this and Lemma 4.3.2(b) and (c) we get that for $s > 3$ we have $\Gamma_{\min,s} = \emptyset$ (this is automatically so if $r = 2$). As we have $\Xi_{(q_1,q_2,q_3)} = -\Xi_{(q_1,q_3,q_2)}$, we get that $\Gamma_{\min,3} \subset \{\Xi_\omega|\omega \in \mathcal{C}^{\text{uni}}_{r,3}\}$. Clearly, the natural $\mathbb{K}$-linear map from $\text{Span}(\{\xi^Y_\omega|\omega \in \mathcal{C}^{\text{uni}}_{r,3}\})$ to

$$\mathbb{K}[\xi^Y_\omega|\omega \in \mathcal{C}_r]/\{[(Y^2_{i,j}|(i,j) \in \mathcal{I}_r) + (Y_{i,j,n,s}|(i,j,n,s) \in \mathcal{J}_r)] \cap \mathbb{K}[\xi^Y_\omega|\omega \in \mathcal{C}_r]\}$$

is injective. From this and Diagram (4.40) we get that $\Gamma_{\min,3} = \{\Xi_\omega | \omega \in \mathcal{C}_{r,3}^{\text{uni}}\}$. So part (a) holds.

Part (b) follows from part (a) and Lemma 4.3.2(c). □

For each $(i, j) \in \mathcal{I}_r$ let $\underline{(i, j)} \in \Delta(2, r)$ be the unique $r + 1$-tuple which has 1 on the $i + 1$ and $j + 1$ positions and let $\Theta^Y_{\underline{(i,j)}}$ be an indeterminate of degree 1. For each $i \in \{0, \ldots, r\}$ let $\underline{i} \in \Delta(2, r)$ be the unique $r + 1$-tuple which has 2 on the $i + 1$ position and let $\Theta^Y_{\underline{i}}$ be an indeterminate of degree 1.

If $\mathbb{H}^r_{2,\text{tot},\mathbb{K}}/(I_{\Delta(2,r)}) \xrightarrow{\varsigma} \mathbb{K}[\xi_\omega | \omega \in \mathcal{C}_r]/(y^2_{i,j} | (i, j) \in \mathcal{I}_r)$ is an isomorphism and $p \neq 2$ (for instance, if $p = 0$), then from Proposition 4.3.3(b) we get that we have a surjective homomorphism

$$\mathfrak{q}_r : \mathbb{K}[\Theta^Y_{\underline{i}} | i \in \{0, \ldots, r\} \cup \mathcal{I}_r] \to \mathbb{H}^r_{2,\mathbb{K}}$$

of $\mathbb{Z}_{\geq 0}$-graded $\mathbb{K}$-algebras that maps each $\Theta^Y_{\underline{i}}$ to $\Theta_{\underline{i}}$. We speak about the Θ^Y-degrees of homogeneous elements of $\mathbb{K}[\Theta^Y_{\underline{i}} | i \in \{0, \ldots, r\} \cup \mathcal{I}_r]$.

4.3.4 On Hilbert Series

We refer to Corollary 4.2.14 for the uniquely determined nondecreasing sequence $(c_p(2, r, l))_{l=1}^{\varrho^-_p(2,r)}$ of integers ≥ 2 with $\varrho^-_p(2, r) \in \mathbb{N} \cup \{0\}$ and for $\varrho_p(2, r) \in \mathbb{N}$. We recall that $\varrho_0(2, 2) = 1$ if $p = 0$, see Theorem 4.2.22(b).

Proposition 4.3.4 *With the notation of Corollary 4.2.14, for all integers $r \geq 3$ the following properties hold for $p = 0$.*

(a) *We have*

$$\varrho_0(2, r) = 2^{r-1}(r-2)^{-1}\binom{2r-3}{r} - 1 - \binom{r-1}{2}$$

(thus $\varrho_0(2, 3) = 2$, $\varrho_0(2, 4) = 16$, and $\varrho_0(2, 5) = 105$).

(b) *We have*

$$\frac{(1 + x^3)[1 + \frac{(r-1)(r-2)x^2}{2} + \sum_{l=1}^{\varrho^-_0(2,r)} x^{2c_0(2,r,l)}]\left[\sum_{a=0}^{\lfloor\frac{r+1}{2}\rfloor} \binom{r+1}{2a} x^a\right]}{(1 - x^2)^{2r-1}}$$

$$\preceq \frac{\frac{(1+x)^{2r-1}}{r-1}\left[\sum_{j=1}^{r-1} \binom{r-1}{j}\binom{r-1}{j-1} x^{j-1}\right]}{(1 - x^2)^{2r-1}}.$$

(c) *We have inequalities*

$$c_0(2, r, \varrho(2, r) - 1) \leq \lfloor \frac{3r-4}{2} \rfloor \quad \text{and} \quad c_0(2, r, \varrho(2, r)) \leq \lfloor \frac{3(r-1)}{2} \rfloor$$

(thus $c_0(2, 3, 1) = 2$, $c_0(2, 3, 2) \in \{2, 3\}$, $c_0(2, 4, 16) \leq 4$, $c_0(2, 5, 104) \leq 5$, *and* $c_0(2, 5, 105) \leq 6$*).*

Proof Let $\mathcal{T}_0$ be as in Theorem 4.2.13(c); so it is a subset with $D(2, r) = 3r$ elements of $\mathbb{H}_2^r(1)_{\mathbb{K}} = \mathrm{Span}(\{\Theta_{m_0,\ldots,m_r} | (m_0, \ldots, m_r) \in \Delta(2, r)\})$ which is algebraically independent and for which the two $\mathbb{K}$-monomorphisms $\mathbb{K}[\mathcal{T}_0] \to \mathbb{H}^r_{2,\mathrm{tot},\mathbb{K}}$ and $\mathbb{K}[\mathcal{T}_0] \to \mathbb{H}^r_{2,\mathbb{K}}$ are integral and flat, i.e., the $\mathbb{Z}_{\geq 0}$-graded quotient $\mathbb{K}$-algebra $\mathbb{H}^r_{2,\mathrm{tot},\mathbb{K}}/(\mathcal{T}_0)$ has dimension 0 (equivalently, it is a local artinian ring). Based on Corollary 4.2.15(b) and the identities

$$D(2, r, 1) - D(2, r) = \binom{r+2}{2} - 3r = \binom{r-1}{2},$$

to prove part (a) it suffices to show that

$$\mathrm{rank}_{\mathbb{K}[\mathcal{T}_0]}(\mathbb{H}^r_{2,\mathrm{tot},\mathbb{K}}) = 2^r (r-2)^{-1} \binom{2r-3}{r}. \tag{4.43}$$

To check this, we first check that we can assume that we have

$$\mathcal{T}_0 \supset \mathcal{D}_0 := \mathrm{Det}_{2,r} = \{\Theta_{\underline{i}} | i \in \{0, \ldots, r\}\} = \{\det(T^{(i)}) | i \in \{0, \ldots, r\}\}.$$

As $p = 0$, $\mathbf{V}_2^{r+1} \to \mathbf{V}_2^{r+1}/\mathbf{SL}_2$ is a universal categorical quotient, hence:

$$\mathrm{Spec}(\mathbb{H}^r_{2,\mathrm{tot},\mathbb{K}}/(\mathcal{D}_0)) = \mathrm{Spec}((\mathrm{Sym}((\mathbf{V}_2^{r+1})^*)/(\mathcal{D}_0))^{\mathbf{SL}_2}) = (S_2^{r+1})^{\mathbf{SL}_2}.$$

Recall that $\dim((S_2^{r+1})^{\mathbf{SL}_2})$ is equal to

$$\dim(\mathbb{K}[u_l^2, u_l v_l, v_l^2 | l \in \{0, \ldots, r\}]^{\mathbf{SL}_2}) = \dim(\mathbb{Y}_r^{\mathbf{SL}_2}) = 2(r+1) - 3 = 2r - 1$$

(one can easily check this based on Lemma 4.1.6(b)). From this and Theorem 4.2.13(a) we get that there exists a subset $\mathcal{E}_0$ with $2r - 1$ elements of $\mathbb{H}_2^r(1)_{\mathbb{K}}$ such that by taking $\mathcal{T}_0 = \mathcal{D}_0 \cup \mathcal{E}_0$, the $\mathbb{Z}_{\geq 0}$-graded $\mathbb{K}$-algebra $\mathbb{H}^r_{2,\mathrm{tot},\mathbb{K}}/(\mathcal{T}_0)$ has dimension 0; so the proof of Theorem 4.2.13(c) applies to give us that the chosen $\mathcal{T}_0$ has the desired properties.

We identify $\mathbb{K}[\mathcal{T}_0]/(\mathcal{D}_0) = \mathbb{K}[\mathcal{E}_0]$. Based on the previous paragraph, to prove that Eq. (4.43) holds it suffices to show that

$$\mathrm{rank}_{\mathbb{K}[\mathcal{E}_0]}(\mathbb{H}^r_{2,\mathrm{tot},\mathbb{K}}/(\mathcal{D}_0)) = 2^r (r-2)^{-1} \binom{2r-3}{r}. \tag{4.44}$$

We consider the reductive group

$$\mathbf{G}_{2,r} := [\mathbf{SL}_2 \times (\mathbb{Z}/2\mathbb{Z})_{\mathbb{K}}^{r+1}]/(\mathbb{Z}/2\mathbb{Z})_{\mathbb{K}}$$

over $\mathrm{Spec}(\mathbb{K})$, where $(\mathbb{Z}/2\mathbb{Z})_{\mathbb{K}}$ is embedded diagonally in the center of the product $\mathbf{SL}_2 \times (\mathbb{Z}/2\mathbb{Z})_{\mathbb{K}}^{r+1}$. It acts naturally on $\mathbb{Y}_r$, with the nontrivial element of the $l+1$-th factor of $(\mathbb{Z}/2\mathbb{Z})_{\mathbb{K}}^{r+1}$ mapping (u_l, v_l) to $(-u_l, -v_l)$ and fixing each u_i and v_i with $i \in \{0, \dots, l-1, l+1, \dots, r\}$. The identity

$$\mathbb{Y}_r^{(\mathbb{Z}/2\mathbb{Z})_{\mathbb{K}}^{r+1}} = \mathbb{K}[u_l^2, u_l v_l, v_l^2 | l \in \{0, \dots, r\}]$$

induces new identifications

$$\begin{aligned} \mathbb{Y}_r^{\mathbf{G}_{2,r}} &= (\mathbb{Y}_r^{\mathbf{SL}_2})^{(\mathbb{Z}/2\mathbb{Z})_{\mathbb{K}}^r} = (\mathbb{Y}_r^{(\mathbb{Z}/2\mathbb{Z})_{\mathbb{K}}^{r+1}})^{\mathbf{PGL}_2} \\ &= (\mathbb{Y}_r^{\text{all-even}})^{\mathbf{SL}_2} = \mathbb{H}^r_{2,\mathrm{tot},\mathbb{K}}/(\mathcal{D}_0), \end{aligned}$$

where we identify $(\mathbb{Z}/2\mathbb{Z})_{\mathbb{K}}^r$ with the quotient of $(\mathbb{Z}/2\mathbb{Z})_{\mathbb{K}}^{r+1}$ by its diagonal $(\mathbb{Z}/2\mathbb{Z})_{\mathbb{K}}$ subgroup. We have a natural direct sum decomposition

$$\mathbb{Y}_r^{\mathbf{SL}_2} = \oplus_{\chi \in \mathrm{Hom}((\mathbb{Z}/2\mathbb{Z})^r, \{-1,1\})} \mathbb{Y}_{r,\chi}^{\mathbf{SL}_2}, \tag{4.45}$$

where the abstract group $(\mathbb{Z}/2\mathbb{Z})^r = (\mathbb{Z}/2\mathbb{Z})_{\mathbb{K}}^r(\mathbb{K})$ acts on $\mathbb{Y}_{r,\chi}^{\mathbf{SL}_2}$ via the character χ. We translate this decomposition in terms of the parity of the entries of partial uv-degrees as follows. For $a \in \{0, \dots, \lfloor \frac{r+1}{2} \rfloor\}$, let $\mathrm{Par}_{r,a}$ be the set of $r+1$-tuples $(o_0, \dots, o_r) \in \{\text{even}, \text{odd}\}^{r+1}$ with the property that the number of times odd shows up in the sequence $o_0, \dots, o_r$ is exactly $2a$; it has $\binom{r+1}{2a}$ elements. The set

$$\mathrm{Par}_r := \bigcup_{a=0}^{\lfloor \frac{r+1}{2} \rfloor} \mathrm{Par}_{r,a}$$

has exactly $\sum_{a=0}^{\lfloor \frac{r+1}{2} \rfloor} \binom{r+1}{2a} = 2^r$ elements. For $\underline{o} = (o_0, \dots, o_r) \in \mathrm{Par}_r$, let

$$\mathbb{Y}_{r,\underline{o}}^{\mathbf{SL}_2}$$

be the direct summand of $\mathbb{Y}^{\mathbf{SL}_2}$ formed by homogeneous polynomials of partial uv-degree $(m_0, \dots, m_r)$ such that for each index $l \in \{0, \dots, r\}$, the parity of m_l is exactly o_l. Then for each $\chi \in \mathrm{Hom}((\mathbb{Z}/2\mathbb{Z})^r, \{-1, 1\})$, there exists a unique element $o_\chi \in \mathrm{Par}_r$ such that we have

$$\mathbb{Y}_{r,\chi}^{\mathbf{SL}_2} = \mathbb{Y}_{r,o_\chi}^{\mathbf{SL}_2}.$$

For instance, if $\chi_0 \in \mathrm{Hom}((\mathbb{Z}/2\mathbb{Z})^r, \{-1, 1\})$ is the trivial character, then we have $o_{\chi_0} = (\text{even}, \text{even}, \text{even}, \dots, \text{even})$ and

$$\mathbb{H}^r_{2,\mathrm{tot},\mathbb{K}}/(\mathcal{D}_0) = \mathbb{Y}_{r,\chi_0}^{\mathbf{SL}_2} = \mathbb{Y}_{r,(\text{even},\text{even},\dots,\text{even})}^{\mathbf{SL}_2}.$$

Each $\mathbb{Y}^{\mathbf{SL}_2}_{r,\chi}$ is a nonzero torsion free $\mathbb{H}^r_{2,\mathrm{tot},\mathbb{K}}/(\mathcal{D}_0)$-module. For instance,

$$0 \neq \prod_{i=0}^{a-1} y_{2i,2i+1} \in \mathbb{Y}^{\mathbf{SL}_2}_{r,(\mathrm{odd},\mathrm{odd},\ldots,\mathrm{odd},\mathrm{even},\ldots,\mathrm{even})},$$

where odd shows up exactly $2a$ times, and there exist no homogeneous elements of $\mathbb{Y}^{\mathbf{SL}_2}_{r,(\mathrm{odd},\mathrm{odd},\ldots,\mathrm{odd},\mathrm{even},\ldots,\mathrm{even})}$ of positive uv-degree less than $2a$ (equivalently, Y-degree less than a). This implies that

$$\dim_{\mathrm{Frac}(\mathbb{H}^r_{2,\mathrm{tot},\mathbb{K}}/(\mathcal{D}_0))}(\mathrm{Frac}(\mathbb{H}^r_{2,\mathrm{tot},\mathbb{K}}/(\mathcal{D}_0)) \otimes_{\mathbb{H}^r_{2,\mathrm{tot},\mathbb{K}}/(\mathcal{D}_0)} \mathbb{Y}^{\mathbf{SL}_2}_{r,\chi}) = 1. \tag{4.46}$$

Quillen–Suslin's theorem implies that each $\mathbb{Y}^{\mathbf{SL}_2}_{r,\chi}$ is a free $\mathbb{K}[\mathcal{E}_0]$-module. From this and Eq. (4.46) we get that for $\chi \in \mathrm{Hom}((\mathbb{Z}/2\mathbb{Z})^r, \{-1, 1\})$ we have

$$\mathrm{rank}_{\mathbb{K}[\mathcal{E}_0]}(\mathbb{H}^r_{2,\mathrm{tot},\mathbb{K}}/(\mathcal{D}_0)) = \mathrm{rank}_{\mathbb{K}[\mathcal{E}_0]}(\mathbb{Y}^{\mathbf{SL}_2}_{r,\chi}). \tag{4.47}$$

Let $q(r) := \mathrm{rank}_{\mathbb{K}[\mathcal{E}_0]}(\mathbb{Y}^{\mathbf{SL}_2}_{r})$. From Eqs. (4.45) and (4.47) we get that

$$q(r) = 2^r \mathrm{rank}_{\mathbb{K}[\mathcal{E}_0]}(\mathbb{H}^r_{2,\mathrm{tot},\mathbb{K}}/(\mathcal{D}_0)).$$

Thus to prove that Eq. (4.44) holds it suffices to show that we have

$$q(r) = 2^{2r}(r-2)^{-1}\binom{2r-3}{r}.$$

To check this we recall (see [GW], Sect. 6 and [Br], Thm. 1) that the Hilbert series of the $\mathbb{Z}_{\geq 0}$-graded $\mathbb{K}$-algebra $\mathbb{Y}^{\mathbf{SL}_2}_{r}$ which is the homogeneous coordinate ring of the Grassmannian $\mathrm{Gr}(2, r+1)$ and in which all $y_{i,j}$s with $0 \leq i < j \leq r$ are of degree 1 (to be called as well Y-degree), is

$$\wp_1(2,r)(x) := \frac{\sum_{j=1}^{r-1}(r-1)^{-1}\binom{r-1}{j}\binom{r-1}{j-1}x^{j-1}}{(1-x)^{2r-1}}. \tag{4.48}$$

The composite $\mathbb{K}$-monomorphism $\mathbb{K}[\mathcal{E}_0] \to \mathbb{H}^r_{2,\mathrm{tot},\mathbb{K}}/(\mathcal{D}_0) \to \mathbb{Y}^{\mathbf{SL}_2}_{r}$ is integral and thus the $\mathbb{K}[\mathcal{E}_0]$-module $\mathbb{Y}^{\mathbf{SL}_2}_{r}$ is free (see [Ke], Cor. 4.2). If we have an ordered $\mathbb{K}[\mathcal{E}_0]$-basis $w_1, \ldots, w_{q(r)}$ of $\mathbb{Y}^{\mathbf{SL}_2}_{r}$ formed by homogeneous elements such that the sequence $(\deg(w_l))_{1\leq l\leq q(r)}$ of Y-degrees is nondecreasing, then we have

$$\wp_1(2,r)(x) = \frac{\sum_{l=1}^{q(r)} x^{\deg(w_l)}}{(1-x^2)^{2r-1}}. \tag{4.49}$$

[We recall that for $(i,j) \in \mathcal{I}_r$ we have $\jmath(\Theta_{\underline{(ij)}}) = y^2_{ij}$, and thus each element of $\mathcal{E}_0$ has Y-degree 2; also, it has T-degree 2.]

From Eqs. (4.48) and (4.49) we get that

$$\sum_{l=1}^{q(r)} x^{\deg(w_l)} = \left[\sum_{j=1}^{r-1}(r-1)^{-1}\binom{r-1}{j}\binom{r-1}{j-1}x^{j-1}\right](1+x)^{2r-1}. \qquad (4.50)$$

By taking $x = 1$ in this equation we get that

$$q(r) = 2^{2r-1}(r-1)^{-1}\left(\sum_{j=1}^{r-1}\binom{r-1}{j}\binom{r-1}{j-1}\right) = 2^{2r}[2(r-1)]^{-1}\binom{2(r-1)}{r},$$

where we used that $\sum_{j=1}^{r-1}\binom{r-1}{j}\binom{r-1}{j-1} = \sum_{j=1}^{r-1}\binom{r-1}{j}\binom{r-1}{r-j}$ is the coefficient of x^r in the product $(1+x)^{r-1}(1+x)^{r-1} = (1+x)^{2(r-1)}$ and hence it is $\binom{2(r-1)}{r}$. Thus $q(r) = 2^{2r}(r-2)^{-1}\binom{2r-3}{r}$, i.e., Eq. (4.43) holds. So part (a) holds.

To check that parts (b) and (c) hold, we first remark that the leading term of the right-hand side of the Eq. (4.50) is $x^{2r-1}x^{r-2} = x^{3r-3}$. Therefore $\deg(w_{q(r)}) = 3r-3$ and $\deg(w_{q(r)-1}) \leq 3r-4$.

As $\mathrm{Perm}(\{0,\ldots,r\})$ acts naturally on $\mathbb{Y}_r$ in such a way that for every element $a \in \{0,\ldots,\lfloor\frac{r+1}{2}\rfloor\}$ it permutes transitively $\mathrm{Par}_{r,a}$ and the set $\{\mathbb{Y}_{r,\chi}^{\mathbf{SL}_2} | o_\chi \in \mathrm{Par}_{r,a}\}$, if $o_\chi \in \mathrm{Par}_{r,a}$, then the Hilbert series $\wp_{1,a}(2,r)(x)$ of $\mathbb{Y}_{r,\chi}^{\mathbf{SL}_2}$ depends, as the notation suggests, only on a. We conclude that:

$$\wp_1(2,r)(x) = \sum_{a=0}^{\lfloor\frac{r+1}{2}\rfloor}\binom{r+1}{2a}\wp_{1,a}(2,r)(x).$$

If $o_\chi \in \mathrm{Par}_{r,a}$, then each element of a homogeneous $\mathbb{K}[\mathcal{E}_0]$-basis of $\mathbb{Y}_{r,\chi}^{\mathbf{SL}_2}$ has Y-degree at least a and therefore, we have

$$x^a\wp_{1,0}(2,r)(x) \preceq \wp_{1,a}(2,r)(x), \qquad (4.51)$$

where $\wp_{1,0}(2,r)(x)$ is defined as follows. The Hilbert series of $\mathbb{H}_{2,\mathbb{K}}^r/(\mathcal{D}_0)$ when viewed as a $2\mathbb{Z}_{\geq 0}$-graded $\mathbb{K}$-algebra (as the $y_{i,j}^2$s have Y-degree 2 or equivalently as the Θs have T-degree 2) is $(1-x^2)^{r+1}\wp_0(2,r)(x^2)$. The Hilbert series $\wp_{1,0}(2,r)(x)$ is the Hilbert series of $\mathbb{H}_{2,\mathrm{tot},\mathbb{K}}^r/(\mathcal{D}_0)$ (viewed with the T-degree grading which is compatible with the one just mentioned for $\mathbb{H}_{2,\mathbb{K}}^r/(\mathcal{D}_0)$) and thus is the sum of $(1-x^2)^{r+1}\wp_0(2,r)(x^2)$ and of the Hilbert series $\wp_{1,0}^{\perp}(2,r)(x)$ of $\mathbb{H}_{2,\mathbb{K}}^{r,\perp}/\mathcal{D}_0H_{2,\mathbb{K}}^{r,\perp}$. Similarly to Inequality (4.51) we argue that

$$x^3(1-x^2)^{r+1}\wp_0(2,r)(x^2) \preceq \wp_{1,0}^{\perp}(2,r)(x). \qquad (4.52)$$

Combining Eq. (4.50) and Inequalities (4.51) and (4.52) with Corollary 4.2.14 we get that part (b) holds.

From part (b) and the two relations $\deg(w_{q(r)}) = 3r - 3$ and $\deg(w_{q(r)-1}) \leq 3r - 4$ we get the inequalities $2c_0(2, r, \varrho_0^-(2, r)) \leq 3r - 3$ and $2c_0(2, r, \varrho_0^-(2, r) - 1) \leq 3r - 4$; so part (c) holds.[11] □

Example 4.3.5 We take $r = 3$ and continue to assume that $p = 0$. As we have $\dim(\mathbb{H}^3_{2,\mathbb{K}}) = 9$ and $\mathbb{H}^3_{2,\mathbb{K}}$ is generated (see Proposition 4.3.3(b)) by the 10 polynomials $\Theta_{(m_0,m_1,m_2,m_3)} \in \mathbb{H}^3_2(1)_{\mathbb{K}}$ indexed by the elements (m_0, m_1, m_2, m_3) of $\Delta(2, 3)$, the $\mathbb{K}$-epimorphism $\mathfrak{q}_r$ induces an identification

$$\mathbb{H}^3_{2,\mathbb{K}} \simeq \mathbb{K}[\Theta^Y_{(m_0,m_1,m_2,m_3)} | (m_0, m_1, m_2, m_3) \in \Delta(2, 3)]/(\Pi_4(0, 1, 2, 3))$$

where $\Pi_4(0, 1, 2, 3)$ is a homogeneous polynomial in the 10 indeterminates of Θ^Y-degree equal to $\operatorname{rank}_{\mathbb{H}^{3,\Theta}_{2,\mathbb{K}}}(\mathbb{H}^3_{2,\mathbb{K}}) = 2 + \varrho_0(2, 3) = 4$, uniquely determined up to multiplication by elements of $\mathbb{K}^\times$. Thus, for $r = 3$ we get that

$$\operatorname{Ker}(\mathfrak{q}_3) = (\Pi_4(0, 1, 2, 3)).$$

The Hilbert series of a hypersurface of degree 4 in $\mathbb{P}^9_{\mathbb{K}}$ is

$$\lambda(2, 3)(x) = \frac{1 - x^4}{(1 - x)^{10}} = \frac{1 + x + x^2 + x^3}{(1 - x)^9}$$

and thus we have $c_0(2, 3, 1) = 2$ and $c_0(2, 3, 2) = 3$ (as suggested by Proposition 4.3.4(b)).

Corollary 4.3.6 *For all integers $r \geq 3$ and $m \in \{4, \ldots, r + 1\}$ the following two types of relations hold for $p = 0$.*

(a) (Ξ cyclic relations) *For all $(\{q_1, \ldots, q_m\}, s) \in \mathcal{P}_{r,m} \times \{3, \ldots, m - 1\}$ we have the following identity*

$$\Xi_{(q_1,q_2,\ldots,q_m)} + \Xi_{(q_1,q_{s+1},q_{s+2},\ldots,q_m)}\Xi_{(q_2,q_3,\ldots,q_s)} = (-1)^s\,\Xi_{(q_1,q_s,q_{s-1},\ldots,q_2,q_{s+1},q_{s+2},\ldots,q_m)}$$

between elements of $H^r_{2,tot,\mathbb{K}}$.

(b) (Ξ Plücker relations) *For $(i, j, n, s) \in \mathcal{J}_r$, let $\Pi^{(2)}_4(i, j, n, s)$ be*

$$(\Theta^Y_{\underline{(i,j)}})^2(\Theta^Y_{\underline{(n,s)}})^2 + (\Theta^Y_{\underline{(i,n)}})^2(\Theta^Y_{\underline{(j,s)}})^2 + (\Theta^Y_{\underline{(i,n)}})^2(\Theta^Y_{\underline{(j,s)}})^2$$

and let $\Pi^{(4)}_4(i, j, n, s)$ be

$$2\Theta^Y_{\underline{(i,j)}}\Theta^Y_{\underline{(i,n)}}\Theta^Y_{\underline{(n,s)}}\Theta^Y_{\underline{(j,s)}} + 2\Theta^Y_{\underline{(i,j)}}\Theta^Y_{\underline{(i,n)}}\Theta^Y_{\underline{(n,s)}}\Theta^Y_{\underline{(j,n)}} + 2\Theta^Y_{\underline{(i,n)}}\Theta^Y_{\underline{(i,s)}}\Theta^Y_{\underline{(j,s)}}\Theta^Y_{\underline{(j,n)}}.$$

---

[11] Due to the presence of the denominator $(1 - x^2)^{2r-1}$ in part (b), one cannot conclude that the inequality $2c_0(2, r, \varrho_0^-(2, r)) + 3 + \lfloor \frac{r+1}{2} \rfloor \leq 3r - 3$ holds, and in fact this inequality is false if $r \in \{3, 4\}$.

Then there exists a homogeneous polynomial

$$\Pi_4^{\det}(i, j, n, s) \in (\Theta_{\underline{i}}^Y, \Theta_{\underline{j}}^Y, \Theta_{\underline{n}}^Y, \Theta_{\underline{s}}^Y)$$

of the same partial degrees as $\Pi_4^{(2)}(i, j, n, s)$ *or* $\Pi_4^{(4)}(i, j, n, s)$ *and such that*

$$\Pi_4(i, j, n, s) := \Pi_4^{(2)}(i, j, n, s) + \Pi_4^{(4)}(i, j, n, s) + \Pi_4^{\det}(i, j, n, s) \in Ker(\mathfrak{q}_r).$$

Proof Part (a) follows from Lemma 4.3.2(a).

Part (b) is a direct consequence of the Plücker relations. More precisely, starting from the Plücker relation $y_{i,j}y_{n,s} - y_{i,n}y_{j,s} + y_{i,s}y_{j,n} = 0$ we first get that

$$y_{i,n}^2 y_{j,s}^2 = y_{i,j}^2 y_{n,s}^2 + 2y_{i,j}y_{n,s}y_{i,s}y_{j,n} + y_{i,s}^2 y_{j,n}^2,$$

and second that

$$4y_{i,j}^2 y_{n,s}^2 y_{i,s}^2 y_{j,n}^2 = (y_{i,n}^2 y_{j,s}^2 - y_{i,j}^2 y_{n,s}^2 - y_{i,s}^2 y_{j,n}^2)^2,$$

and thus finally that

$$y_{i,j}^4 y_{n,s}^4 + y_{i,n}^4 y_{j,s} + y_{i,n}^4 y_{j,s} - 2y_{i,j}^2 y_{i,n}^2 y_{n,s}^2 y_{j,s}^2 - 2y_{i,j}^2 y_{i,n}^2 y_{n,s}^2 y_{j,n} - 2y_{i,n}^2 y_{i,s}^2 y_{j,s}^2 y_{q_1,n} = 0.$$

To prove part (b) we can assume that $r = 3$; so $(i, j, n, s) = (0, 1, 2, 3)$. For $r = 3$, part (b) follows once we recall that $y_{i',j'}^2 = \xi_{(i',j')}$ which implies that the sum $\Pi_4^{(2)}(i, j, n, s) + \Pi_4^{(4)}(i, j, n, s)$ modulo the ideal $(\Theta_{\underline{0}}^Y, \Theta_{\underline{1}}^Y, \Theta_{\underline{2}}^Y, \Theta_{\underline{3}}^Y) + \mathrm{Ker}(\mathfrak{q}_3)$ is 0 (cf. also Eqs. (4.38) and (4.39)). □

Example 4.3.7 We take $r \geq 3$. Let the $\Pi_4(i, j, n, s)$s with $(i, j, n, s) \in \mathcal{J}_r$ be as in Corollary 4.3.6(b): their T-degrees are 8, their partial T-degrees $(m_0, \dots, m_r)$ are such that $m_l \in \{0, 2\}$, and m_l is 0 if and only if $l \in \{0, \dots, r\}\backslash\{i, j, n, s\}$, and their Θ^Y-degrees are 4. As $\mathrm{Ker}(\mathfrak{q}_3) = (\Pi_4(0, 1, 2, 3))$, it follows that $\mathrm{Ker}(\mathfrak{q}_r)$ contains no nonzero homogeneous polynomials of T-degree 4 (so of Θ^Y-degree 2) and all homogeneous polynomials of T-degree 6 or 8 (so of Θ^Y-degrees 3 or 4) must either involve 5 indeterminates among the $T^{(i)}$s with $i \in \{0, \dots, r\}$ or must belong to the $\mathbb{K}$-vector space $\mathrm{Span}(\{\Pi_4(i, j, n, s) | (i, j, n, s) \in \mathcal{J}_r\})$. This implies that for all $r \geq 3$ we have

$$D(2, r, 2) = \binom{1 + \frac{(r+1)(r+2)}{2}}{2},$$

$$D(2, r, 3) \leq \binom{2 + \frac{(r+1)(r+2)}{2}}{3},$$

$$D(2, r, 4) \leq \binom{3 + \frac{(r+1)(r+2)}{2}}{4} - \binom{r+1}{4}.$$

In particular, as $\varrho_0(2,4) = 16$ and $c_0(2,4,16) \le 4$ (see Proposition 4.3.4(c)) and as $D(2,4,1) = 15$, $D(2,4,2) = 120$, $D(2,4,3) = 680 - \alpha$, and moreover $D(2,4,4) = 3,055 - \beta$, with $\alpha, \beta \in \mathbb{N} \cup \{0\}$, we easily get that

$$\wp_0(2,4)(x) = \frac{1 + 3x + 6x^2 + (10-\alpha)x^3 + \alpha x^4}{(1-x)^{12}},$$

and thus $c_0(2,4,1) = \cdots = c_0(2,4,6) = 2$, $c(2,4,7) = \cdots = c_0(2,4,16-\alpha) = 3$, $c_0(2,4,17-\alpha) = \cdots = c_0(2,4,16) = 4$, $\alpha \in \{0,1,\ldots,10\}$ and $\beta = 10 + 11\alpha$.

Representation theory gives an explicit way to compute Hilbert series. For instance, [DK], Cor. 4.6.9 and Rmk. 4.6.10 allow us to compute $\wp_0(2,r)(x^2)$ as the even part of the coefficient of z^0 of the fraction

$$\frac{1 - z^2}{(1 - xz^{-2})^{r+1}(1-x)^{r+1}(1-xz^2)^{r+1}}$$

and thus of the fraction

$$\frac{\sum_{i=0}^{\infty}[x^{2i}\binom{r+i}{i}^2 - x^{2i+1}\binom{r+i}{i}\binom{r+i+1}{i+1}]}{(1-x)^{r+1}}.$$

A simple calculation shows that for $r = 4$ this fraction is a Hilbert series which modulo $x^9\mathbb{Z}_{\ge 0}[[x]]$ is equal to

$$1 + 15x^2 + 10x^3 + 12 - x^4 + 126x^5 + 680x^6 + 855x^7 + 3,045x^8.$$

In other words, we have $\alpha = 0$, $\beta = 10$, and

$$\wp_0(2,4)(x) = \frac{1 + 3x + 6x^2 + 10x^3}{(1-x)^{12}}.$$

As $\beta = 10$, for each $(q_0, q_1; q_2, q_3, q_4)$ with $\{q_0, \ldots, q_4\} = \{0,1,2,3,4\}$, $q_0 < q_1$, $q_2 < q_3 < q_4$, there exists a nonzero homogeneous polynomial

$$\Pi_4(q_0, q_1; q_2, q_3, q_4) \in \mathrm{Ker}(\mathfrak{q}_5)$$

of partial T-degrees $(m_0, m_1, m_2, m_3, m_4)$ such that we have $m_{q_0} = m_{q_1} = 1$ and $m_{q_2} = m_{q_3} = m_{q_4} = 2$.

By combining Corollary 4.2.14 with Proposition 4.3.4(a) and (c), the equality $\varrho_0^-(2,2) = 0$ (see Eq. (4.35)), and Examples 4.3.5 and 4.3.7, we get directly the following consequences.

Corollary 4.3.8 *The following properties hold.*

(a) *For all integers $s \ge 0$ we have $D(2,1,s) = \binom{s+2}{2}$ and*

$$D(2,2,s) = \binom{s+5}{5} = \frac{(s+1)(s+2)(s+3)(s+4)(s+5)}{120}.$$

(b) *For all integers $s \geq 0$ we have $D(2,3,s) = \sum_{i=5}^{8} \binom{s+i}{8}$ and thus*

$$D(2,3,s) = \frac{(s+1)(s+2)(s+3)(s+4)(s+5)(s^3+9s^2+46s+84)}{2(7!)}.$$

(c) *For all integers $s \geq 0$ we have*

$$D(2,4,s) = \binom{s+11}{11} + 3\binom{s+10}{11} + 6\binom{s+9}{11} + 10\binom{s+8}{11}$$

$$= \frac{[\prod_{i=1}^{8}(s+i)](20s^3+103s^2+535s+990)}{11!}.$$

(d) *Let $r \geq 5$ be an integer. Then for all integers $s \geq 0$, $D(2,r,s)$ is*

$$\binom{3r+s-1}{3r-1} + \binom{r-1}{2}\binom{3r+s-2}{3r-1} + \sum_{l=1}^{\varrho^-(2,r)} \binom{3r+s-1-c_l(2,r)}{3r-1}$$

and the rational number

$$\frac{D(2,r,s) - \binom{3r+s-1}{3r-1} - \binom{r-1}{2}\binom{3r+s-2}{3r-1}}{\frac{2^{r-1}}{(r-2)}\binom{2r-3}{r} - 1 - \binom{r-1}{2}}$$

belongs to the interval of the real line

$$\left[\binom{\lfloor\frac{3r}{2}\rfloor+s+1}{3r-1}, \binom{3r+s-3}{3r-1}\right].$$

Thus the unique polynomial $D_{2,r}(x) \in \mathbb{Q}[x]$ such that for all integers $s \geq 0$ we have $D(2,r,s) = D_{2,r}(s)$, is divisible by $\prod_{i=1}^{3r-1-\lfloor\frac{3(r-1)}{2}\rfloor}(x+i)$, has leading coefficient $\frac{2^{r-1}}{(r-2)(3r-1)!}\binom{2r-3}{r}$, and satisfies $D_{2,r}(0) = 1$ and $D_{2,r}(1) = \binom{r+1}{2}$.

4.4 A General Result on σ-Finite Generation

For the sake of future references we generalize Theorem 4.2.17 and Corollary 4.2.19 as follows. Let $d \in \mathbb{N}$ and let

$$\underline{x} := \underline{x}^{(0)} := (x_1 \cdots x_d)^t,\ \underline{x}' := \underline{x}^{(1)} := (x_1' \cdots x_d')^t, \ldots, \underline{x}^{(r)} := (x_1^{(r)} \cdots x_d^{(r)})^t, \ldots$$

be a sequence of column vectors with entries indeterminates of degree 1. We consider the polynomial $\mathbb{Z}_{\geq 0}$-graded $\mathbb{K}$-algebras

$$\mathbb{X} := \cup_{r\geq 0}\mathbb{X}_r, \quad \mathbb{X}_r := \mathbb{K}[x_i^{(l)} | 1 \leq i \leq d, 0 \leq l \leq r].$$

We consider the shift $\mathbb{K}$-algebra monomorphism

$$\sigma : \mathbb{X} \to \mathbb{X}$$

that sends $x_i^{(l)}$ to $x_i^{(l+1)}$ for all $(i, l) \in \{1, \ldots, d\} \times \mathbb{Z}_{\geq 0}$.

If G is a reductive group over Spec($\mathbb{K}$) equipped with a faithful representation $\rho : G \to \mathbf{GL}_{d,\mathbb{K}}$, we let G act on $\mathbb{X}$ via the matrix rules $(h, \underline{x}^{(l)}) \mapsto \rho(h)\underline{x}^{(l)}$ for all $(h, l) \in G(\mathbb{K}) \times \mathbb{Z}_{\geq 0}$; it restricts to an action of G on $\mathbb{X}_r$ for each $r \in \mathbb{Z}_{\geq 0}$. Clearly, if $\vartheta \in \mathbb{X}^G$, then $\sigma(\vartheta) \in \mathbb{X}^G$.

For $j, t \in \mathbb{Z}_{\geq 0}$, we consider the G-invariant $\mathbb{K}$-automorphism

$$\iota_{j;j+t} : \mathbb{X} \to \mathbb{X}$$

that for $i \in \{1, \ldots, d\}$ fixes each $x_i^{(l)}$ with $l \in \mathbb{Z}_{\geq 0}\backslash\{j, j+t\}$ and interchanges $x_i^{(j)} \longleftrightarrow x_i^{(j+t)}$; each $\iota_{j;j}$ is the identity automorphism of $\mathbb{X}$.

Theorem 4.4.1 *We assume there exists $m \in \mathbb{N}$ such that we have a G-separating polynomial on V^m (i.e., the stable locus $(V^m)^s$ is nonempty).*[12] *Let $\mathcal{T}_0$ be an arbitrary transcendence basis of Frac($\mathbb{X}_{m-1}^G$) over $\mathbb{K}$. Let $\mathcal{T} = \{\vartheta_1, \ldots, \vartheta_d\} \subset \mathbb{X}_m^G$ be an arbitrary transcendence basis of Frac($\mathbb{X}_m^G$) over Frac($\mathbb{X}_{m-1}^G$). Then the following properties hold.*

(a) *For each $l \in \mathbb{Z}_{\geq 0}$ the union $\mathcal{T}_0 \cup \left(\cup_{j=0}^{l} \iota_{m;m+j}(\mathcal{T})\right)$ is a transcendence basis of Frac($\mathbb{X}_{m+l}^G$) over $\mathbb{K}$.*

(b) *The set $\mathcal{T}$ is σ-algebraically independent over $\mathbb{K}$, i.e., the set*

$$\{\sigma^j(\vartheta_i) | (i, j) \in \{1, \ldots, d\} \times \mathbb{Z}_{\geq 0}\} = \cup_{j\geq 0}\sigma^j(\mathcal{T})$$

of polynomials of $\mathbb{X}^G$ is algebraically independent over $\mathbb{K}$.

(c) *The field Frac($\mathbb{X}^G$) is σ-generated by a finite subset of Frac($\mathbb{X}_{m+1}^G$).*

(d) *We assume that for each integer $l \geq 0$ we have a direct sum decomposition*

$$\mathbb{X}_{m+l-1}^G = \mathbb{X}_{m+l-1,even}^G \oplus \mathbb{X}_{m+l-1,odd}^G$$

that corresponds to homogeneous degrees in $2a\mathbb{Z}_{\geq 0}$ and $a(2\mathbb{N} - 1)$ (respectively), where $a \in \mathbb{N}$ is fixed. We also assume that for $l >> 1$, $\mathbb{X}_{m+l-1,odd}^G \neq 0$. Then $\mathbb{X}_{m+1,odd}^G \neq 0$ and the field

[12] If the identity component of G is semisimple, then $m \in \{1, \ldots, d\}$ exists by Lemma 4.1.10.

$$Frac(\mathbb{X}^G_{even}) := \cup_{l\geq 0} Frac(\mathbb{X}^G_{m+l-1,even})$$

is σ-generated by a finite subset of $Frac(\mathbb{X}^G_{m+2,even})$.

Proof As the representation $\rho : G \to \mathbf{GL}_{d,\mathbb{K}}$ is faithful, for a generic $m+1$-tuple $(v_0, \ldots, v_m) \in V^{m+1}$ with $(v_0, \ldots, v_{m-1}) \in (V^m)^s$, the stabilizer $\mathrm{Stab}_G((v_0, \ldots, v_m))$ is trivial. Based on this, the proof of the theorem is the same as the proofs of Theorem 4.2.17 and Corollary 4.2.19. We include here only details of the arguments of the last two results which used concrete notation specific to quadratic forms.

As in the proof of Lemma 4.2.16(b) and (c) one checks that for all integers $j > t \geq 0$ we have the following property:

(i) the natural G-morphism

$$\rho_{m+j,m+t} : \mathbf{V}^{m+j} \to \mathbf{V}^{m+t} \times_{(\mathbf{V}^{m+t}/G)} (\mathbf{V}^{m+j}/G)$$

is generically finite. For $t > 0$, $\rho_{m+j,m+t}$ is generically finite of degree 1, the field extension $\mathrm{Frac}(\mathbb{X}^G_{m+t-1}) \to \mathrm{Frac}(\mathbb{X}^G_{m+j-1})$ is purely transcendental of transcendence degree $d(j-t)$, and a subset with $d(j-t)$ elements of $\mathrm{Frac}(\mathbb{X}^G_{m+j-1})$ is a generating transcendence basis (resp. a transcendence basis) of $\mathrm{Frac}(\mathbb{X}^G_{m+j-1})$ over $\mathrm{Frac}(\mathbb{X}^G_{m+t-1})$ if and only if it is a generating transcendence basis (resp. transcendence basis) of $\mathrm{Frac}(\mathbb{X}_{m+j-1})$ over $\mathrm{Frac}(\mathbb{X}_{m+t-1})$.

As in the proof of Theorem 4.2.17 we check that the following properties hold:

(ii) for all integers $l \geq 0$, the set $\iota_{m;m+l}(\mathcal{T})$ is a transcendence basis of

$$\mathrm{Frac}((\mathbb{K}_{m-1}[x_1^{(m+l)}, \ldots, x_d^{(m+l)}])^G)$$

over $\mathrm{Frac}(\mathbb{X}^G_{m-1})$;

(iii) the union $\mathcal{T} \cup (\bigcup_{t\geq 1} \iota_{m;m+t}(\mathcal{T}))$ is an algebraically independent subset of $\mathrm{Frac}(\mathbb{X})$ over $\mathrm{Frac}(\mathbb{X}_{m-1})$; so it is a transcendence basis of $\mathrm{Frac}(\mathbb{X}^G)$ over $\mathrm{Frac}(\mathbb{X}^G_{m-1})$.

For a pair $(t, r) \in \mathbb{N} \times \mathbb{Z}_{\geq 0}$ let $\iota_{0,1,\ldots,t-1;t+r,t+r+1,\ldots,2t-1+r}$ be the G-invariant $\mathbb{K}$-automorphism of $\mathbb{X}$ that for each $i \in \{1, \ldots, d\}$ fixes each $x_i^{(l)}$ with l in the difference set $\mathbb{Z}_{\geq 0} \setminus \{0, 1, \ldots, t-1, t+r, t+r+1, \ldots, 2t-1+r\}$ and interchanges $x_i^{(l)} \longleftrightarrow x_i^{(l+t+r)}$ for all $l \in \{0, \ldots, t-1\}$.

From the property (iii) we first get that part (a) holds and second that (as in the proof of Corollary 4.2.19):

(iv) for each integer $t \geq m$, $\sigma^t(\mathcal{T}) = \iota_{0,1,\ldots,m-1;t,t+1,\ldots,m+t-1}(\iota_{m;m+t}(\mathcal{T}))$ is a transcendence basis of $\mathrm{Frac}(\mathbb{X}^G_{m+t})$ over $\mathrm{Frac}(\mathbb{X}^G_{m+t-1})$;

(v) if $t \in \{1, \ldots, m-1\}$, then $\sigma^t(\mathcal{T}) = \iota_{\pi_{m+t}}(\iota_{m;m+t}(\mathcal{T}))$ is a transcendence basis of $\mathrm{Frac}(\mathbb{X}^G_{m+t})$ over $\mathrm{Frac}(\mathbb{X}^G_{m+t-1})$, where π_{m+t} is the shift by t permutation of

$\{0, \ldots, m+t-1\}$ with the property that $\pi_{m+t}(l)$ is congruent to $l+t$ modulo $m+t$ and where $\iota_{\pi_{m+t}}$ is the G-invariant $\mathbb{K}$-automorphism of $\mathbb{X}$ which for each $i \in \{1, \ldots, d\}$ maps $x_i^{(l)}$ to $x_i^{(\pi_{m+t}(l))}$ for all $l \in \{0, \ldots, m+t-1\}$ and fixes $x_i^{(l)}$ for all integers $l \geq m+t$.

Part (b) follows from the property (iv).

From the second part of the property (i) we get that the proofs of Corollary 4.2.19(b)–(d) apply entirely to give us that parts (c) and (d) hold as well. □

Remark 4.4.2 The fact that the fields in parts (c) and (d) of the theorem above are σ-finitely generated follows, again, from the general fact of difference algebra in [Le], Ch. 4, Thm. 4.4.1. The novel part of parts (c) and (d) is, again, the bounds $m+1$ and $m+2$ (respectively) for the 'order' of the σ-generating set.

Appendix: Abelian Schemes and Hecke Correspondences

In this appendix we review basic definitions pertaining to moduli schemes of principally polarized abelian schemes over $\mathbb{Z}_{(p)}$-schemes and their Hecke correspondences that are being used throughout our monograph. The standard references for this material are [MFK], Ch. 7, [FC], Ch. 1, and [D].

Throughout this appendix we let $g \in \mathbb{N}$, p a prime, $N \geq 3$ an integer prime to p, and R a $\mathbb{Z}_{(p)}$-algebra. Let $L := \mathbb{Z}^{2g}$. Let $\psi : L \otimes_{\mathbb{Z}} L \to \mathbb{Z}$ be a fixed perfect alternating form on L. We denote by ψ_N the reduction of ψ modulo N. For a scheme X over R let $\boldsymbol{\mu}_{N,X}$ be the finite flat étale group scheme over X that represents N-th roots of unity. Let $(L/NL)_X$ and $(\mathbb{Z}/N\mathbb{Z})_X$ be the finite flat étale commutative group scheme over X defined by the finite abelian groups L/NL and $\mathbb{Z}/N\mathbb{Z}$.

For *étale* and *smooth* morphisms of schemes we refer to [BLR], Ch. 2, Sect. 2.2. In particular, a morphism $X \to Y$ of schemes is smooth (resp. étale) iff it is locally of finite presentation and for each Y-scheme Z which is affine and for every closed subscheme Z_0 of Z defined by a sheaf of ideals of $\mathcal{O}_Z$ of square 0, the canonical map $\mathrm{Hom}_Y(Z, X) \to \mathrm{Hom}_Y(Z_0, X)$ is surjective (resp. bijective) (see [BLR], Ch. 2, Sect. 2.2, Prop. 6).

Abelian schemes. An *abelian scheme* over X is a group scheme $A \to X$ which is smooth, proper, with connected geometric fibers (e.g., see [FC], Ch. I, Def. 1.1). So the relative dimension of A over X is well-defined provided X is connected. If $X = \mathrm{Spec}(S)$ we simply say that A is an abelian scheme over S.

The dual of an abelian scheme A (e.g., see [FC], Ch. I, p. 3) is denoted by $\check{A}$.

A principal polarization of A is an isomorphism $\theta : A \to \check{A}$ which locally in the étale topology is defined by some relatively ample invertible sheaf (in the sense of [FC], Ch. I, pp. 3 to 6).

An abelian variety A over an algebraically closed field κ of characteristic p is called *ordinary* if its p-torsion group of κ-valued points $A[p](\kappa)$ is isomorphic to $(\mathbb{Z}/p\mathbb{Z})^{\dim(A)}$. As an extrapolation, an abelian scheme A over X is called *ordinary* if all its geometric fibers over points of $X \times_{\mathrm{Spec}(\mathbb{Z})} \mathrm{Spec}(\mathbb{F}_p)$ are ordinary abelian varieties over fields.

A. Buium and A. Vasiu, *Delta-Invariant Theory for Hecke Correspondences*, Lecture Notes in Mathematics 2396, https://doi.org/10.1007/978-3-032-20571-1

An *isogeny* $u : A_1 \to A_2$ of abelian schemes over X is a finite faithfully flat homomorphism. If X is connected, then A_1 and A_2 have the same relative dimension over X. Assume now that A_1 and A_2 have relative dimension g over X. Let $d \in \mathbb{N}$. We say that u is a *polarized isogeny* with respect to principal polarizations θ_1 and θ_2 on A_1 and A_2, respectively (a polarized isogeny between the polarized abelian schemes (A_1, θ_1) and (A_2, θ_2)), of degree d^g, if the composite endomorphism $A_2 \to \check{A}_2 \to \check{A}_1 \to A_1 \to A_2$ is the multiplication by d, where $\check{u} : \check{A}_2 \to \check{A}_1$ is the isogeny dual to u and for $i \in \{1, 2\}$, the homomorphisms $A_i \to \check{A}_i$ and $\check{A}_i \to A_i$ are θ_i and its inverse; equivalently, if the pull back of θ_2 by u is $d\theta_1$. E.g., see [FC], Ch. VII, Def. 3.1, for the case when d is a power of p.

A *level-N structure* on A is an isomorphism $\mathfrak{l} : (L/NL)_X \to A[N]$ between finite flat group schemes over X (see [MFK], Ch. 7, Def. 7.1). If $\theta : A \to \check{A}$ is a principal polarization of A, then θ restricts to an isomorphism $A[N] \to \check{A}[N]$. Using it and the Weil pairing $A[N] \times_X \check{A}[N] \to \boldsymbol{\mu}_{N,X}$, we get a perfect pairing $\psi_{A,\theta,N} : A[N] \otimes_X A[N] \to \boldsymbol{\mu}_{N,X}$ which is known to be alternating.

One says that $\mathfrak{l}$ is a *symplectic similitude level-N structure*[1] on (A, θ) if locally in the étale topology of X there exists an isomorphism $\zeta_{N,X} : \boldsymbol{\mu}_{N,X} \to (\mathbb{Z}/N\mathbb{Z})_X$ of finite flat group schemes over X such that $\mathfrak{l}$ induces an isomorphism

$$(L/NL, \psi_N)_X \to (A[N], \psi_{A,\theta,N})$$

of symplectic spaces over X, where $\psi_{A,\theta,N} : A[N] \otimes_Y A[N] \to (\mathbb{Z}/N\mathbb{Z})_X$ is the composite of $\psi_{A,\theta,N}$ with $\zeta_{N,X}$.

Moduli schemes of abelian schemes. We denote by $\mathcal{A}_{g,1,N,R,\text{classic}}$ Mumford's moduli scheme of principally polarized abelian schemes of relative dimension g endowed with a level-N structure over R-schemes (see [MFK], Ch. 7, Thm. 7.9).[2] It is a quasi-projective smooth scheme over R.

We denote by $\mathcal{A}_{g,1,N,R}$ the moduli scheme of principally polarized abelian schemes of relative dimension g endowed with a symplectic similitude level-N structure over R-schemes. It is an open closed subscheme of $\mathcal{A}_{g,1,N,R,\text{classic}}$.

Hecke correspondences. For $d \in \mathbb{N}$ relatively prime to p, a *prime to p Hecke correspondence* on $\mathcal{A}_{g,1,N,R}$ defined by polarized isogenies of degree d^g is a smooth scheme $\mathcal{A}_{g,1,N,R,d}$ over R that represents the functor that attaches to every R-algebra S the set of isomorphism classes of polarized isogenies $(A_1, \theta_1) \to (A_2, \theta_2)$ of degree d^g between principally polarized abelian schemes (A_1, θ_1) and (A_2, θ_2) over S of relative dimension g endowed with symplectic similitude level-N structures $\mathfrak{l}_1$ and $\mathfrak{l}_2$ such that u is compatible with $\mathfrak{l}_1$ and $\mathfrak{l}_2$.

---

[1] Often, for the sake of simplicity, it is called in literature shortly as level-N structure. This is not done here as [MFK] is quoted extensively in this monograph and one would like to avoid any terminology conflict with [MFK], Ch. 7, Def. 7.1.

[2] It is Serre's lemma of [Mum], Ch. IV, Sect. 21, Thm. 5 which allows to apply [MFK], Ch. 7, Thm. 7.9 with $N \geq 3$ and not just with $N > 6^g\sqrt{g!}$.

To explain the latter compatibility assume first that S is an algebraically closed field and there exists a prime ℓ different from p such that both N and d are powers of ℓ. If $T_\ell(A_i)$ is the ℓ-adic Tate module of A_i and if isomorphisms

$$\mu_{\ell^m,\mathrm{Spec}(S)} \to (\mathbb{Z}/\ell^m\mathbb{Z})_{\mathrm{Spec}(S)}$$

are chosen for each $m \in \mathbb{N}$ in a compatible way, then, similar to $\psi_{A,\theta,N}$ above, we obtain a perfect alternating form $\psi_i : T_\ell(A_i) \otimes_{\mathbb{Z}_\ell} T_\ell(A_i) \to \mathbb{Z}_\ell$.[3] The $\mathfrak{l}_i$ is defined by an orbit $o(h_i)$ of the natural right action of the group

$$\mathfrak{L}_{\ell,N} := \{M \in \mathrm{GSp}(L,\psi)(\mathbb{Z}_\ell) | M \text{ modulo } N \text{ is the identity matrix}\}$$

on the set of symplectic similitude isomorphisms $(L,\psi) \otimes_{\mathbb{Z}} \mathbb{Z}_\ell \to (T_\ell(A_i), \psi_i)$; so $h_i : (L,\psi) \otimes_{\mathbb{Z}} \mathbb{Z}_\ell \to (T_\ell(A_i), \psi_i)$ is a symplectic similitude isomorphism. The compatibility of u with $\mathfrak{l}_1$ and $\mathfrak{l}_2$ means that the injective $\mathbb{Z}_\ell$-linear map

$$T_\ell(u) : T_\ell(A_1) \to T_\ell(A_2)$$

defined by u induces a symplectic similitude isomorphism

$$T_\ell(u) \otimes 1_{\mathbb{Q}_\ell} : (T_\ell(A_1), \psi_1) \otimes_{\mathbb{Z}_\ell} \mathbb{Q}_\ell \to (T_\ell(A_2), \psi_2) \otimes_{\mathbb{Z}_\ell} \mathbb{Q}_\ell.$$

We emphasize that, if $d > 1$, then the compatibility of u with $\mathfrak{l}_1$ and $\mathfrak{l}_2$ is not expressed by a commutative diagram between finite flat group schemes over $\mathrm{Spec}(S)$.[4] More precisely, it is expressed by a natural compatibility between h_1, h_2 and $T_\ell(u) \otimes 1_{\mathbb{Q}_\ell}$: there exists $h \in \mathrm{GSp}(L,\psi)(\mathbb{Q}_\ell)$ such that the composite $\mathbb{Q}_\ell$-linear isomorphism

$$(T_\ell(u) \otimes 1_{\mathbb{Q}_\ell}) \circ (h_1 \otimes 1_{\mathbb{Q}_\ell}) \circ h$$

maps $L \otimes_{\mathbb{Z}} \mathbb{Z}_\ell$ onto $T_\ell(A_2)$ and defines h_2.

The case when S is an algebraically closed field and N, d are arbitrary requires the usage of the ring of finite adéles with the p-component omitted (or at least of the product of the $\mathbb{Q}_\ell$s indexed by prime divisors ℓ of N) instead of only the $\mathbb{Q}_\ell$ in the previous paragraph. In characteristic 0, this is detailed in [D], Sects. 4.12 and 4.16.

For a general S, the compatibility of a polarized isogeny with symplectic similitude level-N structures means compatibility over all geometric points of $\mathrm{Spec}(S)$.

Taking the disjoint union of all prime to p Hecke correspondences

$$\mathcal{A}'_{g,1,N,R} := \cup_{d \in \mathbb{N} \setminus p\mathbb{N}} \mathcal{A}_{g,1,N,R,d}$$

we obtain a correspondence

---

[3] The language of 'symplectic similitudes' makes the mentioning of such compatible isomorphisms unnecessary and hence they are not mentioned again in this monograph.

[4] In general, this is possible if N and d are relatively prime.

$$\mathcal{A}'_{g,1,N,R} \rightrightarrows \mathcal{A}_{g,1,N,R}$$

in the category of schemes over $\mathrm{Spec}(R)$. The categorical quotient of this correspondence (exists and) is $\mathrm{Spec}(R)$.

Moduli stacks of abelian schemes. We denote by $\mathcal{A}_{g,1,R}$ the stack of principally polarized abelian schemes of relative dimension g over R-schemes (for instance, see [FC], Ch. I, Def. 4.3, p. 19). This is an algebraic stack over $\mathrm{Spec}(R)$ in the sense of [FC], Ch. I, Def. 4.6, p. 20 (see [FC], Ch. I, Sect. 4.11).[5] In particular it has an étale cover that is a smooth scheme over $\mathrm{Spec}(R)$ and hence it has a well-defined relative dimension which equals to $\frac{g(g+1)}{2}$. The stack $\mathcal{A}_{g,1,R}$ induces a functor we still denote by $\mathcal{A}_{g,1,R}$ from the category **Ring** of R-algebras to the category **Set** of sets: for each R-algebra S, $\mathcal{A}_{g,1,R}(S)$ is the set of isomorphism classes of principally polarized abelian schemes over S of relative dimension g. If in the above discussion one takes ordinary abelian schemes instead of abelian schemes, then one obtains a functor $\mathcal{A}_{g,1,\mathrm{ord},R}$ from **Ring** to **Set**.

Similar to the correspondence $\mathcal{A}'_{g,1,N,R} \to \mathcal{A}_{g,1,N,R}$, we have a correspondence $\mathcal{A}'_{g,1,R} \to \mathcal{A}_{g,1,R}$ of stacks, where $\mathcal{A}'_{g,1,R}$ is the union of prime to p Hecke correspondences on $\mathcal{A}_{g,1,R}$. So, for each R-algebra S, $\mathcal{A}'_{g,1,R}(S)$ is the set of isomorphism classes of quintuples $(A_1, \theta_1, A_2, \theta_2, u)$, where (A_1, θ_1) and $A(A_2, \theta_2)$ are principally polarized abelian schemes of relative dimension g over $\mathrm{Spec}(S)$ and $u : A_1 \to A_2$ is a polarized isogeny with respect to principal polarizations θ_1 and θ_2 of degree prime to p. Restricting the correspondence $\mathcal{A}'_{g,1,R} \to \mathcal{A}_{g,1,R}$ to $\mathcal{A}_{g,1,\mathrm{ord},R}$, we obtain its ordinary version, i.e., the correspondence $\mathcal{A}'_{g,1,R,\mathrm{ord}} \to \mathcal{A}_{g,1,R,\mathrm{ord}}$ of stacks.

Some of the definitions in this Appendix are repeated in the body of the monograph for convenience. Moreover, various properties of the objects defined here are being reviewed in Chaps. 1–3 as needed.

[5] In the more modern terminology, it is a Deligne–Mumford stack.

References

[A] A.A. Albert, Symmetric and alternate matrices in an arbitrary field. I, Trans. Am. Math. Soc. **43**(3), 386–436 (1938)

[Bal] G. Baldi, On a conjecture of Buium and Poonen. Ann. Inst. Fourier (Grenoble) **70**(2), 457–477 (2020)

[Bar] M. Barcău, Isogeny covariant differential modular forms and the space of elliptic curves up to isogeny. Compos. Math. **137**(3), 237–273 (2003)

[BB] M. Barcău, A. Buium, Siegel differential modular forms. Int. Math. Res. Not. **28**, 1457–1503 (2002)

[BS] B. Bhatt, P. Scholze, Prisms and prismatic cohomology. Ann. Math. (2) **196**(3), 1135–1275 (2022)

[Bore] A. Borel, Linear Algebraic Groups. Graduate Texts in Mathematics, vol. 26, 2nd edn. (Springer, New York, 1991)

[Borg] J. Borger, Λ-rings and the field with one element, Manuscript dated June 17, 2009, https://arxiv.org/abs/0906.3146

[BG] J. Borger, L. Gurney, Canonical lifts and δ-structures. Selecta Math. (N.S.) **26**(5), Paper No. 67, 48 pp (2020)

[Bos] A.J. Bosch, Note on the factorization of a square matrix into two Hermitian or symmetric matrices. SIAM Rev. **29**(3), 463–468 (1987)

[BLR] S. Bosch, W. Lütkebohmert, M. Raynaud, Néron models. Ergeb. Math. Grenzgeb. (3), **21** [Results in Mathematics and Related Areas (3)] (Springer, Berlin, 1990)

[Br] L. Braun, Hilbert series of the Grassmannian and k-Narayana numbers. Commun. Math. **27**(1), 27–41 (2019)

[Bu95a] A. Buium, Differential characters of abelian varieties over p-adic fields. Invent. Math. **122**(2), 309–340 (1995)

[Bu95b] A. Buium, Geometry of differential polynomial functions, III: Moduli spaces. Am. J. Math. **117**(1), 1–74 (1995)

[Bu97] A. Buium, Differential characters and characteristic polynomial of Frobenius. J. Reine Angew. Math. **485**, 209–219 (1997)

[Bu00] A. Buium, Differential modular forms. J. Reine Angew. Math. **520**, 95–167 (2000)

[Bu03] A. Buium, Differential modular forms on Shimura curves, I. Compos. Math. **139**, 197–237 (2003)

[Bu04] A. Buium, Differential modular forms on Shimura curves, II: Serre operators. Compos. Math. **140**, 1113–1134 (2004)

A. Buium and A. Vasiu, *Delta-Invariant Theory for Hecke Correspondences*, Lecture Notes in Mathematics 2396, https://doi.org/10.1007/978-3-032-20571-1

[Bu05] A. Buium, Arithmetic Differential Equations, Mathematical Surveys and Monographs, vol. 118 (American Mathematical Society, Providence, RI, 2005)

[Bu17] A. Buium, Foundations of Arithmetic Differential Geometry, Mathematical Surveys and Monographs, vol. 222 (American Mathematical Society, Providence, RI, 2017)

[BM] A. Buium, Yu.I. Manin, Arithmetic differential equations of Painlevé VI type, Arithmetic and Geometry, 114–138, London Mathematical Society Lecture Note Series, vol. 420 (Cambridge University Press, Cambridge, 2015)

[BPI] A. Buium, B. Poonen, Independence of points on elliptic curves arising from special points on modular and Shimura curves, I: global results. Duke Math. J. **147**(1), 181–191 (2009)

[BPII] A. Buium, B. Poonen, Independence of points on elliptic curves arising from special points on modular and Shimura curves. II. Local results. Compos. Math. **145**(3), 566–602 (2009)

[Cha] C.-L. Chai, Every ordinary symplectic isogeny class in positive characteristic is dense in the moduli. Invent. Math. **121**(3), 439–479 (1995)

[CCO] C.-L. Chai, B. Conrad, F. Oort, Complex Multiplication and Lifting Problems, Mathematical Surveys and Monographs, vol. 195 (American Mathematical Society, Providence, RI, 2014)

[Cho] T. Chow, Reconciling alternative definitions of Cohen–Macaulay rings, unpublished manuscript, 6 pages (1992), http://timothychow.net/cv.html

[Coh] R. Cohn, *Difference Algebra* (Interscience Publishers John Wiley & Sons, New York-London-Sydeny, 1965)

[Con] A. Connes, *Noncommutative Geometry* (Academic Press Inc., San Diego, CA, 1994)

[D] P. Deligne, Travaux de Shimura, Séminaire Bourbaki, 23ème année (1970/71), Exp. No. 389, pp. 123–165, Lecture Notes in Mathematics, vol. 244 (Springer, 1971)

[Dol] I.V. Dolgachev, *Classical Algebraic Geometry. A Modern View* (Cambridge University Press, Cambridge, 2012)

[Don92] S. Donkin, Invariants of several matrices. Invent. Math. **110**, 389–401 (1992)

[Don93] S. Donkin, Invariant functions on matrices. Math. Proc. Cambridge Philos. Soc. **113**(1), 23–43 (1993)

[DCP] C. De Concini, C. Procesi, *The Invariant Theory of Matrices,* University Lecture Series, vol. 59 (American Mathematical Society, Providence, RI, 2017)

[DK] H. Derksen, G. Kemper, *Computational Invariant Theory. Invariant Theory and Algebraic Transformation Groups, I,* Encyclopaedia of Mathematical Sciences, vol. 130 (Springer, Berlin, 2002)

[E] D. Eisenbud, *Commutative Algebra: With a View Toward Algebraic Geometry,* Graduate Texts in Mathematics, vol. 150 (Springer, New York, 1995)

[FC] G. Faltings, C.-L. Chai, *Degeneration of Abelian Varieties. With an Appendix by David Mumford,* Ergeb. Math. Grenzgeb. (3) [Results in Mathematics and Related Areas (3)], vol. 22 (Springer, Berlin, 1990)

[F] W. Fulton, *Intersection Theory*, 2nd edn. Ergeb. Math. Grenzgeb. (3) [Results in Mathematics and Related Areas (3)], vol. 2 (Springer, Berlin, 1998)

[FH] W. Fulton, J. Harris, *Representation Theory. A First Course,* Graduate Texts in Mathematics, vol. 129, Readings in Mathematics (Springer, New York, 1991)

[G] G.B. Gurevich, *Foundations of the Theory of Algebraic Invariants* (P. Noordhoff Ltd., Groningen, 1964)

[GH] P. Griffiths, J. Harris, *Principles of Algebraic Geometry, Wiley Classics Library* (Wiley, New York, 1994)

[GW] B.H. Gross, M.R. Wallach, On the Hilbert polynomials and Hilbert series of homogeneous projective varieties, Arithmetic geometry and automorphic forms, 253–263, Advanced Lectures in Mathematics (ALM), vol. 19 (International Press, Somerville, MA, 2011)

[H] R. Hartshorne, *Algebraic Geometry,* Graduate Texts in Mathematics, vol. 52 (Springer, New York-Heidelberg, 1977)

[HR] M. Hochster, J.L. Roberts, Rings of invariants of reductive groups acting on regular rings are Cohen-Macaulay. Adv. Math. **13**, 115–175 (1974)

[Ja] J.C. Jantzen, *Representations of Algebraic Groups.* Mathematical Surveys and Monographs, vol. 107, 2nd edn. (American Mathematical Society, Providence, RI, 2003)

[Jo] A. Joyal, δ-anneaux et vecteurs de Witt. C. R. Math. Rep. Acad. Sci. Canada **7**(3), 177–182 (1985)

[Ka73a] N. Katz, p-adic properties of modular schemes and modular forms, *Modular Functions of One Variable, III,* Proceedings of the International Summer School, University of Antwerp, Antwerp, 1972, pp. 69–190, Lecture Notes in Mathematics, vol. 350 (Springer, Berlin, 1973)

[Ka73b] N. Katz, Travaux de Dwork, Séminaire Bourbaki, 24ème année (1971/1972), Exp. No. 409, pp. 167–200, Lecture Notes in Mathematics, vol. 317 (Springer, Berlin, 1973)

[Ka81] N. Katz, Serre–Tate local moduli, Algebraic surfaces (Orsay, 1976–78), pp. 138–202, Lecture Notes in Mathematics, vol. 868 (Springer, Berlin-New York, 1981)

[Ke] G. Kempf, The Hochster-Roberts theorem of invariant theory. Michigan Math. J. **26**(1), 19–32 (1979)

[Ko] E.R. Kolchin, *Differential Algebra and Algebraic Groups,* Pure and Applied Mathematics, vol. 54 (Academic Press, New York, 1973)

[La] S. Lang, *Algebra.* Revised third edition, Graduate Texts in Mathematics, vol. 211 (Springer, New York, 2002)

[Le] A. Levin, *Difference Algebra,* Algebra and Applications, vol. 8 (Springer, New York, 2008)

[Man] Yu.I. Manin, Rational points of algebraic curves over function fields, Izv. AN SSSR Mat. **27**, 1395–1440 (1963), or AMS Transl. **37**, 189–234 (1966)

[Mat] H. Matsumura, *Commutative Ring Theory.* Cambridge Studies in Advanced Mathematics, vol. 8, 2nd edn. (Cambridge University Press, Cambridge, 1989)

[Maz] B. Mazur, Rational points of abelian varieties with values in towers of number fields. Invent. Math. **18**, 183–266 (1972)

[Me] W. Messing, *The Crystals Associated to Barsotti-Tate Groups: With Applications to Abelian Schemes*, Lecture Notes in Mathematics, vol. 264 (Springer, Berlin-New York, 1972)

[Mum] D. Mumford, Abelian varieties. With appendices by C. P. Ramanujam and Yuri Manin. Corrected reprint of the second (1974) edition, Tata Inst. Fundam. Res. Stud. Math., **5**, Published for the Tata Institute of Fundamental Research, Bombay by Hindustan Book Agency, New Delhi (2008)

[MFK] D. Mumford, J. Fogarty, F. Kirwan, *Geometric Invariant Theory*, 3rd edn., Ergeb. Math. Grenzgeb. (2) [Results in Mathematics and Related Areas (2)], **34** (Springer, Berlin, 1994)

[Mur] M.P. Murthy, A note on factorial rings. Arch. Math. (Basel) **15**, 418–420 (1964)

[NS] J. Nekovář, N. Schappacher, On the asymptotic behaviour of Heegner points. Turkish J. Math. **23**, 549–556 (1999)

[PT] J. Pila, J. Tsimerman, Independence of CM points in elliptic curves. J. Eur. Math. Soc. (JEMS) **24**(9), 3161–3182 (2022)

[P] C. Procesi, The invariant theory of $n \times n$ matrices. Adv. Math. **19**(3), 306–381 (1976)

[Sa] P. Samuel, *Lectures on Unique Factorization Domains. Notes by M. Pavman Murthy,* Tata Inst. Fundam. Res. Stud. Math., No. 30, Tata Institute of Fundamental Research, Bombay (1964)

[Se] C.S. Seshadri, Geometric reductivity over arbitrary base. Adv. Math. **26**(3), 225–274 (1977)

[T] H.W. Turnbull, *The Theory of Determinants, Matrices, and Invariants* (Dover Publications Inc., New York, 1960)

[V] A. Vasiu, *Generalized Serre-Tate Ordinary Theory* (International Press, Boston, MA, 2013)

[W] C.A. Weibel, *An Introduction to Homological Algebra,* Cambridge Studies in Advanced Mathematics, vol. 308 (Cambridge University Press, Cambridge, 1994)

LECTURE NOTES IN MATHEMATICS

Springer

Editors in Chief: J.-M. Morel, B. Teissier

Editorial Policy

1. Lecture Notes aim to report new developments in all areas of mathematics and their applications – quickly, informally and at a high level. Mathematical texts analysing new developments in modelling and numerical simulation are welcome. See Lecture Notes in Mathematics for the existing online volumes of LNM.

   Manuscripts should be reasonably self-contained and rounded off. Thus, they may, and often will, present not only results of the author but also related work by other people. They may be based on specialised lecture courses. Furthermore, the manuscripts should provide sufficient motivation, examples and applications. This clearly distinguishes Lecture Notes from journal articles or technical reports which normally are very concise. Articles intended for a journal but too long to be accepted by most journals, usually do not have this "lecture notes" character. For similar reasons it is unusual for doctoral theses to be accepted for the Lecture Notes series, though habilitation theses may be appropriate.

2. Besides monographs, multi-author manuscripts resulting from SUMMER SCHOOLS or similar INTENSIVE COURSES are welcome, provided their objective was held to present an active mathematical topic to an audience at the beginning or intermediate graduate level (a list of participants should be provided).

   The resulting manuscript should not be just a collection of course notes but should require advance planning and coordination among the main lecturers. The subject matter should dictate the structure of the book. This structure should be motivated and explained in a scientific introduction, and the notation, references, index and formulation of results should be, if possible, unified by the editors. Each contribution should have an abstract and an introduction referring to the other contributions. In other words, more preparatory work must go into a multi-authored volume than simply assembling a disparate collection of papers, communicated at the event.

3. Manuscripts should be submitted online either at www.editorialmanager.com/lnm to Springer's mathematics editorial in Heidelberg, or electronically to one of the series editors. Authors should be aware that incomplete or insufficiently close-to-final manuscripts almost always result in longer refereeing times and nevertheless unclear referees' recommendations, making further refereeing of a final draft necessary. The strict minimum amount of material that will be considered should include a detailed outline describing the planned contents of each chapter, a bibliography and several sample chapters. Parallel submission of a manuscript to another publisher while under consideration for LNM is not acceptable and can lead to rejection.

4. The type of material considered for publication includes:

   - Research monographs
   - Lectures on a new field or presentations of a new angle in a classical field
   - Summer schools and intensive courses on topics of current research

In general, **monographs** will be sent out to at least 2 external referees for evaluation.

A final decision to publish can be made only on the basis of the complete manuscript, however a refereeing process leading to a preliminary decision can be based on a pre-final or incomplete manuscript.

Volume Editors of **multi-author works** are expected to arrange for the refereeing, to the usual scientific standards, of the individual contributions. If the resulting reports can be forwarded to the LNM Editorial Board, this is very helpful. If no reports are forwarded or if other questions remain unclear in respect of homogeneity etc., the series editors may wish to consult external referees for an overall evaluation of the volume.

5. Manuscripts should in general be submitted in English. Final manuscripts should contain at least 100 pages of mathematical text and should always include:
   - A table of contents
   - An informative introduction, with adequate motivation and perhaps some historical remarks: it should be accessible to a reader not intimately familiar with the topic treated
   - A subject index: as a rule, this is genuinely helpful for the reader
   - For evaluation purposes, manuscripts should be submitted as pdf files.
6. Careful preparation of the manuscripts will help keep production time short besides ensuring satisfactory appearance of the finished book in print and online. After acceptance of the manuscript authors will be asked to prepare the final LaTeX source files plus the corresponding pdf- or zipped ps-file (Preparing your Manuscript). You can also download our LaTeX package for monographs. The LaTeX source files are essential for producing the full-text online version of the book. For more detailed information, please visit the Book Manuscript Guidelines and Key Style Points. The technical production of a Lecture Notes volume takes approximately 12 weeks. Additional instructions, if necessary, are available on request from lnm@springer.com.
7. Authors receive a total of 30 free copies of their volume and free access to their book on SpringerLink, but no royalties. They are entitled to a discount of 40 % on the price of Springer books purchased for their personal use, if ordering directly from Springer.
8. Commitment to publish is made by a *Publishing Agreement*; contributing authors of multi-author books are requested to sign a *Consent to Publish form*. Springer-Verlag registers the copyright for each volume. Authors are free to reuse material contained in their LNM volumes in later publications: a brief written (or e-mail) request for formal permission is sufficient.

Addresses:

Professor Jean-Michel Morel, City University Hongkong, Kowloon Tong, Hong Kong
E-mail: moreljlmichel@gmail.com

Professor Bernard Teissier, Equipe Géométrie et Dynamique,
Institut de Mathématiques de Jussieu – Paris Rive Gauche, Paris, France
E-mail: bernard.teissier@imj-prg.fr

Springer: Ute McCrory, Editor, Mathematics, Heidelberg, Germany.
E-mail: lnm@springer.com

Index

A. Buium and A. Vasiu, *Delta-Invariant Theory for Hecke Correspondences*, Lecture Notes in Mathematics 2396, https://doi.org/10.1007/978-3-032-20571-1

The manufacturer's authorised representative in the EU is Springer Nature Customer Service Centre GmbH, Europaplatz 3, 69115 Heidelberg, Germany. If you have any concerns regarding our products, please contact ProductSafety@springernature.com

Printed and bound by CPI Group (UK) Ltd, Croydon, CR0 4YY
25/09/2026
02222849-0018